Emerging Technologies in Hazardous Waste Management III

ACS SYMPOSIUM SERIES **518**

Emerging Technologies in Hazardous Waste Management III

D. William Tedder, EDITOR
Georgia Institute of Technology

Frederick G. Pohland, EDITOR
University of Pittsburgh

Developed from a symposium sponsored
by the Division of Industrial and Engineering Chemistry, Inc.,
of the American Chemical Society
at the Industrial and Engineering Chemistry Special Symposium,
Atlanta, Georgia,
October 1–3, 1991

American Chemical Society, Washington, DC 1993

Library of Congress Cataloging-in-Publication Data

Emerging technologies in hazardous waste management III /
   D. William Tedder, editor, Frederick G. Pohland, editor.

   p.   cm.—(ACS symposium series, ISSN 0097–6156; 518)

   "Developed from a symposium sponsored by the Division of Industrial
and Engineering Chemistry, Inc., of the American Chemical Society at
the Industrial and Engineering Chemistry Special Symposium, Atlanta,
Georgia, October 1–3, 1991."

   Includes bibliographical references and index.

   ISBN 0–8412–2530–3

   1. Hazardous   wastes—Management—Congresses.   2. Sewage—
Purification—Congresses.

   I. Tedder, D. W. (Daniel William), 1946–   . II. Pohland, F. G.
(Frederick   George),   1931–   .   III. American   Chemical   Society.
Division of Industrial and Engineering Chemistry. IV. Series.

TD1020.E443   1993
628.4′2—dc20   92–43966
                                                              CIP

1993 Advisory Board

ACS Symposium Series

M. Joan Comstock, *Series Editor*

Foreword

THE ACS SYMPOSIUM SERIES was first published in 1974 to provide a mechanism for publishing symposia quickly in book form. The purpose of this series is to publish comprehensive books developed from symposia, which are usually "snapshots in time" of the current research being done on a topic, plus some review material on the topic. For this reason, it is necessary that the papers be published as quickly as possible.

Before a symposium-based book is put under contract, the proposed table of contents is reviewed for appropriateness to the topic and for comprehensiveness of the collection. Some papers are excluded at this point, and others are added to round out the scope of the volume. In addition, a draft of each paper is peer-reviewed prior to final acceptance or rejection. This anonymous review process is supervised by the organizer(s) of the symposium, who become the editor(s) of the book. The authors then revise their papers according to the recommendations of both the reviewers and the editors, prepare camera-ready copy, and submit the final papers to the editors, who check that all necessary revisions have been made.

As a rule, only original research papers and original review papers are included in the volumes. Verbatim reproductions of previously published papers are not accepted.

M. Joan Comstock
Series Editor

Contents

BIOLOGICAL TREATMENT

SOIL REMEDIATION AND TREATMENT

TREATMENT OF VOLATILE COMPOUNDS

SELECTED MIXED-WASTE TREATMENT APPLICATIONS

INDEXES

Preface

HAZARDOUS WASTES ARE A CONTINUING PROBLEM in today's world, increasing in both quantity and toxicity. At the same time, government regulations restricting their disposal and release are becoming more stringent. Moreover, because wastewater concerns are most prevalent, they will continue to dominate efforts for management and control.

The symposium on which this book is based was created to bring together specialists in emerging technologies for treating and managing gaseous, liquid, and solid wastes. Approximately 150 presentations were given during the meeting. Eighty-four authors presented formal manuscripts for review. The final selection of the 22 chapters included here was based on peer review, scientific merit, the editors' perceptions of lasting value or innovative features, and the general applicability of either the technology itself or the scientific methods and scholarly details provided by the authors.

This volume is a continuation of a theme initiated in 1990. Its predecessors, *Emerging Technologies for Hazardous Waste Management*, ACS Symposium Series No. 422 (1990), and *Emerging Technologies for Hazardous Waste Management II*, ACS Symposium Series No. 468 (1991), are related collections on waste management, but these three volumes are remarkably different. No single volume can do justice to this broad subject; few authors, for example, appear in more than one volume. This volume includes a comprehensive review of supercritical water oxidation technology by investigators at the Massachusetts Institute of Technology and an evaluation of nonlinear group contribution methods for estimating biodegradation kinetics by Environmental Protection Agency researchers. Neither of these topics is discussed in earlier volumes. The use of bioprocessing to achieve separations at low concentrations is another advanced technology that was not previously described. There are also new contributions on mineralization kinetics that will be of value in design trade-off studies.

This book is a useful introduction to hazardous waste treatment for the novice and a valuable reference for the technical expert. No topic is afforded comprehensive treatment, but there is something for everyone. Our authors come from varied backgrounds and interests. They include physicists, chemists, biologists, soil scientists, and an assortment of mechanical, civil, environmental, nuclear, and chemical engineers.

The contributions in this volume are divided into five sections:

physical and chemical wastewater treatment, biological treatment, soil remediation and treatment, treatment of volatile compounds, and selected mixed-waste treatment applications. The first section contains chapters describing photochemical and separations applications. The second section continues with several chapters describing innovative applications of biological treatment. The third section includes several theoretical chapters dealing with soil remediation and treatment. The fourth section includes several innovative technologies for managing volatile species. The fifth section describes two important technology applications for mixed-waste treatment.

The symposium on which this book is based was supported by several organizations that are committed to excellence in solving waste problems and reducing environmental pollution. Their generosity was essential to the overall success of the symposium, and we gratefully recognize it here. Our event sponsors were Eli Lilly and Company, Indianapolis, Indiana; the Hoechst Celanese Corporation, Chatham, New Jersey; and Merck and Company, Merck Chemical Manufacturing Division, Rahway, New Jersey.

D. WILLIAM TEDDER
Georgia Institute of Technology
Atlanta, GA 30332–0100

FREDERICK G. POHLAND
University of Pittsburgh
Pittsburgh, PA 15261–2294

September 3, 1992

Chapter 1

Emerging Technologies for Hazardous Waste Management

An Overview

D. William Tedder[1] and Frederick G. Pohland[2]

[1]School of Chemical Engineering, Georgia Institute of Technology, Atlanta, GA 30332–0100
[2]Department of Civil Engineering, University of Pittsburgh, Pittsburgh, PA 15261–2294

Once exotic, many hazardous wastes are now commonplace. For example, estimates of their production rates in the U.S. were 680 million metric tons/y in 1986 according to Baker and Warren (*1*). Wastewaters comprise about 90% of these estimates. An accompanying summary detailed waste management practices in 1986, and estimated that about 1090 million metric tons of waste were processed in 1986.

Fischer (*2*) presents a study using the data base generated by the Chemical Manufacturers Association. He estimates that in 1989, the U.S. chemical industry produced 6.17 million metric tons of hazardous solid wastes and 987 million metric tons of hazardous wastewaters. Fischer (*2*) also presents cumulative distributions of annual waste production for solids and wastewaters which show that most of the waste generators produce relatively small amounts. About half produced less than 122 metric tons of solid wastes in 1989. Nearly 70% of the 617 chemical plants surveyed did not produce hazardous wastewaters that year, and the average solid waste production rate was 10,000 metric tons/y. On the other hand, the average wastewater rate was 1.6 million metric tons/y, but about 70% of the plants generated virtually no wastewater. Thus, there are a few plants that generate relatively large waste streams, while many more generate much smaller ones.

Based upon Fischer's study for the 1980s, the incineration of hazardous solid wastes appeared to increase steadily over that decade while treatment and disposal decreased. Most of these effects were due to changes at a few plants with large waste streams.

Both studies find over 90 mass% of the wastes currently generated are aqueous, but neither study includes historical waste inventories. These latter quantities include significant amounts of solids (e.g., contaminated soils and sediments).

Manahan (*3*) points out that hazardous wastes also arise from natural processes. Many significant hazards result from various organisms found in na-

0097–6156/93/0518–0001$06.00/0

ture (4). The botulism toxin from *Clostridium botulinum*, for example, is one of the most acutely toxic substance known. Wastes from the food industry may include such hazards as aflatoxin B_1, produced by *Aspergillus niger*, a fungus which grows on moldy food, especially nuts and cereals. Hazardous grain molds include *Cephalosporium, Fusarium, Trichoderma* and *Myrothecium* which produce trichothecenes (5). Among the alkaloids, nicotine is particularly toxic and prevalent in many wastes (6).

Considerable research into biological warfare agents has focused on *Clostridium botulinum* and *Bacillus anthracis* which produce botulism and anthrax, respectively. Although not strictly natural products, biomedical wastes are significant both in the quantities and hazards they represent. They contain many infectious agents that occur naturally (7).

Increasing environmental concerns are forcing waste generators to significantly modify waste management practices. The concentration limits in liquid and gaseous effluents are decreasing. Zero discharge is the goal for many priority pollutants, especially in aqueous wastes. Thus, there are clear needs for improved technologies to achieve these goals—either directly by decontamination or indirectly by recycle and process modification. Zero liquid discharge may become a reality in many instances.

Waste management using separation technologies becomes more difficult and expensive as feed concentrations in the waste stream decrease. This is partly due to the fact that secondary waste production per mol of product recovered is inversely related to the feed concentration. Figure 1 illustrates this effect. It shows data for aqueous secondary waste production using ion exchange and solvent extraction technologies to recover heavy metals and fission products (8–17). Although secondary wastes can usually be concentrated, Figure 1 clearly indicates why the practical returns, and the incentives for using decontamination technologies, rapidly diminish with feed concentrations. As a consequence, this approach is not necessarily the best one. Alternatives include strategies to reduce waste generation rates, to facilitate wastewater recycle and reuse, to permit the implementation of alternative manufacturing chemistry, and to enable sludge reduction and reuse. These latter approaches are sometimes less obvious, but may be preferred from the systems viewpoint.

Brandt (18) describes waste reduction in terms of the four Rs—reduction, reuse, reclamation, and recycle. Wastes may have value if they can be recycled or reused at the generation site, or if they are usable raw materials in another process. On-site recycle is particularly attractive if it also leads to higher product recoveries and waste concentration. In some cases, the waste has little intrinsic value, but recycle improves the overall process efficiency and the production of solid waste forms. Partly because of such complexities, several organizations have developed assistance programs (19, 20).

Off-site disposal necessarily incurs transportation costs, but may be advantageous if the waste can be used as a raw material by another process, promotes energy recovery, or assists in pollution abatement or waste treatment. This latter strategy is often more complex than the first because the supply and demand are less likely to match.

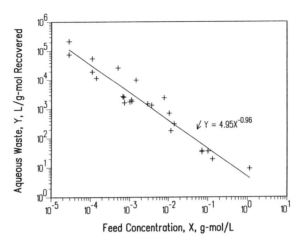

Figure 1. Typical dependence of secondary aqueous waste production on feed concentration using solvent extraction and ion exchange to recover heavy metals and fission products. (Adapted from ref. 8.)

For many solid and liquid wastes, significant reductions are achievable by a variety of means. Simple accounting, inventory control, and more stringent disposal standards often yield significant waste reductions (*21*). In some cases, less hazardous raw materials can be substituted. A current example includes the on-going search for substitutes to chlorofluorocarbons for numerous dry solvent applications. At the other extreme, process modifications may be implemented to reduce waste generation rates and, finally, alternative treatment technologies (e.g., incineration) may be adopted.

Constituents in municipal solids are among the more obvious candidates for recycle (e.g., metals, glass, paper, plastics, and rubber). Some of these constituents are not hazardous, but they are often found in hazardous mixtures (i.e., in a contaminated state). Such wastes may be segregated manually or by methods that exploit differences in physical properties such as density or magnetic susceptibilities. Increasingly, residential generators are being encouraged to segregate bottles and cans. Separation at the point of generation is less expensive, but still labor intensive. This approach can be effective, especially as an interim procedure, but a clear need exists for more efficient segregation technologies that are economical at centralized facilities (*22*).

The contents of this volume focuses on selected waste management technologies that are under development. Since they are emerging technologies, neither process safety nor economic considerations are discussed in detail. Therefore, in most cases, additional research and development are needed before these aspects can be assessed with acceptable levels of confidence.

Physical and Chemical Wastewater Treatment

Improved technologies for managing wastewaters contaminated with toxic and hazardous substances are an important goal for many researchers. This emphasis is appropriate since over 90 vol % of hazardous wastes are primarily aqueous (*1, 2*). The papers in this section primarily review decontamination technologies (either separation and purification or detoxification by chemical reaction). It should be recognized, however, that water reuse and recycle is at least equally important, especially as acceptable discharge limits approach lower concentration levels.

Advanced Oxidation Processes (AOPs) for wastewater treatment have been discussed in earlier volumes in this series and they remain of interest here (*23, 24*). The fundamental chemistry is complex, even in well even for well defined systems. The hydroxyl radical upon which they are based can be generated by many catalytic and photolytic systems, and effectively used to degrade target pollutants. In Chapter 2, Ollis overviews earlier efforts and, by emphasizing photocatalysis, particularly Peyton's work (*25*), concludes that reactor design for these systems is still not well developed. Ollis focuses primarily on the use of ozone, hydrogen peroxide, ultraviolet light, and the catalytic system based on titanium dioxide (*26*).

Supercritical water oxidation (SCWO) can also be used to generate the hydroxyl radical and oxidize pollutants. Tester et al. provide an in depth review of this technology in Chapter 3. It is of interest because many hazardous substances, including polychlorinated hydrocarbons, can be effectively oxidized under conditions that avoid the production of nitrogen oxides. Their contribution reviews critical technology components and operations as required for commercial-scale development. They also summarize fundamental research activities in this area. Although other uses of supercritical fluids have been studied [e.g., supercritical extraction with CO_2 (27–30)], SCWO has clear advantages. These authors conclude that the basic idea is sound.

Fenton's reaction can also be used to generate the hydroxyl radical through the reaction:

$$Fe^{2+} + H_2O_2 \rightarrow Fe^{3+} + OH\cdot + HO^- \tag{1}$$

The use of Fe^{3+} has not been as widely studied, but it can also catalyze the oxidation of organics. In Chapter 4, Pignatello and Sun compare herbicide mineralization in several related systems (Fe^{3+}/H_2O_2, and the photocatalyzed systems $Fe^{3+}h\nu$, $H_2O_2/h\nu$, and $Fe^{3+}/H_2O_2/h\nu$). Initial rate studies indicated that a favorable synergism occured when Fe^{3+} and H_2O_2 mixtures were irradiated.

In Chapter 5, Vella and Munder compare potassium permanganate, Fenton's reagent, and ClO_2 to oxidize substituted phenols. They find that chlorine dioxide is the most flexible of these oxidants, since it is less susceptible to interferences and more effective over a wide pH range. Potassium permanganate is similar to ClO_2, but it exhibits slightly greater interferences. Fenton's reagent only works well in a narrow pH range and is very susceptible to inhibitors and common scavengers for $OH\cdot$ (e.g., phosphate and carbonate). On the other hand, it is effective at the lowest concentrations of the three. They find that $KMNO_4$ is less effective than Fenton's reagent, which is much less effective than ClO_2 in reducing toxicity.

The use of solid catalysts to enhance degradation rates is complicated by mass transfer considerations. Reactants must transfer from the bulk liquid phase to the catalytic site. In general, this involves diffusion through a boundary layer surrounding the solid particle, pore diffusion to the catalytic site, and the adsorption of reactants (*31, 32*). The opposite path is followed by products, and either may become limited due to mass transfer rates. In situations where mass transfer controls, ultrasound irradiation may be helpful to reduce boundary layer thickness and enhance intrapore diffusion rates. Ultrasound may also enhance reaction rates if it increases the effective catalyst surface area (e.g., through pitting or fragmentation). In Chapter 6, Johnston and Hocking report accelerated photocatalytic treatment rates with ultrasound and TiO_2 and UV to degrade chlorophenols. They find ultrasound enhances reaction rates in this system and suggest possible explanations.

Water purification by adsorption and chemical treatment (e.g., flocculation) is well established technology, but its effectiveness in treating hazardous wastes is not as well defined. Reactive dye discharges, for example, are potential problems for the textile industry. When such discharges occur, they typically involve large volumes and present color problems in addition to their toxicities. Typical reactive dyes also have low vapor pressures and are, therefore, candidates for adsorption-based removal systems. In Chapter 7, Michelsen et al. evaluate this technology for the control of Navy 106 reactive wash waters from pad dying operations, and the use of chemical reduction to pretreat the more concentrated dye streams in the process. They find that it is more attractive to treat the concentrated wastes, and compare several alternatives.

Biological Treatment

Biological treatment is a traditional technology for maintaining water quality that is finding new applications in the treatment of hazardous wastes occurring as either gases or solids. The chemistry of biological systems is complex, however, and has the concomitant problem of biomass maintenance. This latter challenge may be particularly severe whenever biological agents are used to destroy or concentrate hazardous toxins.

In Chapter 8, Eckenfelder and Norris provide an overview of the applicability of biological processes for the treatment of soils. Since the biodegradation of organic wastes (or "land farming") has been used in some form for many years, the paramount issues associated with its application to hazardous wastes are basically those that have been defined from years of wastewater treatment (i.e., temperature, pH, moisture, nutrient availability, oxygen levels, etc.). The primary focus of most aerobic bioremediation processes is the provision of adequate nutrients and oxygen, but additional considerations may arise in hazardous waste applications (e.g., in situ vs ex situ treatment, its use in conjunction with other remediation technologies, and the control of toxic effluents). Thus, the extent of bioremediation at any given site must be evaluated in light of overall needs and constraints.

The effectiveness of bioremediation clearly depends on the susceptibility of the target pollutants. This vulnerability can be measured for individual species (e.g., using respirometry), but a combinatorial problem exists because of the numbers of pollutants, biological systems, and substrate matrices that need evaluation. There is, therefore, a need to develop predictive methods for estimating degradation kinetics using generalized methods. One approach, described by Tabak and Govind in Chapter 9, uses molecular structure and biodegradation relationships to define group contribution effects and thus estimate degradation rates. They conclude that this approach can be used to predict first order kinetics and Monod rate constants.

Hydrogen cyanide and its salts have numerous industrial applications. Cyanide-bearing wastes are toxic and must be managed. One approach, discussed by Shah and Aust in Chapter 10, uses *Phanerochaete chrysosporium* to mineralize potassium cyanide and salts of Fe, Cd, Cu, Cr and Cr^{6+} in soil and liquid cultures. They show that a pure lignin peroxidase from *P. chrysosporium* oxidizes cyanide using a pathway involving the cyanyl radical. This microorganism has the advantage of being relatively non-specific; it also degrades lignin and munition wastes, for example (*33*). Ground corn cobs can be used as the nutrient.

Anaerobic treatment processes frequently involve toxic and hazardous substances. However, treated effluents may still contain unacceptable pollutant concentrations, and the biological factors which influence degradation rates are often poorly defined. In Chapter 11, Sathish et al. describe their efforts to develop better methods of quickly obtaining reliable kinetic measurements. They propose the use of a fixed-growth environment and well-defined media in which acclimated cultures are developed along with selected co-substrates.

Goodloe et al. describe the anaerobic digestion of industrial activated sludge in Chapter 12. These sludges may be hazardous if they contain residual chemicals (e.g., phenols or heavy metals). Volume reduction and sludge disposal is often based on incineration, but anaerobic treatment offers a possible alternative that also produces a methane-bearing biogas that has fuel value. They conclude that polymer hydrolysis was not the rate-limiting step in their system, and were able to achieve about a 35% reduction in volatile solids with a hydraulic residence time of 15 days. Their minimum hydraulic residence time for stable operation was about five to seven days.

Soil Remediation and Treatment

The widespread use of hydrocarbon fuels has resulted in significant contamination problems both in soils and sediments. Typically, waste oils are recovered by decantation, extraction, or evaporation with condensation. The most attractive means depends upon the waste matrix, oil concentrations, and their properties. However, the conventional means for oil recovery are generally not applicable for cleaning contaminated soils (*34*). When petroleum products are released to the environment, they tend to become adsorbed and chemisorbed onto soil particles. The primary factors are the relative permeabilities of the oil contaminants in the soil, their viscosities, and interfacial tensions. Secondary factors include the extent of oily contamination in the soil field and the areal distribution of oily pools.

There are many alternatives for contaminated soil cleanup. As might be expected, each has strengths and weaknesses. Ex situ thermal treatment can be effective (*35*), for example, but also expensive and destructive to the soil. Ex situ chemical (*36*) and bioremediation (*37*) methods are often helpful for the remediation of specific pollutants, or classes of pollutants, but they too have disadvantages. In situ methods are often less traumatic to the soil ecology, but also much slower and less complete (*38, 39*). As an added complication, the most effective technologies are typically site specific and require extensive experimental testing to identify. Thus, real needs exist for better generic capabilities for soil cleanup that are more economical, easier to implement and have reduced environmental impacts.

Just and Stockwell compare the effectiveness of in situ and ex situ technologies in Chapter 13. They focus on issues surrounding the cleanup of solvent-contaminated soils, but their perspective is useful beyond that immediate problem. They compare low-temperature thermal treatment, radio frequency heating, steam stripping, vacuum extraction, aeration, bioremediation, soil flushing and washing. Both advantages and disadvantages of the technologies are discussed along with summaries of specific applications. Just and Stockwell conclude that a clear treatment choice often does not exist a priori.

Greater progress will be made in a priori selection if a better and more fundamental understanding of soil chemistry and its interactions with pollutants exists. Toward that end, Goldberg presents a soil model in Chapter 14 that appears useful for describing metal adsorption on oxide minerals, clays, and soils.

Their constant capacitance model provides a molecular description of adsorption phenomena using an equilibrium approach. It is based on a rational soil model that requires the definition of surface species, chemical reactions, equilibrium constants, surface activity coefficients, and mass and charge balances. This model is an extension of work by Schindler and Stumm (40–42) that is based on four essential assumptions: (a) all surface complexation occurs within the inner spheres, (b) anion adsorption occurs via a ligand exchange mechanism, (c) the aqueous species activity coefficients are defined using a constant ionic medium reference state, and (d) a linear relationship exists between surface charge and potential. The predictions from this model are encouraging.

In Chapter 15, Marsi and Evangelou present a model describing the effects of brackish solutions on chemical and physical behavior in temperate region soils. They study the effects of three variables: (a) ionic strength, (b) the sodium adsorption ratio and (c) pH on the Vanselow exchange coefficient, the adsorbed ion activity coefficients, and the dispersion and saturated hydraulic conductivity relationships. In studying two soils, they find that these variables regulate saturated hydraulic conductivity by clay dispersion and swelling.

Chapter 16 is a contribution by Sato et al. that examines the applicability of Fenton's reagent for the treatment of soils contaminated with perchloroethylene and polychlorinated biphenyls. It complements the discussions of Fenton chemistry found in Chapters 4 and 5 for aqueous systems. Their laboratory results suggest that Fenton's reagent may be useful for ex situ decontamination. Under their conditions they were able to achieve reasonable mineralization rates.

Treatment of Volatile Compounds

In situ soil wash solutions and groundwater may become contaminated with volatile organics. These waters are easily treated by air stripping to remove the volatiles, but the pollutants still require management to avoid an air pollution problem. In addition, many industrial processes emit volatile pollutants that require abatement. Air venting with activated carbon adsorption, for example, is one commonly-used method for controlling worker exposure and volatile emissions. However, it suffers from the disadvantage that adsorption beds must either be regenerated or periodically replaced. In the former case, the volatiles must still be managed on site. In the latter, volatiles must be controlled off site.

Chemical reaction, particularly oxidation, represents one possible avenue for detoxifying many airborne volatiles. In Chapter 17, Shaw et al. describe the use of Pt, PdO, and MnO_2 for the destruction of methylene chloride and trichloroethylene. A 1.5% Pt on γ-alumina catalyst, either on a cordierite monolith or as a powder, completely oxides 200 ppm of trichloroethylene in air at 500 °C and space velocities of 30,000 v/v-h. A PdO catalyst is about equally effective at 600 °C, but produces highly chlorinated byproducts whose generation rates are reduced by the presence of hydrogen sources (e.g., water or methane).

Alternatively, volatile organics may be photo-oxidized in air using ultraviolet light. In Chapter 18, Blystone et al. discuss this option by using a xenon flashlamp as the radiation source. They report apparent quantum yields and first order rate constants for several chlorinated species (e.g., trichloroethylene, perchloroethylene, chloroform and methylene chloride). Species with higher absorbances have higher destruction rates. The destruction rates in some mixtures may exhibit favorable synergistic interactions. They conclude that the application of this technology to groundwater treatment via air stripping is possible.

A third option for the destruction of volatile chlorinated organics in air is discussed by Krause and Helt in Chapter 19. They examine the use of a microwave discharge plasma reactor at atmospheric pressure to destroy trichloroethylene and 1,1,1-trichloroethane. They find that either species can be destroyed using either oxygen, water, or O_2 and H_2O vapor mixtures as co-reactants using an argon carrier. Both the extent of reaction and the product distribution depend upon the power input to the reactor; the primary products are CO, CO_2, HCl, Cl_2, and H_2.

Apel et al. discuss a fourth method for promoting the destruction of toxic vapors in Chapter 20. Their approach uses fixed-bed bioreactors in which a suitable microorganism is immobilized on an inert support. They studied the degradation of methane, trichloroethylene, and p-xylene using methanotrophic bacteria, and find that gas-phase bioreactors appear to offer significant potential for the cleanup of selected gaseous effluents. They also point out that this technology is in competition with conventional aqueous-phase bioreactors, and that solutes with higher aqueous solubilities may be treated more effectively as liquids.

Selected Mixed Waste Treatment Applications

Mixed wastes can possess both radiological and chemical hazards. Mixed wastes containing organic species may undergo radiolysis and the continuous or intermittent release of flammable gases (e.g., CH_4 or H_2). Examples include high-level liquids in the Hanford Underground Storage Tanks and solid transuranic wastes at Idaho and Rocky Flats. In such cases, there is an added incentive to stabilize the wastes (e.g., by destroying the organic species or separating them from the radioactive components). As a further complication, the treatment of mixed wastes invariably requires extensive precautions to protect workers.

In Chapter 21, Hickman et al. describe an electrochemical process for treating mixed wastes. Variations using this or similar redox chemistry have been studied for many years (43–47), but their applicability and utility are still often overlooked. In the case of mixed wastes, it is particularly interesting in that: (a) carbonaceous species (either solids or liquids) are oxidized more readily than refractory metal oxides, (b) refractory metal oxides (e.g., PuO_2) can be ionized and dissolved, and (c) the corrosive properties of higher valence species can be eliminated by the addition of excess reductant. The first property enables the selective removal of many organics, the second provides a way to recover and concentrate the metals, and the last allows greater freedom to use less expensive materials of construction. Hickman et al. use a semipermeable membrane to partition the anolyte and catholyte solutions. In some cases, however, a membrane is not required and this is another advantage in waste handling systems. In cerium and nitric acid, for example, Ce^{6+} exists as $H_2Ce(NO_3)_6$ and is not attracted to the cathode.

Chapter 22 provides an evaluation of polyethylene encapsulation of low-level wastes. Kalb et al. describe their findings based on the investigation of specific failure mechanisms (e.g., biodegradation, radiation, chemical attack, flammability, etc.) Polyethylene was found to be extremely resistant to each of these potential failure modes under anticipated storage and disposal conditions. Polyethylene is highly resistant to microbial degradation and attack by aggressive chemicals. Radiation doses through 10^8 rad increase crosslinking and improve strength and other physical properties. They conclude that polyethylene waste forms exceed minimum performance standards established by the Nuclear Regulatory Commission and the Environmental Protection Agency for commercial low-level wastes and hazardous wastes, respectively.

Summary

Hazardous wastes are a continuing problem in today's world, increasing in both quantity and toxicity. At the same time, government regulations to restrict their disposal and release are becoming more stringent. Moreover, since wastewater concerns are most prevalent, they will continue to dominate efforts for management and control.

Separation and destruction technologies can be helpful in decontaminating waste streams. On the other hand, decontamination and discharge may not be the best long-term strategy, since many of the traditional technologies are inefficient for managing trace contaminants. Waste stream recycle and pollution prevention may be more attractive, especially as release limits continue to drop and design criteria change and become more stringent. As a consequence, there are still many research opportunities to develop innovative technologies that are more effective. MacNeil, for example, reviews membrane applications in detail (48). Those which enable the regeneration and recycle of reagents are of particular interest. In some cases, the incremental waste management costs can be offset by reduced net reagent use in waste-generating process, and Thornburg et al. provide one example for recovering nitric and hydrofluoric acids from spent pickling liquors (*49*).

Biological waste treament processes exhibit considerable potential because of their low cost and relatively benign characteristics. Microorganisms will continue to play important roles in water purification through such techniques as immobilization on beds, and the exploitation of their naturally occurring bioaccumulative properties. Their abilities to oxidize and reduce various species should not be overlooked, as these abilities can be exploited to either liberate or precipitate metals. They also have useful applications in soil treatment and air pollution control.

Waste recycle and process integration will become increasingly important, and they often offer the greatest opportunities for utilizing existing equipment and capabilities. While specialized unit operations can provide new and improved alternatives, it is the overall system performance that determines the effectiveness of any given waste management plan. Process synthesis and integration are particularly important in the area of pollution prevention, but it is essential to have an overall strategy and a clear vision for each application. In general, the best approaches will convert wastes into assets, reduce resource consumption and production costs, and provide more robust and competitive production and marketing networks. Improvements in waste management can also improve product recovery and promote reagent reuse, and these factors should always be considered in the development of an overall plan.

Literature Cited

1. Baker, R. D. and Warren, J. L. Generation and management of hazardous waste in the United States. In *Preprints: AIChE 2nd Topical Pollution Prevention Conference, August 20-21, 1991, Pittsburgh, PA*, pages 163–166, American Institute of Chemical Engineers, New York, NY, 1991.
2. Fischer, L. M. The chemical manufacturers association hazardous waste database. In *Preprints: AIChE 2nd Topical Pollution Prevention Conference, August 20-21, 1991, Pittsburgh, PA*, pages 167–171, American Institute of Chemical Engineers, New York, NY, 1991.
3. Manahan, S. E. *Hazardous Waste Chemistry, Toxicology and Treatment.* Lewis Publishers, Chelsea, Michigan, 1990.
4. Harris, J. B., Ed. *Natural Toxins: Animal, Plant, Microbial.* Oxford University Press, New York, 1987.
5. Cross, Jr., F. L. and Robinson, R. Infectious waste. In Freeman, H. M., Ed., *Standard Handbook of Hazardous Waste Treatment and Disposal,* pages 4.35–4.45, McGraw-Hill, New York, 1989.

6. Gosselin, R. E., Smith, R. P., and Hodge, H. C. *Nicotine*, pages III–311–III–314. Williams and Wilkins, Baltimore/London, 5th edition, 1984.
7. Brunner, C. R. and Brown, C. H. Hospital waste disposal by incineration. *J Air Poll Cont Fed*, 38:1297–1309, **1988**.
8. Tedder, D. W. A review of separations in hazardous waste management. *Sep Purif Methods*, 21(1):23–74, **1992**.
9. Perona, J. J., Blomeke, J. O., Bradshaw, R. L., and Roberts, J. T. *Evaluation of Ultimate Disposal Methods for Liquid and Solid Radioactive Wastes. V. Effects of Fission Product Removal on Costs of Waste Management*. Technical Report ORNL-3357, Oak Ridge National Laboratory, Oak Ridge, TN, 1963.
10. Bray, L. A., LaBorde, C. G., and Richardson, G. L. One step extraction yields radioisotopes. *C&E News*, 41(21):46–47, May 27 **1963**.
11. LaRiviere et al., J. R. *The Hanford Isotopes Production Plant Engineering Study*. Technical Report HW-77770, Hanford Atomic Products Operation, Richland, WA, 1963.
12. Bond, W. D. and Leuze, R. E. *Feasibility Studies of the Partitioning of Commercial High-Level Wastes Generated in Spent Nuclear Fuel Reprocessing*. Technical Report ORNL-5012, Oak Ridge National Laboratory, Oak Ridge, TN, 1975.
13. Oak Ridge National Laboratory. *Chemical Technology Division, Annual Progress Report.* , Oak Ridge, TN, March 31 1975.
14. Fullam, H. T. and VanTuyl, H. H. Promethium technology: a review. *Isotopes Radiation Tech*, 7(2), **1969**.
15. Wheelwright et al., E. J. *Flowsheet for Recovery of Curium*. Technical Report BNWL-1831, Battelle Northwest Laboratory, Richland, WA, 1974.
16. Wheelwright, E. J. *Ion Exchange—A Generic Nuclear Industry Process for the Recovery and Final Purification of Am, Cm, Pm, Sr, Pu, Np, Cs, Tc, P, Rh, and Pd*. Technical Report BNWL-SA-1945, Battelle Northwest Laboratory, Richland, WA, 1968.
17. Schulz, W. W. and Benedict, G. E. *Neptunium-237 Production and Recovery*. AEC Critical Review Series, National Technical Information Service, Springfield, VA, 1972.
18. Brandt, A. S. Canadian perspectives. In *Proc Thirty-First Ontario Waste Conf, Ontario, Canada*, pages 3–20, Ontario Ministry of the Environment, Toronto, Ontario, 1984.
19. Thompson, F. M. and McComas, C. A. Technical assistance for hazardous waste reduction. *Environ Sci Technol*, 21:1154–1158, **1987**.
20. Bishop, J. Waste reduction. *HazMat World*, 56–61, Oct **1988**.
21. Hunt, G. E. and Schecter, R. N. Minimization of hazardous waste generation. In Freeman, H. M., Ed., *Standard Handbook of Hazardous Waste Treatment and Disposal*, chapter 5.1, pages 5.3–5.27, McGraw-Hill, New York, 1989.

22. Saltzberg, E. R. and Cushnie, G. C., Eds. *Centralized Waste Treatment of Industrial Wastewater.* Noyes Publishers, Park Ridge, NJ, 1985.

23. Tedder, D. W. and Pohland, F. G., Eds. *Emerging Technologies in Hazardous Waste Management.* Volume 422 of *ACS Symposium Series*, American Chemical Society, Washington, DC, 1990.

24. Tedder, D. W. and Pohland, F. G., Eds. *Emerging Technologies in Hazardous Waste Management II.* Volume 468 of *ACS Symposium Series*, American Chemical Society, Washington, DC, 1991.

25. Peyton, G. R. Modeling advanced oxidation processes for water treatment. In Tedder, D. W. and Pohland, F. G., Eds., *Emerging Technologies in Hazardous Waste Management*, chapter 7, pages 100–118, Volume 422 of *ACS Symposium Series*, American Chemical Society, Washington, DC, 1990.

26. Pacheco, J. E. and Holmes, J. T. Falling-film and glass-tube solar photocatalytic reactors for treating contaminated water. In Tedder, D. W. and Pohland, F. G., Eds., *Emerging Technologies in Hazardous Waste Management*, chapter 3, pages 40–51, Volume 422 of *ACS Symposium Series*, American Chemical Society, Washington, DC, 1990.

27. Andrews, A. T., Ahlert, R. C., and Kooson, D. S. Supercritical fluid extraction of aromatic contaminants from a sandy loam soil. *Environ Progress*, 9(4):204–210, Nov **1990**.

28. Dooley, K. M., Ghonasgi, D., and Knopf, F. C. Supercritical CO_2– cosolvent extraction of contaminated soils and sediments. *Environ Progress*, 9(4):197–203, Nov **1990**.

29. Groves, F. R., Brady, B. O., and Knopf, F. C. State of the art on the supercritical extraction of organics from hazardous wastes. *CRC Crit Rev Environ Control*, 15:237, **1985**.

30. Modell, M. *Supercritical Fluid Technology Hazardous Waste Treatment.* Vienna, March 1987.

31. Hougen, O. A. and Watson, K. M. *Chemical Process Principles, Part Three, Kinetics and Catalysis.* John Wiley & Sons, Inc., New York, 1947.

32. Ruthven, D. M. *Principles of Adsorption and Adsorption Processes.* John Wiley & Sons, New York, 1984.

33. Fernando, T. and Aust, S. D. Biodegradation of munition waste, tnt (2,4,6-trinitrotoluene) and rdx (hexahydro-1,3,5-trinitro-1,3,5-triazine by *Phanerochaete chrysosporium.* In Tedder, D. W. and Pohland, F. G., Eds., *Emerging Technologies for Hazardous Waste Management II*, chapter 11, pages 214–232, *ACS Symposium Series No. 468*, American Chemical Society, Washington, D.C., 1991.

34. Testa, S. M. and Winegardner, D. L. *Restoration of Petroleum-Contaminated Aquifers.* Lewis Publishers, Inc, Chelsea, Michigan, 1991.

35. Flytzani-Stephanopoulos, M., Sarofim, A. F., Tognotti, L., Kopsinis, H., and Stoukides, M. Fundamental studies of incineration of contaminated

soils in an electrodynamic balance. In Tedder, D. W. and Pohland, F. G., Eds., *Emerging Technologies for Hazardous Waste Management II*, chapter 3, pages 29–49, ACS Symposium Series No. 468, American Chemical Society, Washington, D.C., 1991.

36. Tiernan, T. O., Wagel, D. J., VanNess, G. F., Garrett, J. H., Solch, J. G., and Rogers, C. Dechlorination of organic compounds contained in hazardous wastes: potassium hydroxide with polyethylene glycol reagent. In Tedder, D. W. and Pohland, F. G., Eds., *Emerging Technologies in Hazardous Waste Management*, chapter 14, pages 236–251, Volume 422 of *ACS Symposium Series*, American Chemical Society, Washington, DC, 1990.

37. Borazjani, H., Ferguson, B. J., McFarland, L. K., McGinnis, G. D., Pope, D. F., Strobel, D. A., and Wagner, J. L. Evaluation of wood-treating plant sites for land treatment of creosote- and petachlorophenol-contaminated soils. In Tedder, D. W. and Pohland, F. G., Eds., *Emerging Technologies in Hazardous Waste Management*, chapter 15, pages 252–266, Volume 422 of *ACS Symposium Series*, American Chemical Society, Washington, DC, 1990.

38. Counce, R. M., Thomas, C. O., Wilson, J. H., Singh, S. P., Ashworth, R. A., and Elliott, M. G. An economic model for air stripping of vocs from groundwater with emission controls. In Tedder, D. W. and Pohland, F. G., Eds., *Emerging Technologies for Hazardous Waste Management II*, chapter 10, pages 177–212, ACS Symposium Series No. 468, American Chemical Society, Washington, D.C., 1991.

39. Chawla, R. C., Porzucek, C., Cannon, J. N., and Johnson, Jr., J. H. Importance of soil-contaminant-surfactant interactions for in situ soil washing. In Tedder, D. W. and Pohland, F. G., Eds., *Emerging Technologies in Hazardous Waste Management II*, chapter 16, pages 316–341, Volume 468 of *ACS Symposium Series*, American Chemical Society, Washington, DC, 1991.

40. Schindler, P. and Gamsjäger, H. *Kolloid-Z. Z. Polymere*, 250:759–763, **1972**.

41. Schindler, P. In Anderson, M. and Rubin, A., Eds., *Adsorption of Inorganics and organic Ligands at Solid-Liquid Interfaces*, pages 1–49, Ann Arbor Science, Ann Arbor, MI, 1981.

42. Stumm, W., Kummert, R., and Sigg, L. *L. Croatica Chem Acta*, 53:291–312, **1980**.

43. Bourges, J., Madic, C., Koehly, G., and Lecomte, M. Plutonium dissolution in nitric acid by electrogenerated silver. *J Less Common Met*, 121:303–311, Jul-Aug **1986**.

44. Bray, L., Ryan, J., and Wheelwright, E. Electrochemical process for dissolving plutonium dioxide and leaching plutonium from scrap or wastes. pages 120–127, Volume 83 of *AIChE Symposium Series No. 254*, American Institute of Chemical Engineers, New York, 1987.

45. Dukas, S., Black, D., McClure, L., and Offutt, G. Electrolytic dissolution experience at the idaho chemical processing plant. pages 128–134, Volume 83 of *AIChE Symposium Series No. 254*, American Institute of Chemical Engineers, New York, 1987.

46. Chander, K., Marathe, S., and Jain, H. Applications of electrochemical techniques in nuclear fuel cycle. *Trans Soc Advan Electrochem Sci Tech*, 22(4):185–188, Oct–Dec **1987**.

47. Sakurai, S., Tachimori, S., Akatsu, J., Kimura, T., Yoshida, Z., Mutoh, H., Yamashita, T., and Ohuchi, K. Dissolution of plutonium dioxide by electrolytic oxidation method. I. Determination of dissolution conditions and preliminary test for scaling-up. *J Atomic Energy Soc Japan*, 31(11):1243–50, Nov **1989**.

48. MacNeil, J. C. Membrane separation technologies for treatment of hazardous wastes. *CRC Crit Rev Environ Control*, 18(12):91–131, **1988**.

49. Thornburg, G., McArdle, J. C., Piccari, J. A., and Byszewski, C. H. Recovery and recycle of valuable constituents in spent pickling acids. *Environ Progress*, 9(4):N10–N11, Nov **1990**.

RECEIVED September 3, 1992

PHYSICAL AND CHEMICAL WASTEWATER TREATMENT

Chapter 2

Comparative Aspects of Advanced Oxidation Processes

David F. Ollis

Chemical Engineering Department, North Carolina State University, Raleigh, NC 27695

Photo-oxidation processes for purification and decontamination treatment of water include UV/ozone, UV/peroxide, and near-UV/photocatalysis. This paper extendss Peyton's(1) preliminary study of the strengths and weaknesses of these three AOPs as well as the ozone/peroxide process.

Advanced Oxidation Processes (AOPs) have been defined broadly as those aqueous phase oxidation processes which are based primarily on the intermediacy of the hydroxyl radical in the mechanism(s) leading to destruction of the target pollutant or xenobiotic or contaminant compound. The AOPs of concern here are water treatment operations using ozone in combination with ultraviolet light (O_3/UV), ozone plus hydrogen peroxide (O_3/H_2O_2), hydrogen peroxide and ultraviolet light (UV/H_2O_2), and photocatalysis, which uses titanium dioxide (TiO_2) in combination with light (UV) and molecular oxygen (O_2).

Potential application of such AOP processes include decontamination and deodorization. In integration with other water treatment operations, which seems hardly to have been examined, these AOPs may serve either as a polishing operation, or as an initiation step, wherein a recalcitrant compound is partially degraded but the remaining destruction of partial oxidation by-products is left to a less expensive technology such as biological treatment.

A comparative assessment of various AOP technologies is only beginning; a recent brief example is that of Peyton(1). The present

0097–6156/93/0518–0018$06.00/0

paper summarizes Peyton's and updates it , in particular with regard to aspects of the newest AOP, photocatalysis. With such early stage overviews, as many interesting questions arise as are solutions found. A primary purpose of such a review is then as much to guide future research and to help establish the vocabulary and issues of the conversation as to reach any definitive conclusions.

AOP PROCESS CHEMISTRIES

All AOP chemistries have appreciable similarities due to the participation of hydroxyl radicals in some, most, or all of the main mechanisms which are operative during reaction.   It is thus, as Peyton implies, the differences in efficiencies of central steps and the susceptibilities to penalties due to possible side reactions, that the AOP comparisons may be usefully organized (1).   We outline these process chemistries briefly, with the intent to indicate oxidant and photon usage requirements, and to identify possible points of physical or chemical interferences in operation.

Ozone oxidation of dissolved organics may occur (Figure 1)throught direct oxidation of the organic, or indirectly, through conversion of ozone into ozonide ion (O_3^-) and subsequent hydroxyl radicals (OH·) which in turn may attack the organics directly (Staehlin and Hoigne(2)). As hydroxyl radicals are strong oxidants vs the milder ozone, we may expect that simple ozonation dominated by the direct ozone-organic reactions may produce reactant degradation but not complete mineralization to carbon dioxide, whereas use of circumstances which enhance conversion of ozone along hydroxyl radical routes will also lead to mineralization. An example of such altered behavior is phenol destruction (Figure 2) in aqueous phase as demonstrated by Gurol and Vatistos (3), wherein both ozone and ozone/UV provide simlar rates of phenol disapparance, but the latter also leads simultaneously to mineralization (complete disappearance of total organic carbon) due to the stronger oxidizing power of the photo-produced hydroxyl radicals.   With ozonation only (no UV), the initial reaction products are both less reactive than the reactant, leading to slower conversion of remaining organic carbon (Fig. 2b), and more absorbant of light, leading to slower phenol disappearance (Fig. 2a). The benefit of the UV-ozone combination is thus continued attack of partial oxidation products.

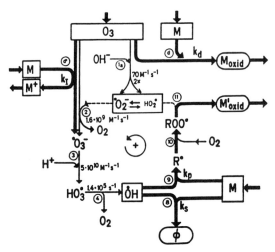

Figure 1.Pathways for ozone reactions with organics (M) (Reproduced with permission from ref. 2. Copyright 1985.)

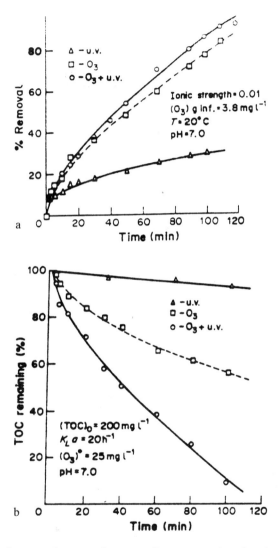

Figure 2(a) Comparison of overall removal of phenols by UV radiation, ozone, and ozone + UV radiation at pH 7.0; (b) comparison of total organic carbon removal by UV radiation, ozone, and ozone/UV at pH 7.0   (Reproduced with permission from ref. 2. Copyright 1985.)

Ozone-UV provides an appreciable enhancement of not only the rate of complete destruction achievable, and but also of the range of compounds attacked, due to the greater oxidation power of hydroxyl radical vs. ozone per se. The mechanism of ozone-UV is argued to involve a "wheel" of reactive intermediates (Figure 3)(4,5). The original oxidant, ozone, is seen to enter the wheel either directly as an ozonide ion, or indirectly via hydroxyl radical, produced through the photodecomposition of ozone to give hydrogen peroxide and the subsequent photolysis of the peroxide.

A central kinetic feature of this wheel is that it can also propagate without addition of peroxide or light, if initiators of ozone decomposition are naturally present, and if the particular pool of organics available includes a good concentration of "promoters" which yield sufficient $^{\bullet}O_2RH$ to regenerate hydrogen peroxide to continue the ozone decomposition. Absent such initiators and promoters, or in the later stages of an oxidative degradation when these rate enhancers will have been consumed, the addition of hydrogen peroxide and/or light is required to continue ozone utilization.

The UV/peroxide mechanism may be drawn to involve a similar circular oxidant wheel (Figure 4)(6). The absence of ozone is attractive of course, because of the elimination of both capital and operating costs associated with ozone generation. Two disadvantages arise vs. the UV/O3/ (peroxide) wheel of the previous Figure. First, the photon consumption is now fixed stoichiometrically, since every passage arond the wheel requires (at least) one photon. A second is that the UV absorbance of hydrogen peroxide (about 20 $mol^{-1}\cdot cm^{-1}$ at 254 nm) is only (20/3000) or about .7% that of dissolved ozone. Thus, a clear oxidation initiation rate penalty is implied.

Ozone/peroxide (no photons) is the only AOP not involving use of light. As the initial product of ozone photolysis is hydrogen peroxide which in turn can directly aid ozone decomposition, also (7), the direct chemical addition of peroxide can initiate ozone conversion into ozonide ion (see wheel), Fig. 3), and the light requirement disappears. The process economics for this process appear quite attractive at low contaminant levels, as illustrated by Aieta et al (14) for TCE. The potential disadvantages of a light-free process include (1): (i) any direct photolysis or photoactivation of organic is lost (e.g., lose R-> R* + O$_2$ --> RO$_2\cdot$ --> products, but this is normally only a small rate contribution); (ii) peroxide photolysis

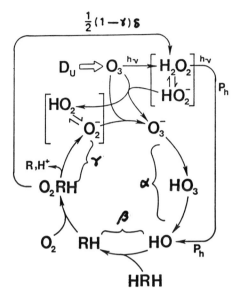

Figure 3. Important reaction pathways of the ozone/peroxide/UV advanced oxidation process.(Reproduced with permission from ref. 5. Copyright 1988.)

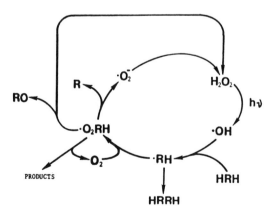

Figure 4. Reaction pathways for the UV-hydrogen peroxide advanced oxidation process.

(Reproduced with permission from ref. 13. Copyright 1990.)

is also lost; though inefficient per se, it can reinitiate the "wheel", and (iii) light allows a compensatory rate enhancement since, as the wheel slows down, dissolved ozone begin to accumulate, and the hydrogen peroxide formation (from ozone) and photolysis increase, thereby reinitiating the "wheel".

Heterogeneous photocatalysis, or its more cumbersome equivalent of semiconductor mediated photo-oxidation, is an example of indirect photochemistry wherein a light-requiring catalyst is continuously photoactivated and therby serves as a source of active centers, again including hydroxyl radicals. The possible mechanisms consistent with the observed saturations kinetics have been discussed elsewhere (8). Here, as for UV/peroxide, the photon activation step indicates a stoichiometric requirement for photons.

The reverse reactions of photoactivation provide appreciable penalties for both UV/peroxide and photocatalysis. The former is penalized by a cage effect, whereing the photoproduced hydroxyl radicals may immediately recombine to give back H_2O_2; this is typically a fifty percent penalty to give about one radical produced per photon absorbed, rather than the two arising from an irreversible activation. In photocatalysis, the bulk or surface of the catalyst allows for the direct recombination of the photoproduced hole (h^+) and electron (e^-), again wasting a photon. This recombination may be catalyzed by impurities, as is known to occur in light activated semiconductor devices at the semiconductor junctions (9). It is also a strong function of the photoactivation rate: at low light levels, the desired photocatalyzed rate may occur with quantum yields approaching unity, whereas the second order recombination rate routinely means that the overall quantum yield for reaction varies as $(intensity)^{-0.5}$ at higher ilumination levels. (10a, b, c).

COMPARATIVE COMMENTS

Peyton's brief comparative essay provides a useful beginning point for the conversation. He summarized these processes in simple tabular form on the "basis of effectivness, cost, ease of implementation, and susceptability to interference by various water components" (1). We summarize several of his forms here, adding materially to the photocatalysis data, to provide modified tables for each circumstance.

(1) pH Sensitivity. The ozone decomposition via hydrogen peroxide is strongly pH dependent, due to the equilibrium $HO_2^- + H^+$ H_2O_2 (Figure 1). This sensitivity will influence both the rates of UV/O_3 (H_2O_2) and the O_3/H_2O_2 AOP processes. The pH sensitivity of photocatalysis is modest, as exemplified by the photocatalyzed reaction rate data for (Figure 5) phenol destruction (Augugliaro et al (1988)) (11a). For easily dissociated molecules, however, the pH dependence can be strong, as exemplified by the variation in rates of photocatalyzed degradation of trichloroacetate and chloroethylammonium ions (Figure 6) (11b). This influence was argued by the authors as due to a combination of ionization equilibria involving the titanium dioxide surface, i.e.,

$$(\text{surf}) - TiO\,H_2^+ \longleftrightarrow (\text{surf}) - TiOH + H^+$$
$$K_{a1}$$

$$(\text{surf}) - TiOH \longleftrightarrow (\text{surf}) - TiO^- + H^+$$
$$K_{a2}$$

where $pK_{a1}^s = 2.4$ and $pK_{a2}^s = 8.0$. Calculated values for surface concentrations of $TiO_2\ CCl_3$ and $TiN^+\ H_2(CH_2)_2\ Cl$ paralleled the observed rate data of Figure 6. The modified Peyton pH table is thus that of Table 1.

(2) Bicarbonate inhibition. Bicarbonate is expected to diminish the potency of the •OH radical by the reaction leading to formation of the less reactive carbonate radical anion ($\cdot CO_3^-$) ($\cdot OH + CO_3^= \longrightarrow OH^- + \cdot CO_3^-$). A similar negative penalty in photocatalysis has been recently noted: approximately a 70% rate reduction was measured when bicarbonate increased from near 0 ppm to 500 ppm (12); this is noted in Table 1, last column. (The bicarbonate question is appreciably more complex: thus the initial reagent such as phenol may be easily attacked by both hydroxyl and carbonate radical anion, whereas later partial oxidation products such as methanol or

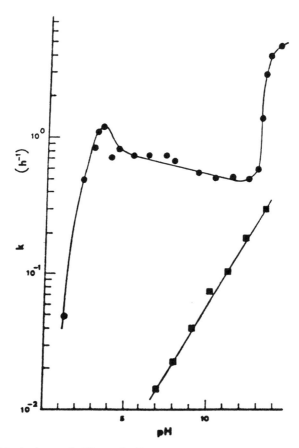

Figure 5.   Variation of Phenol disappearance rate constant with pH (squares: presence of photocatalyst TiO$_2$; circles; absence of photocatalyst). (Reproduced with permission from ref. 11a. Copyright 1988.)

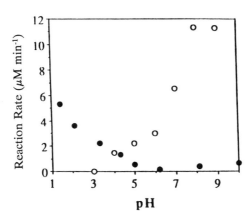

Figure 6. pH dependence of degradation rate of trichloroacetate (•) and chloroethylammonium (O) ions where $[CCl_3 CO_2^-]_0$ = 10 mM = $[C(CH_2)_2 NH_2]_0$; $[TiO_2]$ = 0.5 g/L, $[O_2]_0$ = 0.25 mM and I_a = 2.5 $\cdot 10^{-4}$ einstein-L^{-1}-min^{-1} (310-380 nm).
(Reproduced with permission from ref. 11b. Copyright 1991.)

TABLE 1

pH SENSITIVITY OF ADVANCED OXIDATION PROCESSES (pH 4-10)

Reaction

Process	(Initiation)	(Promotion[1])	Bicarbonate (Sparging)/ (Scavenging)
O_3/UV	no/yes	somewhat[2]	yes/yes
O_3/H_2O_2	yes	somewhat[2]	yes/yes
H_2O_2/UV	no	N/A	?/yes
Photocatalysis	no/yes	N/A	yes

(1) Assuming promoters present.
(2) Intermediate reactions

(Adapted from ref. 1.)

formaldehyde may only be attacked by hydroxyl; calculations supporting this notion of Peyton's are shown in Figure 7) (13).

Combining items (1) and (2) gives Table 1.

(3) High organics concentrations. Organics per se are not a problem for the kinetics of any oxidation process, although the stoichiometric requirements for oxidant(s) and light (where applicable) will clearly increase with organic content.

(4) UV absorbance/scattering The light-based processes will all be severely penalized by light absorption by other than ozone, peroxide or photocatalyst unless the resultant photoactivated organic is reactive, i.e., R-> R* -> •RO$_2$ so that a contribution to the wheel (Fig. 1) is made. This comment may apply to photocatalysis as well, since dissolved molecular oxygen is the ultimate oxidant, and many reactions of R* with O$_2$ are known.

Combining items (3) and (4) leads to Table 2

(5) Cost of oxidant The generation of ozone represents often about half of both the capital and operating costs of an ozone utilizing AOP process. (e.g., Aieta et al (14), and Hackman (15)). Ozone cost is high, peroxide is intermediate, and oxygen is low.

(6) Cost of UV depends on the quantum yield for the desired reaction. If UV (254 nm) photons can be generated from electricity ($.05/kw-hr) at a 30% efficiency (lamp), then photons are produced for power costs of $.02 per mole equivalent (an Einstein). The losses due to non-productive events involving photons are the following:
    *ozone/UV: good oxidant molar absorbance (about 3000 per mol-cm); possible light scattering due to bubbles (of air/oxygen/ozone)
    *peroxide/UV: weak oxidant absorbance (about 20 mol.$^{-1}$cm^{-1}) (excess transmissivity)
    *photocatalysis: strong catalyst absorbance; possible losses due to back-scattering

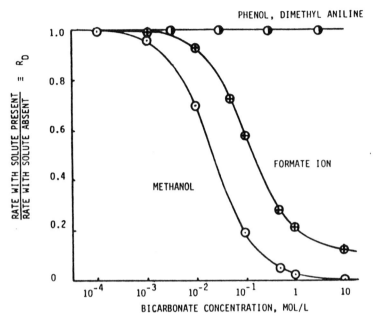

Figure 7. Bicarbonate scavenging . Ratio of rates with/without added bicarbonate; rate penalty depends on competition between reactant and scavenger. (Reproduced with permission from ref. 13. Copyright 1990.)

TABLE 2

COMPARISON OF AOPs FOR DIFFERENT SITUATIONS

Process	High Organics	High[1] Alkalinity	High UV Absorbance/ Scattering	No Promoters Present
O_3/UV	+	++	-	-
O_3/H_2O_2	++	+	++	+
H_2O_2/UV	+	-	-	++
Photocatalysis	+	+ / -	-	++

(1) Consider sparging.

(Adapted from ref. 1.)

The quantum efficiency of these processes often is intensity dependent.   For example,Peyton, Huang, Burleson and Glaze (15) report that the rate constants for trichloroethylene photolysis and photolytic ozonation varied respectively as the 0.67 and 0.42 powers of intensity when lake water was used, leading to quantum efficiencies which vary as (k/I), or $I^{-.33}$ and $I^{-.58}$ respectively. Similarly, photocatalytic quantum efficiencies vary as $I^{-.5}$ at appreciable to high intensities. (10a, b, c).

The cost of solar photons is not included here; it varies from very high if collector concentrators are used (due both to equipment costs and to exaggerated quantum inefficiencies) (17) to low if use of an on-site, already depreciated waste treatment pond is envisaged.

(7) _Ease of implementation_ is unknown for photocatalysis, since no large scale, long term operating experience exists.

These comments lead to Table 3.

Reactor   design is not well developed for any of these technologies.   Moreover, comparison between these potentially competing technologies is less clear because of the different reactor design challenges in each circumstance:

(1) _ozone/peroxide   and ozone/UV_ require both destruction of any ozone in the gas phase exhaust. Such destruction is handled by a gas phase   added   catalyst,   which,   adventitiously,   appears   to simultaneously oxidize some or most of any volatile, stripped contaminant (This is an amusing accidental example of an integrated treatment scheme, where two different destruction chemistries (one liquid, one gas phase) are used).

(2) _UV/ozone , UV/peroxide, and UV/photocatalysis_ all require lamps which are inexpensive and long lived.   As the same lamps can be used to drive all three processes, their operating problems are unimportant among the three processes, but important relative to the ozone/peroxide process. A particular problem of importance is the ease of scale-up as the light-based technologies are considered for larger applications. UV/peroxide is easy for design because the low optical density (weak absorbance) allows easy distribution of

TABLE 3

COMPARISON OF AOPs FOR DIFFERENT SITUATIONS

Process	Cost of Oxidant	Cost of UV	Ease of Implementation (maintenance)	Sensitive to Geographic Location
O_3/UV	high	med.	+(-)	no
O_3/H_2O_2	high	0	+	no
H_2O_2/UV	med.	high	++(-)	no
Photocatalysis	very low	med-high	+(-)	no

(Adapted from ref. 1)

light.   UV/ozonation and UV/ozone-peroxide provide higher optical density solutions and thus better photon utilization; however, the higher optical density means poorer distribution of light in the reactor, necessitating either better liquid mixing or use of more lamps to provide a more uniform volumetric photon dose to each element of treated fluid.

Photocatalysis introduces an additional, solid phase.   If it is present as a slurry, then provisions for catalyst recovery and recyle are needed, or a fluidized bed might be considered, or an immobilized bed configuration could be tried via immobilization on meshes, beads, walls, etc.

The design and modelling of heterogeneous photoreactor systems represents a notable lack in the literature; a fine counter example is the studies of Santarelli et al (1978) (18).   A good starting point for the interested researcher is the review of radiation fields in heterogeneous systems by Alfano, Cassano and co-workers(19).

The general literature topic of photoreactor design has been organized very well by Cassano's group; two particularly complete examples are (19,20).

(3) Mass transfer limitations have been shown to arise under several circumstances:

(a) In UV/ozone, increasing UV consumption rates per unit volume can lead to rate enhancement sufficient to encounter the mass transfer limited regime for ozone dissolution (Prengle et al (21)). Such mass transfer limits imply decreased quantum efficiencies and increased operating costs.

(b) In UV/photocatalysis, immobilization of the photocatalyst in order to avoid recovery and recycle of a slurry photocatalyst can lead to mass transfer limited reaction. For example, immobilization of TiO_2 inside 6 mm ID tubing leads to rate variations with liquid flowrate which are consistent with mass transfer limited behavior (22).

REFERENCES

(1)   Peyton, G. "A Comparison of Advanced Oxidation Processes", Paper presented at Symposium on Advanced Oxidation Process, June 4-5, 1990, Toronto, Canada.

(2)   Staehlin, J. and Hoigne, J., Environ. Science Technology, 19, 1206 (1985).

(3)   Gurol, M. and Vatistos, Water Research, 21, 895 (1987).

(4)   Peyton, G. R., Smith, M. A. and Peyton, B. M., Univ. Ill., Water Resources Center, Research Report 206, (1987).

(5)   Peyton, G. R. and Glaze, W. H., Environ. Sci. Technol., 22, 761 (1988).

(6)   Peyton, G. R. and Smith, M. A., unpublished work (1989) (referenced in (1)).

(7)   Peyton, G. R. and Glaze, W. H., Environ. Sci. Technol., 22, 761 (1988).

(8)   Turchi , C. and Ollis, D. F., J. Catalysis, 122, 178, (1990).

(9)   Morrison, S. R., Semiconductor Oxide Surfaces.

(10) (a) Egerton and King, I. Oil Paint Chem. Technol., (1978), (b) Okamoto, K., et al, Bull. Chem. Soc. Japan, 58, 2023 (1985), (c) Kormann, et al, Environ. Sci. Technol., (1990).

(11) (a) Augugliaro, V., et al, Toxicol. Environ. Chem., 16, 89 (1988), (b) Kormann, C., Bahnemann, D. W. and Hoffman, M. R., Environ. Sci. Technol., 25, 494 (1991).

(12) Blake, D., Webb, J., Turchi, C. and Magrini, K., Solar Energy Materials (in press), (1991)

(13) Peyton, G. R. "Oxidative Treatment Methods for Removal of Organic Compounds from Drinking Water Supplies", in Significance and Treatment of Volatile Organic Componds in Water Supplies, Ram, N. M., Christman, R. F., and Cantor, K. P. (eds), Lewis Publishers, Chelsea, Michigan, 1990,p. 313-362.

(14) Aieta, E. M., et al, J. Am. Water Wks. Assoc., May, 1988, p. 64.

(15) Hackman, E. E., Toxic Organic Chemicals: Destruction and Water Treatment, Noyes Data Corp., Park Ridge, NJ (1978).

(16) Peyton, G. R., Huang, F. Y., Burleson,J. L., and Glaze, W. H., Environ. Sci. Technol. , 16,, 448 (1982).

(17) Ollis, D. F., "Solar-Assisted Photocatalysis" in Photochemical Conversion and Storage of Energy, M. Schiavello (ed), Kluwer Publ., 1990, pp. 593-622.

(18) Stramigioli, C. ,F. Santarelli, and F. P. Foroboschi, Appl. Sci. Res. 33, 23 (1977).

(19) Alfano, R. M., Romero, R. L., and Cassano, A. E. , Chem. Eng. Sci. 41, 1137 (1986).

(20) Cassano, A. E. and Alfano, O. M., "Photoreactor Design", in Handbook of Heat and Mass Transfer, vol 3. Catalysis, Kinetics and Reactor Engineering, N. Cheremisinoff(ed), Gulf Publishing, Houston, 1989, pp. 593-669.

(21) Garrison, R. L., Mauk, C. F. and H. W. Prengle, First Int'l. Symp., Int'l. Ozone Inst., Washington, DC 1973. See also Prengle, H. W. et al, Hydrocarbon Processing, October 1975.

(22) Turchi, C. and Ollis, D. J. Phys Chem, 92, 6852, (1988).

(23) De Bernadez, E. R., Claria, M. A., and Casano, A. E., "Analysis and Design of Photoreactors", in Chemical Reaction and Reactor Engineering, Carberry, J. J. and Varma, A. (eds), Marcel Dekker, Inc. New York, 1987, pp. 839-921.

RECEIVED September 3, 1992

Chapter 3

Supercritical Water Oxidation Technology
Process Development and Fundamental Research

Jefferson W. Tester[1], H. Richard Holgate[1], Fred J. Armellini[1],
Paul A. Webley[1], William R. Killilea[2], Glenn T. Hong[2],
and Herbert E. Barner[3]

[1]Chemical Engineering Department and Energy Laboratory, Massachusetts Institute of Technology, 77 Massachusetts Avenue, Room E40–455, Cambridge, MA 02139
[2]MODAR, Inc., Natick, MA 01760
[3]ABB Lummus Crest, Inc., Bloomfield, NJ 07003

Hazardous organic wastes, including chlorinated hydrocarbons, in aqueous media containing salts can be effectively oxidized by treatment above the critical point of pure water (374°C, 221 bar). High destruction efficiencies may be achieved at low reactor residence times (approximately 1 minute or less) for temperatures above 550°C. Under these conditions, no NO_x compounds are produced. The high solubility of organics and oxygen and the low solubility of salts in supercritical water make it an attractive medium for both oxidation and salt separation. This paper reviews critical technology components and operations required for commercial-scale development of supercritical water oxidation (SCWO). These include fluid handling and compression, heat exchange and recuperative heat recovery, reactor and salt separator design and materials considerations. In addition, a summary of fundamental research activities in the areas of oxidation reaction kinetics and mechanisms and solid salt nucleation and deposition is presented in the context of its impact on SCWO process development.

Oxidation carried out in a supercritical water environment at temperatures above 374°C and pressures above 221 bar (22.1 MPa) provides a viable method for the efficient destruction of organic wastes in a fully contained system. This patented process has been termed supercritical water oxidation, SCWO, and has been under commercial development for about the past 11 years by MODAR, Inc. (1-7). The process has also proven suitable for treatment of human metabolic wastes, and is being considered for use in life support systems on long-term spaceflights (8-11). A relatively recent variation of SCWO involves carrying out the process in a deep well utilizing the hydrostatic head of fluid in the wellbore to help provide the necessary pressure level (see, e.g., Gloyna (12)).

The main objectives of this paper are twofold: (1) to review the status of aboveground SCWO technology from a process engineering perspective, and (2) to review fundamentally focused research on reaction kinetics, phase equilibria and solid salt separation critical to understanding the chemistry and phase behavior of reactions carried out in the SCWO process. The effectiveness of SCWO for treating a wide range of waste stream compositions will be reviewed. The status of development of SCWO as a commercial-scale process along with a discussion of remaining technical issues and critical sub-process elements will also be presented.

0097–6156/93/0518–0035$11.50/0
© 1993 American Chemical Society

Economic predictions for SCWO have appeared in the literature (for example, see Thomason and Modell (6); Stone and Webster Engineering Corp. (13); and Modell (3,14)). Because of the evolving nature of the technology and the inherent uncertainties associated with orders-of-magnitude scale-up from pilot-sized units to commercial-sized systems, these studies should be viewed with reservation. For these same reasons, we have decided not to include economic projections in this paper, but rather to focus on engineering technology and basic supporting research.

Earlier reviews of SCWO technology by Freeman (15), Modell (3) and Thomason et al. (5), while providing important basic process engineering material, need to be updated with respect to current process development and fundamental research on SCWO. A more recent paper by Shaw et al. (16) introduces several aspects of research related to reaction chemistry in supercritical water. It is our intent in this review to provide readers with a comprehensive listing of published papers, patents, and reports on various aspects of the technology.

The supercritical water oxidation (SCWO) process brings together water, organics, and oxygen at moderate temperatures (400°C and above) and high pressures (about 25 MPa). Under these conditions, a single fluid phase reaction system exists where many of the inherent transport limitations of multi-phase contacting are absent. The temperatures are high enough to induce spontaneous oxidation of the organics, and the heat of reaction raises the mixture temperature to levels as high as 650°C. Above about 550°C, organics are oxidized rapidly and completely to conversions greater than 99.99% for reactor residence times of 1 minute or less. Heteroatomic groups, such as chlorine, sulfur and phosphorus, are oxidized to acids which can be neutralized and precipitated as salts by adding a base to the feed. For aqueous solutions containing between 1 and 20 wt% organics, supercritical water oxidation could be more economical than controlled incineration or activated carbon treatment and more efficient than wet oxidation (6). With appropriate temperatures, pressures, and residence times, organics are oxidized completely to carbon dioxide, water, and molecular nitrogen, without formation of NO_x compounds or other toxic products of incomplete combustion. Modell and co-workers (17) also destroyed problematic polychlorinated biphenyls (PCBs) and DDT at greater than 99.99% efficiency without formation of dioxins. Cunningham et al. (18) and Johnston et al. (19) oxidized synthetic pharmaceutical and biopharmaceutical wastes and achieved >99.99% destruction of total organic carbon and >99.9999% destruction of thermophilic bacteria.

Oxidation in supercritical water is not limited to aqueous organics. Biomass, sewage, and soil can be slurried and fed to the reactor as a two-phase mixture. Wastes with large particles or high solids content may need to be homogenized prior to treatment. In principle, any mixture of organic waste that can be pumped to high pressure is treatable by SCWO. Precipitated salts and other solids can be removed from the product stream, providing a clean, high-temperature, high-pressure energy source consisting almost entirely of CO_2, H_2O, and N_2. Heat can be recovered from the treated process stream and used to preheat the feed, or the stream can be used to generate steam for direct process use or electric power generation in a high-efficiency Rankine cycle. The heat of the oxidation reaction is thus partially recoverable and can be used to offset the cost of treatment.

One of the main attributes of SCWO is its ability to rapidly and completely oxidize a wide range of organic compounds in a reaction system that meets the concept of a "Totally Enclosed Treatment Facility". Dilute aqueous wastes ranging from 1 to 20% organic content are particularly well-suited to SCWO processing. The scale of application is also adaptable, from portable benchtop or trailer-mounted units treating 1 to up to several thousands gallons of toxic waste per day, to large stationary plants capable of processing 10,000 to 100,000 gallons per day of waste with 8 to 10% organic content. In the present designs, additional air pollution abatement equipment is not required. Nonetheless, solids removal and storage are necessary, and water effluent polishing units incorporating ion exchange may be needed to remove small concentrations of dissolved metal ions. Major drawbacks of SCWO are the high pressures required (>230 bar) and the fact that corrosion at certain points in the process is significant. For some wastes, solids handling can also pose difficulties.

Properties of Supercritical Water

In the region near the critical point, the density of water (shown in Figure 1) changes rapidly with both temperature and pressure, and is intermediate between that of liquid water (1 g/cm^3) and low-pressure water vapor (<0.001 g/cm^3). At typical SCWO conditions, the water density is approximately 0.1 g/cm^3. Consequently, the properties of supercritical water are quite different from those of liquid water at ambient conditions, particularly those properties related to solvation.

Insight into the solvation characteristics and molecular structure of the aqueous system can be obtained by experimental measurements of dielectric constant, ionic dissociation constant, and Raman spectral emissions in passing from the subcritical to supercritical regions. Figure 2 shows that the static dielectric constant of water at 25 MPa drops from a room-temperature value of around 80 to about 5 to 10 in the near-critical region, and finally to around 2 at 450°C and above (*21-23*). Along the same isobar, as shown in Figure 2, the ionic dissociation constant falls from 10^{-14} at room temperature to 10^{-18} in the near-critical regime and to 10^{-23} under supercritical conditions (*24*). Furthermore, Raman spectra of deuterated water in the supercritical region show only a small residual amount of hydrogen bonding (*25,26*). As a result, supercritical water acts as a non-polar dense gas, and its solvation properties resemble those of a low-polarity organic. Hydrocarbons exhibit generally high solubility in supercritical water.

Near the critical point, the solubility of an organic compound in water correlates strongly with density and is thus very pressure dependent in this region. Benzene solubility in water is a good example (*27,28*). At 25°C, benzene is sparingly soluble in water (0.07 wt%). At 260°C, the solubility is about 7 to 8 wt% and fairly independent of pressure. At 287°C, the solubility is somewhat pressure dependent, with a maximum of 18 wt% at 20 to 25 MPa. In this pressure range, the solubility rises to 35 wt% at 295°C, and at 300°C, the critical point of the benzene-water mixture is surpassed. When the mixture becomes supercritical, by definition, there is only a single phase. Thus, the components are miscible in all proportions.

Other hydrocarbons exhibit similar solubility behavior near 25 MPa: Binary mixtures of n-alkanes (C_2-C_7) and water become supercritical (and, therefore, completely miscible) at temperatures below about 370°C (*27,29,30*). Supercritical water also shows complete miscibility with "permanent" gases such as nitrogen (*31*), oxygen and air (*32*), hydrogen (*33*), and carbon dioxide (*34,35*), in addition to small organics like methane (*36*). The loci of critical points of the binary mixtures containing water have also been reported by Franck and co-workers for the aforementioned organic substances and gases. Alwani and Schneider (*37*) have reported the phase behavior of aromatic hydrocarbons (including 1,3,5-trimethyl-benzene and naphthalene) with supercritical water. All of the compounds studied to date are completely miscible with water above 400°C at 25 MPa. Franck and his research group have also extended earlier phase behavior information for binary systems to include phase equilibrium data for ternary systems containing salts such as NaCl, CaCl$_2$, and NaBr (*36,38,39*).

In contrast to the high solubility of organics, the solubility of inorganic salts in supercritical water, as shown in Figure 3, is very low. Many salts that have high solubilities in liquid water have extremely low solubilities in supercritical water. For example, NaCl solubility is about 37 wt% at 300°C and about 120 ppm at 550°C and 25 MPa (*44*); CaCl$_2$ has a maximum solubility of 70 wt% at subcritical temperatures,

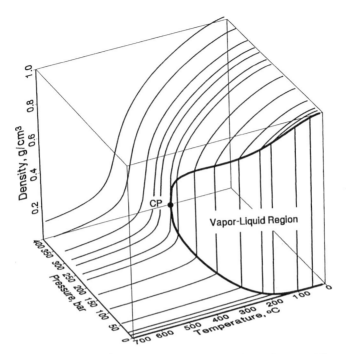

Figure 1.   Pressure-Temperature-Density Behavior of Pure Water.   CP denotes the critical point; the vapor-liquid region indicates conditions under which two phases are present.   Density calculated from the equation of state of Haar *et al. (20)*.

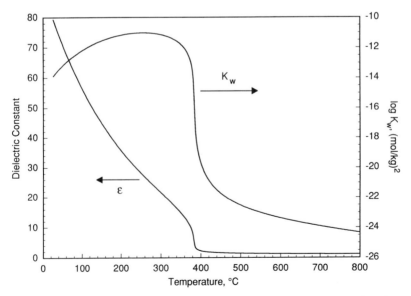

Figure 2. Solvation Properties of Pure Water at 25 MPa.   Properties shown are the dielectric constant ε *(21)* and the ionic dissociation constant K_w *(24)*.

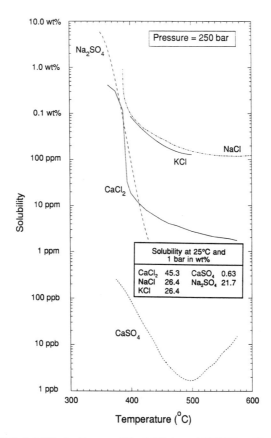

Figure 3. Salt Solubility in Supercritical Water at 25 MPa. Data interpolated from: KCl—Jasmund (*40*); Na$_2$SO$_4$—Martynova (*41*), Ravich and Borovaya (*42*); CaCl$_2$ and CaSO$_4$—Martynova (*41*); NaCl—Bischoff and Pitzer (*43*), Pitzer and Pabalan (*44*); room temperature solubilities—Linke (*45*). At temperatures below 450°C, a dense NaCl brine appears (KCl and CaCl$_2$ may also form dense brines at some temperatures).

which drops to 3 ppm at 500°C and 25 MPa (*41*). The fact that inorganics are practically insoluble is consistent with a low dielectric constant for water of about 2 and an extremely low ionic dissociation constant K_w of less than 10^{-22} at 500°C and 25 MPa. At the same time, supercritical water possesses high diffusivity (*46,47*) and low viscosity (*46,48*).

The combination of these solvation and physical properties makes supercritical water an ideal medium for oxidation of organics. When organic compounds and oxygen are dissolved in water above the critical point, they are immediately brought into intimate molecular contact in a single homogeneous phase at high temperature. With no interphase transport limitations and for sufficiently high temperatures, kinetics are fast and the oxidation reaction proceeds rapidly to completion. The products of hydrocarbon oxidation are CO_2 and H_2O. Heteroatoms are converted to inorganic compounds, usually acids, salts, or oxides in high oxidation states, which can be precipitated from the mixture along with other unwanted inorganics that may be present in the feed. Phosphorus is converted to phosphate and sulfur to sulfate; nitrogen-containing compounds are oxidized to N_2 with some N_2O. Because of the relatively low reactor temperature, neither NO_x nor SO_2 is formed (*49*).

Process Flow Sheet Description

There are basically two concepts proposed for SCWO: (1) an above-ground system (MODAR process), and (2) the below-ground system (sometimes referred to as the Oxidyne process). Both process concepts react mixtures of oxidant (oxygen, air, or hydrogen peroxide) with organic wastes with or without dissolved salts. The inherent difference between the methods is the technique for achieving high pressure. The surface system requires high pressure pumps or compressors, while the underground system utilizes the natural hydraulic head of the fluid contained in a deep well to provide a substantial part of the pressurization (*50,51*).

The critical steps of the above-ground MODAR process are depicted in the flowsheet given in Figure 4. While certain components and processing steps may vary somewhat depending on waste composition and treatment objectives, the general characteristics of the SCWO process can be subdivided into seven major steps:

(1)   Feed preparation and pressurization
(2)   Preheating
(3)   Reaction
(4)   Salt formation and separation
(5)   Quenching, cooling and energy/heat recovery
(6)   Pressure letdown and phase disengagement
(7)   Effluent water polishing

Each of these is described in somewhat more detail below.

(1)   **Feed Preparation and Pressurization:** Organic waste materials in an aqueous medium are pumped from atmospheric pressure to the pressure in the reaction vessel. Air can be compressed to system pressure and metered into the reaction vessel. Alternatively, oxygen, stored as a liquid, can be pumped to the pressure of the reaction vessel and then vaporized. In some cases, it may be advantageous to use oxidants such as hydrogen peroxide; however, the commercial utility of this oxidant is limited due to its higher cost (approximately 30 to 40 times higher than oxygen).

Feed to the process is controlled to an appropriate upper limit of heating value of 4200 kJ/kg (1800 Btu/lb) by adding dilution water or blending higher-heating-value waste material with lower-heating-value waste material prior to feeding it to the reactor. When the aqueous waste has too low a heating value,

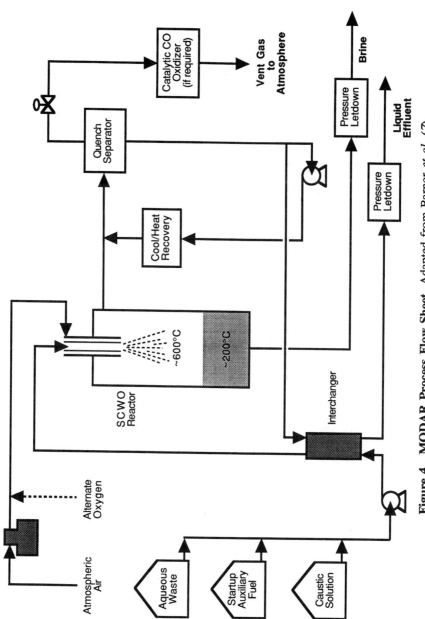

Figure 4. MODAR Process Flow Sheet. Adapted from Barner *et al.* (7).

fuel such as natural gas or fuel oil may be added. Optionally, a combination of preheat by exchange with reactor effluent and fuel addition or preheat alone may be used. One should keep in mind that other flow sheet designs are possible where higher heating values may be desirable, such as for energy recovery.

When organic wastes contain heteroatoms such as Cl, F, P, or S which produce mineral acids upon oxidation, and it is desired to neutralize these acids and form appropriate salts, caustic is injected as part of the feed stream.

(2) **Preheating:** The feed stream can be preheated by cross-exchange with the hot discharge effluent. This promotes rapid initiation of the oxidation reaction and helps to optimize the overall plant energy balance by better heat integration. Other preheating options are possible (e.g., see Thomason *et al.* (*5*)).

(3) **Reaction:** Mixing of the oxidant and organic waste streams with the hot reactor contents initiates the exothermic oxidation reaction which heats the reacting mixture to temperatures of 550°C to 650°C, accelerating reaction rates and reducing residence times for complete destruction. A second stage reactor may be included in other designs to meet destruction requirements for highly toxic materials or to ensure complete oxidation of CO and NH_3 (see Thomason *et al.* (*5*)).

(4) **Salt Formation and Separation:** Because salts have such a low solubility in supercritical water, their precipitation is rapid under almost shock-like conditions. The higher-density solid salts separate from the reacting phase and fall to the bottom of the reaction vessel where they can be redissolved and removed as a concentrated brine or collected as solids and removed periodically. A small fraction of salt is entrained with the hot reactor overhead effluent.

(5) **Quenching, Cooling and Energy/Heat Recovery:** The gaseous products of reaction, along with the supercritical water, leave the reactor at its top. In other design flow sheets a portion of the supercritical fluid may be recycled to the oxidizer by a high-temperature, higher-pressure pump. The reactor effluent (other than that recycled), consisting of supercritical water, carbon dioxide, a small amount of entrained salt and possibly nitrogen, is first mixed with cold recycle fluid to redissolve the salt and then is further cooled to be discharged at atmospheric conditions.

Excess thermal energy contained in the effluent can be used to generate steam for external consumption to produce electricity at high efficiency or for high-temperature industrial process heating needs. For larger-scale systems, energy recovery may potentially take the form of power generation by direct expansion of the reactor products through a supercritical steam turbine. Such a system would be capable of generating significant power in excess of that required for air compression or oxygen pumping and feed pumping. For very dilute aqueous wastes, it can be more economical to use a regenerative heat exchanger to preheat the waste than to add supplemental fuel.

(6) **Pressure Letdown and Phase Disengagement:** The cooled effluent from the process separates into a liquid water phase and a gaseous phase, the latter containing primarily carbon dioxide along with oxygen, which is in excess of the stoichiometric requirements (and nitrogen when air is the oxidant). The separation is carried out in multiple stages in order to minimize erosion of valves as well as to maximize the separation due to phase equilibrium constraints.

(7) **Effluent Water Polishing:** Because of the corrosive nature of supercritical brines and the fact that heavy metals are present in many waste streams, trace

metal concentrations (e.g., Cr, Ni, Zn, Hg) will appear in aqueous effluent streams from SCWO (*9,52*). Consequently, a polishing step involving ion exchange or selective adsorption may be needed. This would be particularly important in applications where recycled process or potable water is required.

For a discussion of the below-ground SCWO process, readers are referred to the following papers: Gloyna (*12*); Gulf Coast Waste Disposal Authority (*53*); Stanford and Gloyna (*54*); Smith and Raptis (*55*); Smith (*56*); and Smith *et al.* (*57*). One of the most significant conclusions of the Stanford and Gloyna (*54*) paper on the performance of the below-ground process is that the energy savings due to avoidance of high-pressure pumping is more than offset by other losses, including heat loss to the surrounding rock formation and energy requirements during startup. Such losses suggest that well depths should be minimized.

Application of SCWO to Waste Treatment

The SCWO process can be applied to a wide range of wastes containing oxidizable components. Details are provided by Thomason *et al.* (*5*) and Modell (*3*) and are only outlined here. Although initial development has been for aqueous wastes with 1-20 wt% organics, the process can successfully treat wastes in which organic concentrations are as low as 0% or as high as 100%. Concentrated organic wastes and dilute aqueous wastes are frequently available at the same site and can be blended to provide a feed with the appropriate heating value in the 1-20 wt% concentration range.

With proper high-pressure pumping systems, sludges and slurries become good candidates for treatment and can be oxidized with high destruction efficiencies. While many organic solids will also dissolve in supercritical water, the initial partial oxidation to fragmented, lower-molecular-weight compounds is normally very fast at typical SCWO processing temperatures, so the need for an intervening solvation step is preempted.

Another application involves remediation of contaminated soils, as found at many federal Superfund sites. Bench-scale testing has shown that hazardous and toxic organics can be extracted from the soil and simultaneously oxidized in a SCWO reactor (*5*). The process would continuously feed a slurry of the soil in water (contaminated groundwater or recycled process effluent water) to the SCWO reactor system. The effluents are clean water, CO_2 and N_2, and soil that is both sterilized and free of organics.

For very dilute wastes (i.e., below 1 wt% organics), activated carbon adsorption or biological oxidation is often an effective treatment method. The major cost of carbon treatment is the cost of regenerating the carbon. The carbon loading, in turn, is directly related to the concentration of organic contaminant. Hence, the cost of waste treatment is nearly proportional to the organic concentration and usually becomes prohibitive for wastes containing more than 1% organics. Biological treatment systems, on the other hand, become poisoned and often cannot be sustained for many complex waste mixtures or wastes with organic concentrations of 1% or more.

Incineration, on the other hand, is usually restricted for economic reasons to waste streams of relatively high organic concentrations. To achieve high destruction efficiencies for hazardous and toxic wastes, incineration is performed at temperatures as high as 900-1300°C (1700-2400°F) and often with excess air as high as 100%. With aqueous wastes, the energy required to vaporize and heat water to these temperatures is substantial. If the waste contains 25% organics or more, there is sufficient heating value in the waste to sustain the incineration process. With decreasing organic content, the supplemental fuel required to satisfy the energy balance becomes a major cost. Furthermore, incineration is also being regulated to restrict stack gas emissions to the atmosphere. Extensive equipment must now be used downstream of the reaction system to remove NO_x, acid gases, and particulates (fly ash) from the stack gases before discharge. The cost of this equipment often exceeds that of the incinerator itself.

In the range of concentration of 1 to 20 wt% organics, both wet air oxidation and supercritical water oxidation have certain practical and potential economic advantages over controlled incineration or activated carbon treatment (5). In wet air oxidation, carried out typically at temperatures ranging from 200 to 300°C, destruction of toxic organic chemicals (e.g., chlorophenols, nitrotoluenes) can be as high as 99.9% with adequate residence time but many materials are more resistant (e.g., chlorobenzenes, PCBs). Total COD (chemical oxygen demand) reduction is usually only 75-95% or lower (58), indicating that while the toxic compounds may undergo satisfactory destruction, certain intermediate products remain unoxidized. Because the wet air oxidation is not complete, the effluent from the process can contain appreciable concentrations of volatile organics and may require additional treatment such as bio-oxidation. Supercritical water oxidation typically achieves greater than 99.99% reduction in total organic carbon (5), so SCWO offers a much more thorough treatment option than wet air oxidation. Thus, for aqueous streams which require high destruction efficiency and complete oxidation, SCWO may well be the method of choice.

Given current trends toward source reduction and waste minimization, there may be unique applications for SCWO as an integrated component in the manufacture of chemicals. For example, SCWO could be regarded in the same context as distillation with chemical reaction, where in this case residual organics are chemically stripped out of a water solvent stream that is recycled continuously within the process.

History and Status of Development

The supercritical water oxidation (SCWO) process has been patented for above-ground and for underground applications. Table I gives a listing of most of the relevant patents for both applications (related high-pressure wet air oxidation patents have not been included; see, e.g., Barton et al. (69) and Thiel et al. (70)). Most of the work published on SCWO is related to the above-ground surface system, and this consequently is the focus of our review. The earliest patent in this field is due to Dickinson (59). In his process, the oxidation reaction commences at conventional wet oxidation conditions, but the heat released by the exothermic reaction increases the process stream temperature to a supercritical level. Nonetheless, the above-ground MODAR process actually has its roots at MIT with an invention cited by Dr. Michael Modell and co-workers (71) on thermal reforming of glucose under oxygen-free conditions in supercritical water. In 1980, Dr. Modell left MIT to develop SCWO technology. He founded MODAR, Inc. in Natick, Massachusetts as a commercial venture to develop and market the technology. MODAR continues as a commercial leader in above-ground SCWO technology. Since 1986, Dr. Modell has pursued special applications of the technology, laboratory testing, and development services with a second venture called the Modell Development Corporation (or MODEC) located in Framingham, Massachusetts.

At the time of writing this review, an Austin, Texas based subsidiary of RPC Energy Services, Inc. of Atlanta, Georgia, Eco Waste Technologies (EWT), is conducting research at the University of Texas Balcones Research Center (UTBRC) on SCWO technology. The work initially centered on the use of a shallow-shaft concept with Dr. Earnest F. Gloyna serving as principal investigator. The RPC grant to UT was redirected to a surface study of SCWO and, in cooperation with UTBRC, EWT designed and constructed a 40 gph SCWO reactor that is currently being utilized as a test unit. The unit has been successfully tested on a variety of wastes and surrogate materials (Lyon, D., RPC Energy Services, Inc., personal communication, 1992).

Table I. U.S. Patents Related to Supercritical Water Oxidation Technology

Patent No.	Date	Type*	Inventor	Assignee	Title and Reference
4,292,953	10/6/81	A	Dickinson	Inventor	Pollutant-Free Low Temperature Combustion Process Utilizing the Supercritical State (59)
4,338,199	7/6/82	A	Modell	MODAR, Inc.	Processing Methods for the Oxidation of Organics in Supercritical Water (1)
4,377,066	3/22/83	A	Dickinson	Inventor	Pollution-Free Pressurized Fluidized Bed Combustion Utilizing a High Concentration of Water Vapor (60)
4,380,960	4/26/83	A	Dickinson	Dipac Associates	Pollutant-Free Low Temperature Combustion Process Utilizing the Supercritical State (61)
4,543,190	9/24/85	A	Modell	MODAR, Inc.	Processing Methods for the Oxidation of Organics in Supercritical Water (2)
4,564,458	1/14/86	U	Burleson	Inventor	Method and Apparatus for Disposal of a Broad Spectrum of Waste Featuring Oxidation of Waste (62)
4,593,202	6/3/86	A	Dickinson	Dipac Associates	Combination of Supercritical Wet Combustion and Compressed Air Energy Storage (63)
4,594,164	6/10/86	U	Titmas	J.A. Titmas Associates	Method and Apparatus for Conducting Chemical Reactions at Supercritical Conditions (50)
4,792,408	12/20/88	U	Titmas	J.A. Titmas Associates	Method and Apparatus for Enhancing Chemical Reactions (51)
4,822,394	4/18/89	U	Zeigler & Peterscheck	Vertech Treatment Systems	Method and Apparatus for the Production and Liquefaction of Gases (64)
4,822,497	4/18/89	A	Hong et al.	MODAR, Inc.	Method for Solids Separation in a Wet Oxidation Type Process (65)
4,861,497	8/29/89	A	Welch & Siegwarth	Inventors	Method for the Processing of Organic Compounds (66)
4,891,139	1/2/90	U	Zeigler & Peterscheck	Inventors	Method for Wet Oxidation Treatment (67)
5,075,017	12/24/91	A	Hossain and Blaney	Kimberly-Clark Corp.	Method for Removing Polychlorinated Dibenzodioxins and Polychlorinated Dibenzofurans from Paper Mill Sludge (68)

* type A denotes above-ground, type U denotes underground.

In the decade since its conception, supercritical water oxidation has undergone extensive testing, both at the laboratory (bench) scale and at the pilot-plant scale, to determine its ability to treat a wide variety of waste compounds. The MODAR pilot-scale unit is a modularized fiftyfold scale-up from the bench-scale apparatus, capable of processing 190 L/day of organic (or 1900 L/day of 10% aqueous organic). The pilot unit is compact and skid-mounted, measuring approximately 12 m x 3 m x 2.6 m (4). Typical steady-state operating temperatures for both the bench-scale and the pilot-scale units are in the range of 600° to 640°C.

Table II lists most of the compounds oxidized successfully with the MODAR process. The destruction efficiencies listed are based on the disappearance of the particular compounds. Even for those cases where destruction efficiencies are not specifically listed, the wastes were typically oxidized to greater than 99.99% destruction. In many cases, effluent organic concentrations were below detectable levels. Particularly notable in Table II are the high destruction efficiencies observed for chlorinated organics and aromatics, including polychlorinated biphenyls (PCBs) and dibenzo-p-dioxins. Dioxins fed to the reactor were reduced to levels below the analytical detection limit of 1 μg/kg. In addition, as contrasted to incineration or low-pressure combustion, dioxins are not formed during oxidation. Figure 5 demonstrates the thorough, indiscriminate destruction possible in the MODAR process: the broad range of organic components present in the reactor feed has been completely eliminated at a detection limit of 1 ppm (7). In addition to the specific organic compounds listed in Table II, SCWO has successfully destroyed a variety of mixed waste streams, ranging from dioxin-contaminated soil (5) to fermentation wastes (19). Work by others has included testing on forest products and pulp mill sludges (72,14), municipal sludge (73,74), volatile acids (75), model life support system wastes (76), industrial sludge (77), acetamide and acetic acid (78), propellants (79), and other mixed wastes (80).

Today the MODAR supercritical water oxidation process is nearing commercialization and opportunities for a full-scale demonstration plant are being actively pursued. Most recently MODAR and ABB Lummus Crest have joined forces to address design issues in directing their development efforts (see Barner et al. (7) for details). While the SCWO process has clearly proved viable for a broad range of wastes, and development of process equipment and operations has advanced tremendously in the past decade, there are several important engineering and design areas where improvements will help ensure the long-term success of the technology.

Critical Engineering Design and Performance Issues

Because of the higher temperatures and pressures employed in SCWO relative to wet air oxidation, more restrictive design constraints apply. Laboratory bench-scale and pilot-scale testing of the SCWO process over the last decade with a number of different waste types have clearly established several important issues. These include:

1. Oxidation reaction operating conditions.
2. Materials of construction.
3. Control and removal of precipitating (sticky) salts.
4. Mechanical design of the reactor, salt separator, letdown valves, recycle pump (if needed), and other high pressure equipment.
5. Design of turbo-expanders for processing supercritical mixtures of CO_2, H_2O and N_2 (if energy recovery is envisioned).
6. Process integration to provide a safe, environmentally acceptable operating system.

The tradeoff that exists between reactor residence time and temperature to achieve high destruction efficiencies needs to be weighed against the size and cost of the

Table II. Chemicals Successfully Treated by Supercritical Water
Oxidation and Typical Destruction Efficiencies[a]

Organic Compound	Bench-Scale	Pilot-Scale	Destruction Efficiency[b], %
Acetic Acid	x		
Acetylsalicylic Acid (Aspirin)	x		
Ammonia		x	>99.71
Aroclors (PCBs)	x	x	>99.995[c]
Benzene	x		
Biphenyl	x		99.97
Butanol	x		
Carbon Tetrachloride		x	>96.53[c]
Carboxylic Acids	x		
Carboxymethyl Cellulose	x		
Cellulose	x		
Chlorinated Dibenzo-p-dioxins	x		>99.9999
Chlorobenzene		x	
Chloroform		x	>98.83[c]
2-Chlorophenol		x	>99.997[c]
o-Chlorotoluene	x	x	>99.998[c]
Cyanide		x	
Cyclohexane	x		99.97
DDT	x		99.997
Decachlorobiphenyl	x		
Dextrose	x		99.6
Dibenzofurans	x		
3,5-Dibromo-N-cyclohexyl-N- methyltoluene-α,2-diamine	x		
Dibutyl Phosphate	x		
Dichloroacetic Acid	x		
Dichloroanisole	x		
Dichlorobenzene	x		
4,4´-Dichlorobiphenyl	x		99.993
1,2-Dichloroethylene	x		99.99
Dichlorophenol	x		
Dimethyl Sulfoxide		x	
Dimethylformamide		x	
4,6-Dinitro-o-cresol	x		
2,4-Dinitrotoluene	x		99.9998
Dipyridamole	x		
Ethanol	x		
Ethyl Acetate		x	
Ethylene Chlorohydrin	x		
Ethylene Glycol	x		>99.9998[c]
Ethylenediamine Tetraacetic Acid	x		
Fluorescein	x	x	>99.9992[c]
Hexachlorobenzene	x		
Hexachlorocyclohexane	x	x	>99.9993[c]
Hexachlorocyclopentadiene	x		99.99
Isooctane	x		
Isopropanol	x	x	

Continued on next page

Table II. Chemicals Successfully Treated by Supercritical Water Oxidation and Typical Destruction Efficiencies (continued)

Organic Compound	Bench-Scale	Pilot-Scale	Destruction Efficiency[b], %
Mercaptans	x		
Methanol	x	x	
Methyl Cellosolve	x		
Methylene Chloride	x	x	
Methyl Ethyl Ketone	x		99.993
Nitrobenzene		x	>99.998[c]
2-Nitrophenol	x		
4-Nitrophenol	x		
Nitrotoluene	x		
Octachlorostyrene	x		
Octadecanoic Acid Magnesium Salt	x		
Pentachlorobenzene	x		
Pentachlorobenzonitrile	x		
Pentachloropyridine	x		
Phenol	x		
Sodium Hexanoate	x		
Sodium Propionate	x		
Sucrose	x		
Tetrachlorobenzene	x		
Tetrachloroethylene	x	x	99.99
Tetrapropylene H	x		
Toluene	x		
Tributyl Phosphate	x		
Trichlorobenzenes	x		99.99
1,1,1-Trichloroethane	x	x	>99.99997[c]
1,1,2-Trichloroethane		x	>99.981[c]
Trichloroethylene	x		
Trichlorophenol	x		
Trifluoroacetic Acid	x		
1,3,7-Trimethylxanthine	x		
Urea	x		
o-Xylene	x		99.93

Continued on next page

**Table II. Chemicals Successfully Treated by Supercritical Water
Oxidation and Typical Destruction Efficiencies (continued)**

Complex Mixed Wastes/Products (Bench-Scale Tests)

Adumbran	Human Waste
Bacillus Stearothermophilus	Ion Exchange Resins
(Heat-Resistant Spores)	(Styrene-Divinyl Benzene)
Bran Cereal	Malaria Antigen
Carbohydrates	Olive Oil
Casein	Paper
Cellulosics	Protein
Coal	Sewage Sludge
Coal Waste	Soybean Plants
Corn Starch	*Sulfolobus Acidocaldarius*
Diesel Fuel	Surfactants
E. Coli	Transformer Oil[d]
Endotoxin (Pyrogen)	Yeast

Inorganic Compounds (Bench-Scale Tests)[e]

Alumina	Magnesium Phosphate
Ammonium Chloride[d]	Magnesium Sulfate
Ammonium Sulfate	Mercuric Chloride
Boric Acid	Potassium Bicarbonate[d]
Bromides	Potassium Carbonate[d]
Calcium Carbonate[d]	Potassium Chloride[d]
Calcium Chloride[d]	Potassium Sulfate[d]
Calcium Oxide[d]	Silica
Calcium Phosphate[d]	Sodium Carbonate[d]
Calcium Sulfate[d]	Sodium Chloride[d]
Fluorides	Sodium Hydroxide[d]
Heavy Metal Oxides	Sodium Nitrate
Hydrochloric Acid[d]	Sodium Nitrite
Iron	Sodium Sulfate[d]
Iron Oxide[d]	Soil
Lithium Sulfate	Sulfur, Elemental
Magnesium Oxide	Titanium Dioxide

[a]Sources: Thomason *et al.* (*5*), Thomason and Modell (*6*), Modell (*2,3*), and
   unpublished data from MODAR, Inc.
[b]No entry for destruction efficiency indicates that a quantitative determination was not
   reported.
[c]Compound undetectable in effluent; quoted efficiency is based on analytical detection
   limit.
[d]Pilot-scale tests were also performed successfully.
[e]Inorganic compounds were not destroyed but the process was operated successfully
   with those compounds present.

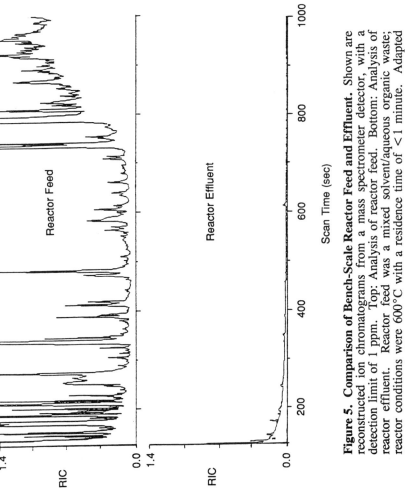

Figure 5. Comparison of Bench-Scale Reactor Feed and Effluent. Shown are reconstructed ion chromatograms from a mass spectrometer detector, with a detection limit of 1 ppm. Top: Analysis of reactor feed. Bottom: Analysis of reactor effluent. Reactor feed was a mixed solvent/aqueous organic waste; reactor conditions were 600°C with a residence time of <1 minute. Adapted from Barner et al. (7).

reaction vessel. High temperatures, of course, lead to faster kinetics and therefore shorter required residence times to achieve the same destruction efficiency. Most organics and gases are miscible with water at pressures near the critical point of water, below the frequently cited operating pressure of 235 to 250 bar. Successful operation of the process at lower pressures could have significant advantages for scale-up, since the larger-scale reactors would not need to be designed to contain such high pressures. Process pressures would nevertheless still be considerable, and economical mechanical containment of the process will be a vital concern. ABB Lummus Crest has completed proprietary designs in which the containment material is kept cool by separating it from the reactor fluid with a suitably inert material (7).

The designs of many high-pressure components used in SCWO processing are evolving, with considerable progress made, but optimal systems will require more research and development. Of particular importance in all process flow sheets is the reactor/salt separator where both proper mechanical design and choice of materials of construction are essential to guarantee safe, reliable operation. In some concepts, product effluent recycle for heat recovery and effluent expansion in turbines is envisioned. In these cases the complete removal of suspended particles from the fluid will be required to minimize abrasion and erosion of valves and rotating components.

Supercritical water can lead to severe corrosion, particularly when aggressive ions such as Na^+, H^+, Cl^-, F^-, and NH_4^+ are present. This is also true during the preheating and cooldown parts of the process where temperatures in the range of 200 to 400°C are encountered. Thus, not only must the reactor and downstream equipment have high-temperature strength with thermal shock resistance, but they must also be corrosion resistant. Stainless steel alloys are very susceptible to corrosion and are unsuitable for long-term SCWO service (81,82). Current practical systems require high-strength nickel superalloys such as Inconel 625 (in wt%: 55-68 Ni, 20-23 Cr, 8-10 Mo, 3-4 Nb and <5 Fe) or Hastelloy C276, or the use of suitable ceramic materials (7). Even high-nickel alloys are not completely corrosion resistant (82), and recent studies report on the corrosion of these alloys during SCWO service (7,83-85).

Effective and efficient salt separation remains a challenge. In a current process concept, salts formed during oxidation are redissolved in a cooler liquid brine region maintained at the bottom of the reactor; however, salt agglomeration on the walls of the reactor remains a problem. In all cases, methods of proper pressure letdown to remove brine solutions or salt-water slurries from the reactor/salt separator are required. ABB Lummus Crest has developed several concepts for controlling salt accumulation on the reactor walls (7). One system is based on using a falling film of cold (approximately 300°C) brine covering the critical height of reactor where salt build-up is anticipated. The brine film serves to collect and redissolve salt impinging on the reactor wall at supercritical conditions. The brine can then be removed as a product stream from the reactor. While this "water wall" concept is promising, it has yet to undergo testing at process conditions. In advanced mechanical designs, ABB Lummus Crest has developed two proprietary hybrid-reactor design concepts based on using a cold-wall pressure-containment vessel. In both of these designs, reaction fluid is isolated from the pressure containment vessel by means of suitable materials of construction to help reduce corrosion rates.

The long-term durability of process equipment must be ensured. A cold-wall reactor concept is a major step in this direction, as durability of the reactor itself is certainly a concern, but the lifespan of other process equipment must also be addressed. In particular, heat-exchange equipment used to heat or cool the reactor streams may undergo significant corrosion as the fluid passes through the critical region (80,83).

While practical engineering solutions currently exist to meet these critical design constraints, the basic research being conducted to understand the fundamentals of SCWO should eventually allow significant process refinements to be achieved.

Supporting Basic Research

Although SCWO has been successfully applied to the destruction of organic compounds for more than a decade, only limited fundamental data on important physicochemical processes are available. As shown in Figure 6, relevant work in three basic research areas is linked to major process steps in SCWO. Materials research on corrosion has been carried out primarily by companies and for the most part only limited results have been reported in the open literature (7,80,83-85). In addition, effects of supercritical water on heat transfer rates have been addressed by some researchers (86-91). Basic work on reaction kinetics and mechanisms, phase equilibria and stability, and solid nucleation and precipitation in supercritical fluids has been extensive both in industry and universities and has been openly reported (92-95). However, only a small portion of this vast literature is directly relevant to SCWO; the majority is focused on extraction or reaction in supercritical fluids such as carbon dioxide and ethylene. The discussion that follows reviews basic research directly related to reaction and salt formation in supercritical water systems.

Basic research is focused on several fundamental processes that make up the complex suite of coupled reaction and phase transition phenomena that occur during the oxidation of organic compounds in the SCWO process. Especially important is the impact of the high fluid density and temperature of SCW on reaction rates. The ability to quantitatively characterize reaction kinetics, fluid mixing rates, and salt nucleation and deposition rates is vital to developing efficient process designs for SCWO that maximize destruction efficiencies while minimizing cost. In order to achieve this ultimate goal, researchers have attempted to decouple the problems by studying individual processes first, such as the oxidation kinetics of single molecular species in SCW and the phase behavior and precipitation kinetics of salts in SCW.

Compounds have been selected as models because they represent rate limiting steps in the oxidation of more complex H/C/N/O compounds (e.g., CH_3OH, C_2H_5OH, NH_3, and CO), because they are themselves characteristic waste compounds (e.g., phenol, acetic acid) or because they represent neutralized products of heteroatom oxidation reactions (e.g., $-Cl \rightarrow HCl \rightarrow NaCl$).

Reaction Kinetics and Mechanisms. Most early studies of SCWO (see, e.g., Modell *et al.* (*17*); Timberlake *et al.* (*8*); Hong *et al.* (*9*); Johnston *et al.* (*19*); Staszak *et al.* (*96*)) were concerned primarily with achieving high destruction efficiencies, rather than elucidating reaction mechanisms and rate dependencies under well-defined temperature, pressure, and composition conditions. Furthermore, many studies of reactions in supercritical water have been concerned with specific chemical processes, such as coal liquefaction or the pyrolytic reaction of organics in and with the water itself, rather than waste detoxification or the reaction of organics with oxygen in a supercritical water environment.

The published literature on oxidation and other reaction studies carried out in supercritical water is summarized in Table III. Four university groups, at MIT (Tester and co-workers), Illinois/Georgia Tech (Eckert and co-workers), Texas (Gloyna and co-workers) and Michigan (Savage and co-workers), have reported results on the oxidation of individual model compounds in SCW. Another related study of catalytic oxidation of 1,4 dichlorobenzene in SCW was conducted by Jin *et al.* (*154*). Furthermore, Rofer and Streit (*155,156*) at Los Alamos National Laboratory reported some early experimental data taken by the MIT group on methane and methanol oxidation. They also describe modeling work using elementary reaction networks. Franck and co-workers at Karlsruhe have studied pyrolysis, hydrolysis, and oxidation reactions in supercritical hydrothermal flames and under flameless conditions (*148,149*).

Reaction rate data in SCW are typically correlated using global rate expressions where rates are expressed with a power-law dependence on concentration and an

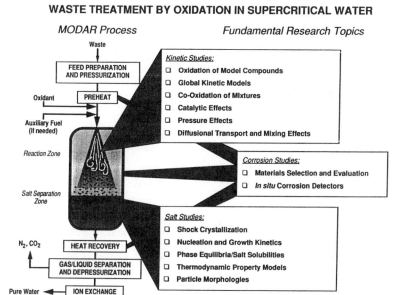

WASTE TREATMENT BY OXIDATION IN SUPERCRITICAL WATER

Figure 6. **Research and Development Linkage to the MODAR SCWO Process.**

Table III.   Selected Experimental Studies of Specific Reactions and Salt Formation in Supercritical Water

---

Oxidation Reactions and Kinetics of Model Compounds

Phenol, Acetic Acid, and Others: Wightman (*97*)
Carbon Monoxide: Helling and Tester (*98-100*), Holgate *et al.* (*90*)
Ethanol: Helling and Tester (*100*)
Methane: Webley and Tester (*101,102*)
p-Chlorophenol: Yang and Eckert (*103*)
Ammonia: Helling and Tester (*100*), Webley *et al.* (*104,105*)
Methanol: Webley and Tester (*106*), Webley *et al.* (*104,105*)
2,4-Dichlorophenol, Acetamide, and Acetic Acid: Lee and Gloyna (*107*), Lee *et al.* (*78*), Wilmanns and Gloyna (*75*)
Phenol: Thornton and Savage (*108*), Thornton *et al.* (*109*)
Hydrogen: Holgate and Tester (*110*)

Other Reaction Studies

Reforming, Decomposition, and Hydrolysis of Sugars and Cellulose
    Amin *et al.* (*111*)   Antal and co-workers (*112-115*)

Dehydration of Alcohols
    Antal and co-workers (*116-120*)

Hydrolysis and Pyrolysis
    Whitlock (*121*)   Antal and co-workers (*129,130*)
    Klein and co-workers (*122-128*)

Reactions in Coal Liquefaction and Extraction
    Deshpande *et al.* (*131*)   Greenkorn and Chao (*139*)
    Houser and co-workers (*132-135*)   Ross *et al.* (*140-142*)
    Amestica and Wolf (*136*)   Penninger and co-workers (*143,144*)
    Fish and co-workers (*137-138*)   Abraham and Klein (*145*)

Electrochemistry
    Flarsheim *et al.* (*146,147*)

Hydrothermal Flames
    Franck and co-workers (*148,149*)

Nucleation and Growth Phenomena of Salts in Supercritical Water Oxidation

---

NaCl and Na₂SO₄: Armellini and Tester (*150,151*)

Salt Separator Design

Killilea *et al.* (*152*)
Dell'Orco and Gloyna (*153*)

Arrhenius temperature dependence. Assuming a first-order dependence on organic substrate concentration alone, which is frequently borne out experimentally, an Arrhenius plot for rate constants can be developed as shown in Figure 7, where a summary of the MIT results is given. Global rate expressions are tabulated in Table IV. One notes the significant range of activation energies and pre-exponential factors observed. For more complex organic molecules, partial oxidation to lower molecular weight fragmentation products will occur rapidly in SCWO even at temperatures near the critical point. Further oxidation of these fragmentation products is slower and eventually one finds that oxidation of simple compounds like CO and NH_3 becomes rate limiting. Li *et al.* (*158*) have developed a generalized kinetic model for wet air oxidation and SCWO based on reported global kinetic expressions for oxidation of model wastes or waste surrogates.

Although global correlations are useful and can greatly augment process design, the eventual *a priori* optimization of SCWO as a waste destruction method depends critically on a better understanding of the elementary processes occurring. Especially important is the impact of the high fluid density of SCW on oxidation rates. Using current combustion elementary reaction sets as a starting point, a concerted effort is underway by several groups at MIT (see, for example, Helling and Tester (*99*), Webley and Tester (*102*), and Holgate and Tester (*110*)), Los Alamos National Laboratory (*155,156*), and Sandia National Laboratories and Lawrence Livermore Laboratory (*159-161*) to adapt such free-radical mechanisms to supercritical reaction conditions by applying pressure and non-ideality corrections. Global and elementary reaction models have been developed for the oxidation of carbon monoxide, methanol, methane, hydrogen, ammonia, and a number of other model compounds which are important in limiting complete oxidation. In addition, Margolis and Johnston (*162*) have reported the development of a coupled hydrodynamic-kinetic model of SCWO.

To supplement the listing of compounds studied given in Table II, a few of the important findings are described below:

Carbon Monoxide: The oxidation kinetics of carbon monoxide over the temperature range 420-570°C at a pressure of 246 bar (24.6 MPa) were originally reported by Helling and Tester (*98,99*) and have been recently reexamined by Holgate *et al.* (*90*). In addition to direct oxidation with oxygen, Helling and Tester found that reaction of carbon monoxide with water (the water-gas shift reaction) was significant.

Direct oxidation	$CO + {}^1/_2 O_2 \rightarrow CO_2$
Water-gas shift	$CO + H_2O \rightarrow CO_2 + H_2$

At low temperatures, over half of the carbon monoxide was oxidized by water and not oxygen. This fraction decreased with increasing temperature, to about 10% at 540°C. The effects of temperature and concentration on direct and indirect oxidation kinetics of carbon monoxide were correlated with global models. The oxidation of carbon monoxide was found by Helling and Tester (*98*) to be globally first order in carbon monoxide and independent of oxygen concentration over the range investigated. Elementary reaction models were also used in an attempt to model the complex chemistry of the reaction pathways, with only some success.

Holgate *et al.* (*90*) largely succeeded in reproducing the earlier data, and accounted for the role of the water-gas shift reaction during preheating of the carbon monoxide feed. O_2/CO feed ratios were extended to the substoich-iometric regime, revealing a slight dependence of the direct-oxidation rate on oxygen not observed in the earlier data but consistent with gas-phase studies of CO oxidation. The direct-oxidation data shown in Figure 7 are a compilation of both sets of CO oxidation data, with the contribution of the water-gas shift

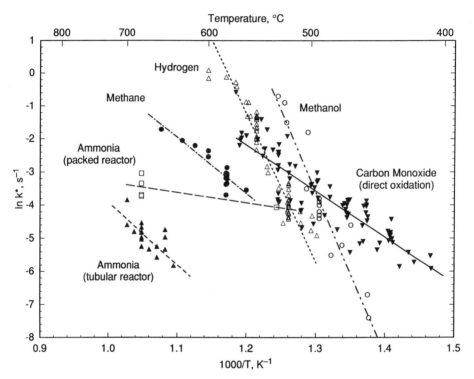

Figure 7. Apparent First-Order Arrhenius Plot for Oxidation of Model Compounds in Supercritical Water at 24.6 MPa. Data from Helling (*157*); Helling and Tester (*99,100*); Webley (*91*); Webley *et al.* (*104,105*); Webley and Tester (*102*); Holgate and Tester (*110*); Holgate *et al.* (*90*). Unless otherwise noted, only tubular reactor experiments with oxygen present are shown.

Table IV. Global Rate Expressions for Oxidation of Organics in Supercritical Water at 24.6 MPa

$$-\frac{d[C]}{dt} = A \exp(-E_a/RT) [C]^a [O_2]^b$$

Rates in mol/L-s; Concentrations [ ] in mol/L

Organic (C)	A $(mol/L)^{1-a-b}/s$	E_a kJ/mol	a	b	Number of Runs	Temperature Range, °C	Comments
Carbon Monoxide (direct oxidation)	$10^{8.5 \pm 3.3}$*	134. ± 32.	0.96	0.34	82	420–570	Water-gas shift contribution removed. Source: Helling and Tester (99), Holgate et al. (90)
Ethanol	$10^{21.8 \pm 2.7}$	340. ± 41.	1	0	8	480–540	Reaction orders assumed. Source: Helling and Tester (100)
Methane	$10^{11.4 \pm 1.1}$	179.1 ± 18.0	0.99	0.66	18	560–650	1 packed reactor run, 2 pyrolysis runs. Source: Webley and Tester (102)
Methanol	$10^{26.2 \pm 5.8}$	408.8 ± 85.4	1	0	21	450–550	1 packed reactor run, 1 pyrolysis run; data from 2 reactor systems. Source: Webley (91), Webley et al. (104)
Ammonia (tubular)**	$10^{6.5 \pm 3.6}$	156.9 ± 64.8	1	0	17	650–700	1 pyrolysis run; reaction orders assumed. Source: Webley et al. (104,105)
Ammonia (packed bed)**	$10^{0.1 \pm 2.8}$	30. ± 49.	1	0	6	530–680	1 pyrolysis run; reaction orders assumed. Source: Webley et al. (105)
Hydrogen	$10^{22.8 \pm 2.2}$	372. ± 34.	1	0	57	495–600	2 pyrolysis runs. Source: Holgate and Tester (110)

*All errors are at the 95% confidence level.

**Ammonia oxidation data did not fit the power-law rate form well. A catalytic model provided the best fit to the tubular reactor data for ammonia (105).

pathway removed. There is a high degree of scatter in the data because the oxygen dependence is not accounted for in the first-order plot. Preliminary experiments in a packed bed reactor, packed with particles of the reactor material (Inconel 625) and with a surface area-to-volume ratio 30 times that of the tubular reactor, yielded rates somewhat lower than those obtained in the tubular reactor.

Ethanol: Helling and Tester (*99*) determined Arrhenius parameters for ethanol oxidation over the temperature range 480°-540°C, assuming the reaction was first order in ethanol and zero order in oxygen. The major products of the reaction were carbon monoxide, carbon dioxide, and acetaldehyde. The reaction exhibited an apparent activation energy of 340 kJ/mol. The ethanol data points have been omitted from Figure 7 because they overlap the methanol points and would be obscured from view.

Methane: The oxidation kinetics of methane over the temperature range 560°-650°C were determined by Webley and Tester (*102*) at a pressure of 246 bar. The major products of the oxidation were carbon monoxide and carbon dioxide. No methanol was detected in the effluent. The oxidation was first order in methane and 0.66-order in oxygen over the range of concentrations investigated. The dependence of the reaction on oxygen concentration accounts for much of the scatter in the methane data in Figure 7.

The activation energy was 179.1 kJ/mol over the experimental temperature range. Attempted pyrolysis of methane gave no conversion at 650°C and 15 seconds residence time. Oxidation of methane in the packed bed reactor showed no significant rate increase over the tubular reactor. Elementary reaction models were used to model methane oxidation, but the models showed limited ability to reproduce experimental results.

Methanol: The oxidation kinetics of methanol over the temperature range 450°-550°C were determined at a pressure of 246 bar (*91,105*). The oxidation was found to be highly activated, with an apparent activation energy of 408.8 kJ/mol. The products of oxidation were carbon monoxide, carbon dioxide, and hydrogen. The oxidation rate was found to be approximately first order in methanol and zero order in oxygen over the concentration range investigated. Attempted pyrolysis of methanol gave a conversion of 2.2% at 544°C and 6.6 seconds residence time. A single oxidation experiment in the packed bed reactor showed no rate increase over the tubular reactor. Elementary reaction models of methanol oxidation correctly predicted the experimental conversions, although the predicted activation energy was too high, no hydrogen was predicted in the effluent, and the product CO/CO_2 ratio was too high.

Ammonia: Webley *et al.* (*104,105*) determined the oxidation kinetics of ammonia over the temperature range 650°-700°C at a pressure of 246 bar and residence times of 10-15 seconds. Over this range conversions ran from 0 to 15%. The product of ammonia oxidation is exclusively nitrogen; no NO or NO_2 was detected. The oxidation was roughly first order with respect to ammonia, and was weakly dependent on oxygen concentration, although the typical power-law rate form did not fit the data well. Instead, a catalytic model yielded a much better fit. Duplicate runs in the packed bed reactor, in the temperature range 530°-680°C, showed that the oxidation was significantly affected by the increased presence of Inconel 625. Conversions up to four times higher were observed in the packed bed reactor at operating conditions similar to those in the tubular reactor, demonstrating that ammonia oxidation is at least partly catalytic. Small amounts of N_2O were detected in the effluent of the packed reactor. A pyrolysis experiment at 700°C yielded no decomposition of ammonia to nitrogen and hydrogen.

Ammonia-Methanol Mixtures: The oxidation kinetics of mixtures of ammonia and methanol were examined by Webley *et al.* (*104,105*) in a tubular and a packed reactor over the temperature range 480°-520°C. The conversion of methanol was slightly enhanced by the presence of ammonia. The conversion of ammonia was unaffected by the presence of methanol in the tubular reactor, although in the packed reactor the rate of ammonia oxidation was slightly retarded by the addition of methanol.

Hydrogen: Holgate and Tester (*110*) examined the oxidation kinetics of hydrogen in a tubular reactor in the temperature range of 495°-600°C and at a pressure of 246 bar. Over the range of experimental conditions, the reaction was independent of oxygen concentration and first-order in hydrogen concentration, with an activation energy of 372 kJ/mol. Figure 7 also shows that hydrogen is more refractory than carbon monoxide up to a temperature of about 550°C, indicating that hydrogen oxidation was probably unimportant in the carbon monoxide oxidation experiments in which hydrogen was formed as a product. A series of experiments at 550°C demonstrated a pronounced induction time of about 2 s in the oxidation reaction. An elementary reaction model for hydrogen oxidation was able to adequately reproduce the experimental data, including the induction time, and demonstrated nearly the same global rate parameters as the data, including reaction orders and overall activation energy.

Other Model Wastes (Phenolics, Acetic Acid, etc.): Several authors have studied the oxidation of various waste compounds in supercritical water. Most of these studies have not produced rigorous kinetic expressions for waste oxidation, but rather have concerned themselves with destruction efficiencies and/or product distributions. Among the first of these studies was that of Wightman (*97*), who studied phenol, acetic acid, and other compounds, including chlorobenzene, carbon tetrachloride, pyridine, aniline, ammonia, and acetonitrile, at temperatures to 430°C and pressures to 5500 psig in a limited series of experiments. Yang and Eckert (*103*) examined the oxidation of p-chlorophenol at 340 and 400°C and 140 and 240 atmospheres, and the reaction was found to be independent of oxygen concentration and first- to second-order in chlorophenol. The reaction was moderately accelerated by copper and manganese salts, and was greatly accelerated by an increased surface-to-volume ratio in the reactor. More recently, Savage and co-workers (*108,109*) have studied phenol oxidation near the critical point of water; they operated at relatively low temperatures to observe the wide spectrum of products obtainable under those conditions. Finally, Lee *et al.* (*78*) studied the relative efficiencies of oxygen and hydrogen peroxide in the oxidation of acetic acid and 2,4-dichlorophenol, at 400-500°C and 240-350 bar. They found that destruction of the wastes was significantly enhanced by the use of H_2O_2 as the oxidant.

Reaction Modeling: Elementary reaction models have been applied to several of the above wastes, with varying degrees of success. The modeling results from Tester and co-workers at MIT for three of these wastes (methane, methanol, and hydrogen) are summarized in Figure 8 as a reaction pathway diagram (*91,102,110*). Major fluxes of each reactive species are shown, with flux arrow thicknesses proportional to the fraction of the total species flux. While some of these models cannot currently reproduce quantitatively the experimental data, many of the predicted overall reaction rates are within an order of magnitude of those observed experimentally, and it is likely that the basic mechanisms are essentially correct. The close agreement of the hydrogen oxidation model with the data is particularly encouraging.

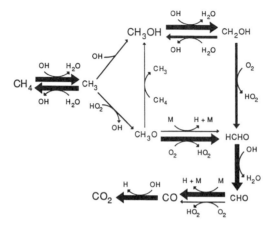

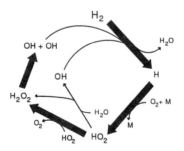

Figure 8. Predicted Elementary Reaction Pathways for Oxidation of Model Compounds in Supercritical Water. Major reaction pathways for the oxidation of methane, methanol, and hydrogen, as predicted by elementary reaction models, are shown. Arrow thicknesses are proportional to fraction of species fluxes; thickest arrows = 100% of species flux. M = third body species (taken to be water). Sources: Webley and Tester (*102*); Webley (*91*); Holgate and Tester (*110*).

The pathways shown in Figure 8 demonstrate several important points. First, the mechanisms of methane and methanol destruction are intimately linked, and would themselves be sub-mechanisms for the oxidation of larger hydrocarbons. The final steps of destruction are common to all hydrocarbons, with the last step being the oxidation of carbon monoxide. Second, the hydroxyl (OH) radical plays an extremely important role in the oxidation process, as it is the oxidant in many of the destruction steps. Molecular oxygen, on the other hand, plays only a limited role. Third, the mechanism for hydrogen oxidation demonstrates how the "macroscopic" oxidant O_2 is converted in a sequence of steps to the "microscopic" oxidant OH; it is this sequence of steps which apparently masks the role of oxygen in the overall oxidation process and leads to the experimental observation of an oxidation reaction whose rate is paradoxically independent of oxygen concentration.

Inorganic Salt Formation and Separation Studies. As discussed earlier, solid salt separation is a critical engineering design issue in the SCWO process. Much of the research results in this area remain proprietary, but general salt separator design concepts can be found in patents and published process descriptions (see, e.g., Hong *et al.* (*65*) and Barner *et al.* (*7*)). Though considerable progress in SCWO salt separation techniques has been made in industry, a fundamental understanding of solid salt formation is required to develop next-generation salt separators. An experimental program was started at MIT in 1988 to study the kinetics and mechanisms of salt nucleation and growth from supercritical water, and recently, Dr. Earnest F. Gloyna and co-workers at the University of Texas have begun a research project to study particulate and salt separation from SCWO effluent streams.

Basic supporting research on solids formation can be divided into three areas of study: phase equilibrium, salt nucleation and growth, and separator design. A few of the important results in these areas are described below:

Phase Equilibrium: A first step in understanding solids formation during the SCWO process is to establish phase relationships under typical operating conditions. An apparatus has been constructed at MIT for high temperature and pressure studies of aqueous salt solutions. Salt studies with NaCl, Na_2SO_4, and NaCl/Na_2SO_4 mixtures have been carried out in an optically accessible cell to examine phase behavior and solids nucleation in salt-water systems near and above the critical point of pure water. A cross-sectional diagram of the optical cell is shown in Figure 9. The cell was constructed from a high-nickel, corrosion-resistant alloy, Inconel 625, and is capable of operating up to 600°C and 400 bar.

An experimental procedure was developed for examining phase relationships in salt-water systems at constant pressure. Isobaric phase equilibrium information is desirable, since the oxidation and separation steps in the SCWO process are essentially isobaric, and salt formation mechanisms may depend on phase behavior. To evaluate this technique, static isobaric experiments were performed on the NaCl-H_2O system at a typical process pressure of 250 bar. Figure 10 shows the results of the isobaric experiments of Armellini and Tester (*151*) plotted on the NaCl-H_2O temperature-composition phase diagram. The results agree well with the interpolated data of the other researchers. Currently, work on additional binary and ternary systems is being performed.

Data on salt solubility in supercritical water under typical SCWO conditions are required to model precipitation kinetics and to calculate maximum attainable separation efficiencies. Very few solubility data exist, and there are large discrepancies in the published results. Armellini and Tester (*168*) have reviewed the available literature values for both NaCl and Na_2SO_4 solubility, and have also performed solubility experiments with both salts.

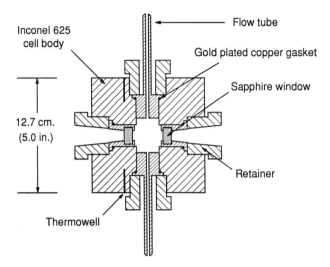

Figure 9. Cross-Sectional View of MIT's High-Temperature and High-Pressure Optically Accessible Cell. Adapted from Armellini and Tester (*151*).

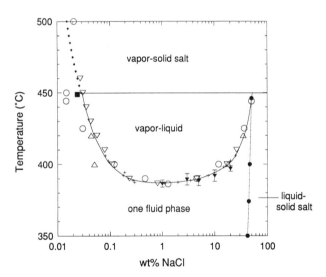

Figure 10. NaCl-H₂O Temperature-Composition Phase Diagram at 25 MPa. Data from: (▼) Armellini and Tester (*151*); (—) compilation of Bischoff and Pitzer (*43*); (○) Sourirajan and Kennedy (*163*); (■) Bischoff *et al.* (*164*); (●) estimated from Linke (*45*); (△) Parisod and Plattner (*165*); (♦) prediction of Pitzer and Pabalan (*44*); (▽) Ölander and Liander (*166*); (+) Khaibullin and Borisov (*167*). Adapted from Armellini and Tester (*151*).

Once reliable phase equilibrium data are obtained, thermodynamic models can be used to correlate and extend the results. Thermodynamic relationships valid for the wide range of conditions encountered in the SCWO process exist only for the NaCl-H$_2$O system (see, e.g., Pitzer and Pabalan (*44*); Pitzer and Tanger (*169*)). Characteristic of most of the models is an extreme temperature and density dependence of the fitted parameters. This is unavoidable due to the highly non-ideal nature of the system. The MIT group is attempting to extend conventional modeling approaches to typical SCWO systems and conditions. So far, they have established that while cubic equations of state, such as the Peng-Robinson equation, and conventional electrolyte activity coefficient models can be used to properly represent the NaCl-H$_2$O system in certain regions of temperature, pressure, and composition, neither can be used over a wide range of conditions to capture the high degree of non-ideality in this system (*150*).

Salt Nucleation and Growth: Flow experiments by Armellini and Tester (*151*) have for the first time allowed observation of salt formation at temperatures, pressures, and supersaturation values similar to those used in the MODAR SCWO process. From these experiments, a better understanding of the salt formation mechanisms involved in the process is being achieved, and salt particle sizes are being estimated. As a first approximation, binary salt-water systems have been used to simulate the SCWO of a waste containing a single dissolved salt. In preliminary runs, transitions through the critical region of pure water as well as the nucleation of solid salt from a supercritical brine have been successfully observed.

Figure 11 shows photos of two different brine jets tested. The jet flow rate was 0.5 g/min with an exit temperature of approximately 150°C, and the external supercritical water stream flowed at 10 g/min at a temperature of around 550°C. The pressure was constant at 250 bar. The 3.0 wt% NaCl jet appears fairly turbulent due to the density differences in the two feeds, very similar in appearance to a pure water jet. The 3.0 wt% Na$_2$SO$_4$ jet and surrounding solution, on the other hand, appear much darker. This was caused by small particles which nucleated from the solution, scattering the light.

Figure 12 shows SEM photomicrographs of the solids collected under conditions similar to the two runs described above. The NaCl solids shown in Figure 12a were grouped in clusters of highly amorphous kernel-shaped particles with lengths between 10 and 100 μm. At higher magnification, the particles appeared to have many hollow inner regions. The Na$_2$SO$_4$ solid shown in Figure 12b comprised many small particles which were fused together. Most of the primary particles were between 1 and 2 μm long.

The phase behavior of two salt-water systems provides a possible explanation of the different visual appearance of the mixed streams and the morphology of the solids formed. Unlike the NaCl-H$_2$O system (see Figure 10), the Na$_2$SO$_4$-H$_2$O system does not contain a vapor-liquid region at a pressure of 250 bar and temperatures from ambient to around 700°C. Instead, a continuous drop in Na$_2$SO$_4$ solubility is observed as temperatures approach and surpass the critical point of water. This solubility behavior most likely resulted in direct nucleation of many small Na$_2$SO$_4$ particles from the injected jet, while the vapor-liquid region in the NaCl-H$_2$O system most likely first caused the formation of concentrated liquid droplets and then larger particles, which could not be visually detected (*151*).

Separator Design: Killilea and co-workers at MODAR, Inc. have reported results of a project supported by NASA for studying the separation of inorganic salts from a SCWO reactor under microgravity conditions for possible use in treating waste water on long-term space missions (*152*). MODAR tested both an

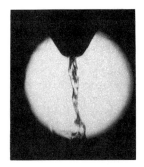 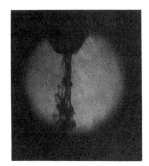

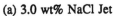

(a) 3.0 wt% NaCl Jet   (b) 3.0 wt% Na$_2$SO$_4$ Jet

Figure 11.   Photographs of Salt Jets Mixing with a Coaxially Flowing Supercritical Water Stream.   Window view diameter is 1.27 cm (0.5 in.). Adapted from Armellini and Tester (*151*).

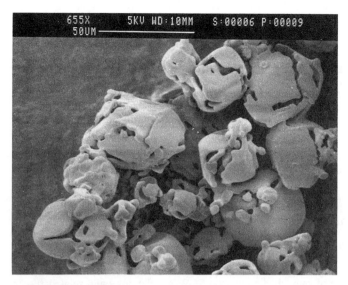

(a) NaCl particle cluster

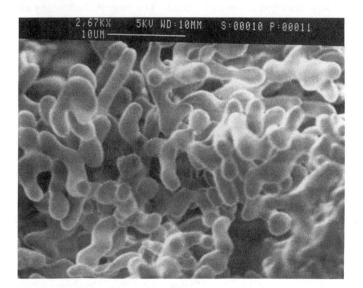

(b) Na$_2$SO$_4$ agglomerated solid

Figure 12. SEM Photomicrographs of Salts Formed from Rapid Mixing with Supercritical Water. Adapted from Armellini and Tester (*151*).

impingement/filtration reactor/separator and a cyclone reactor/separator on a simulated metabolic waste stream. The impingement method, where solids were collected on a replaceable canister, was found to be more effective than cyclone separation, which was hindered by solid salt adhesion to the vessel walls.

Gloyna and co-workers (153) tested a tubular design for solids separation on simulated waste streams containing dissolved sodium salts. For some runs, separation efficiencies greater than 95% (based on net mass of salt removed) were achieved. Due to the simple design of the system, a large fraction of the solid salts remained accumulated on the reactor walls, which resulted in reactor plugging during some runs. The actual MODAR industrial design for the reactor/separator (7,152) overcomes this problem by directing most of the precipitated solids to a cool reactor zone where they are redissolved and removed in a concentrated brine solution.

Conclusions

In this review, we have discussed the status of SCWO technology in the context of several critical process development issues related to reactor operating conditions (temperature, pressure, and residence time), materials selection, and salt formation and removal. We have also attempted to show how basic research into the fundamentals of reaction chemistry and solid salt solubility and nucleation in supercritical water provides support important to refining process development and design. In addition, one of our major objectives in preparing the review was to provide a comprehensive bibliography of published papers on SCWO technology and related research.

The basic idea of using supercritical water as a reaction medium for the efficient destruction of aqueous organic wastes is sound. The process is attracting attention and gaining a large following at the commercial level and in the university research community. Undoubtedly, further practical improvements to the technology will occur as larger plants are built and operated. The potential for SCWO to provide an economically competitive control technology for many hazardous wastes, particularly dilute aqueous waste streams, is very high. Bench- and pilot-scale SCWO tests have yielded positive results in terms of high destruction efficiency, lack of NO_x formation, and successful salt removal. What remains is the resolution of a few process development issues related to scale-up and long-term, reliable operation.

Acknowledgments

The authors would like to thank all our colleagues for their contributions to the general area of SCWO technology. It would be impossible to recognize everyone, but we would like to especially thank Professors Jack Howard and Adel Sarofim, and the late Herman P. Meissner at MIT, and Michael Modell at MODEC, as well as other colleagues at MODAR including K. C. Swallow, Ralph Morgen, Terry Thomason, David Ordway, and Alain Bourhis, and Richard Helling at Dow Chemical. Professors Michael Antal at the University of Hawaii, Michael Klein at the University of Delaware, Keith Johnston at the University of Texas, Charles Eckert at Georgia Tech, E. Ulrich Franck at Karlsruhe, and Johannes Penninger at Eindhoven have also provided thoughtful comments and discussions. Sheridan Johnston and Nina Bergan at Sandia National Laboratories, Ted Wydeven and Ravi Upadhye of NASA-Ames, and Dick Lyon of RPC Energy Services, Inc. contributed helpful comments on this manuscript. Funding for the MIT portions of the SCWO research discussed in this paper was provided in part by NASA, the USDOE, and the Army Research Office. We are particularly grateful to Donald Price and Albert Behrend of NASA and Robert Shaw of the Army Research Office for their interest in our work. We would also like to express our gratitude to Anne Carbone for her help in preparing the manuscript for publication.

Literature Cited

1. Modell, M.; "Processing Methods for the Oxidation of Organics in Supercritical Water"; U.S. Patent 4,338,199, July 6, **1982**.
2. Modell, M.; "Processing Methods for the Oxidation of Organics in Supercritical Water"; U.S. Patent 4,543,190, September 24, **1985**.
3. Modell, M.; "Supercritical Water Oxidation"; In *Standard Handbook of Hazardous Waste Treatment and Disposal*; Freeman, H.M., Ed.; McGraw-Hill: New York, **1989**, pp. 8.153-8.168.
4. Swallow, K.C.; Killilea, W.R.; Malinowski, K.C.; Staszak, C.N.; "The MODAR Process for the Destruction of Hazardous Organic Wastes—Field Test of a Pilot-Scale Unit"; *Waste Management* **1989**, *9*, 19.
5. Thomason, T.B.; Hong, G.T.; Swallow, K.C.; Killilea, W.R.; "The MODAR Supercritical Water Oxidation Process"; In *Innovative Hazardous Waste Treatment Technology Series, Volume 1: Thermal Processes*; Freeman, H.M., Ed.; Technomic Publishing Co.: Lancaster, PA, **1990**, pp. 31-42.
6. Thomason, T.B.; Modell, M.; "Supercritical Water Destruction of Aqueous Wastes"; *Haz. Waste* **1984**, *1*(4), 453.
7. Barner, H.E.; Huang, C.Y.; Johnson, T.; Jacobs, G.; Martch, M.A.; W.R. Killilea; "Supercritical Water Oxidation: An Emerging Technology"; Presented at ACHEMA, June 9, **1991**, and accepted for publication in *J. Haz. Mat.*
8. Timberlake, S.H.; Hong, G.T.; Simson, M.; Modell, M.; "Supercritical Water Oxidation for Wastewater Treatment: Preliminary Study of Urea Destruction"; Presented at the 12th Intersociety Conference on Environmental Systems, San Diego, CA, July 19-21, **1982**; SAE Technical Paper Series No. 820872.
9. Hong, G.T.; Fowler, P.K.; Killilea, W.R.; Swallow, K.C.; "Supercritical Water Oxidation: Treatment of Human Waste and System Configuration Tradeoff Study"; Presented at the 17th Intersociety Conference on Environmental Systems, Seattle, WA, July **1987**; SAE Technical Paper Series No. 871444.
10. Hong, G.T.; Killilea, W.R.; Thomason, T.B.; "Supercritical Water Oxidation: Space Applications"; ASCE Space '88 Proceedings, Albuquerque, NM, August 29-31, **1988**.
11. Hong, G.T.; "Supercritical Water Reactor for Space Applications"; Final Report, Phase I, NASA Contract No. NAS9-18473, **1991**.
12. Gloyna, E.F.; "Supercritical Water Oxidation—Deep-Well Technology for Toxic Wastewaters and Sludges"; Technical Report No. W-89-1; University of Texas at Austin, **1989**.
13. Stone and Webster Engineering Corporation; "Assessment and Development of an Industrial Wet Oxidation System for Burning Waste and Low-Grade Fuels"; Final Report, DOE Contract No. DE-FC07-88ID12711, **1989**.
14. Modell, M.; "Treatment of Pulp Mill Sludges by Supercritical Water Oxidation"; Final Report, DOE Contract No. FG05-90CE40914, **1990**.
15. Freeman, H.; "Supercritical Water Oxidation"; In *Innovative Thermal Hazardous Organic Waste Treatment Processes*; Pollution Technology Review No. 125; Noyes Publications: Park Ridge, NJ, **1985**, pp. 25-29.
16. Shaw, R.W.; Brill, T.R.; Clifford, A.A.; Eckert, C.A.; Franck, E.U.; "Supercritical Water: A Medium for Chemistry"; *Chem. Eng. News* **1991**, *69*(51), 26.
17. Modell, M.; Gaudet, G.G.; Simson, M.; Hong, G.T.; Biemann, K.; "Supercritical Water: Testing Reveals New Process Holds Promise," *Solids Waste Management* **1982**, August.
18. Cunningham, V.L.; Burk, P.L.; Johnston, J.B.; Hannah, R.E.; "The MODAR Process: An Effective Oxidation Process for Destruction of Biopharmaceutical By-Products"; Presented at the AIChE Summer National Meeting, Boston, MA, August **1986**; Paper 45c.

19. Johnston, J.B.; Hannah, R.E.; Cunningham, V.L.; Daggy, B.P.; Sturm, F.; Kelly, R.M.; "Destruction of Pharmaceutical and Biopharmaceutical Wastes by the MODAR Supercritical Water Oxidation Process"; *Bio/Technology* **1988**, *6*, 1423.
20. Haar, L.; Gallagher, J.S.; Kell, G.S.; *NBS/NRC Steam Tables*; Hemisphere Publishing Corp.: New York, **1984**.
21. Uematsu, M.; Franck, E.U.; "Static Dielectric Constant of Water and Steam"; *J. Phys. Chem. Ref. Data* **1980**, *9*(4), 1291.
22. Franck, E.U.; Rosenzweig, S.; Christoforakos, M.; "Calculation of the Dielectric Constant of Water to 1000°C and Very High Pressures"; *Ber. Bunsenges. Phys. Chem.* **1990**, *94*, 199.
23. Heger, K.; Uematsu, M.; Franck, E.U.; "The Static Dielectric Constant of Water at High Pressures and Temperatures to 500 MPa and 550°C"; *Ber. Bunsenges. Phys. Chem.* **1980**, *84*, 758.
24. Marshall, W.L.; Franck, E.U.; "Ion Product of Water Substance, 0-1000°C, 1-10,000 bars. New International Formulation and Its Background"; *J. Phys. Chem. Ref. Data* **1981**, *10*(2), 295.
25. Franck, E.U.; "Properties of Water"; In *High Temperature, High Pressure Electrochemistry in Aqueous Solutions*; Jones, D. de G., and Staehle, R.W., Eds.; National Association of Corrosion Engineers: Houston, TX, **1976**, pp. 109-116.
26. Kohl, W.; Lindner, H.A.; Franck, E.U.; "Raman Spectra of Water to 400°C and 3000 bar"; *Ber. Bunsenges. Phys. Chem.* **1991**, *95*(12), 1586.
27. Connolly, J.; "Solubility of Hydrocarbons in Water Near the Critical Solution Temperature"; *J. Chem. Eng. Data* **1966**, *11*(1), 13.
28. Rebert, C.J.; Kay, W.B.; "The Phase Behavior and Solubility Relations of the Benzene-Water System"; *AIChE J.* **1959**, *5*, 285.
29. Yiling, T.; Michelberger, T.; Franck, E.U.; "High-Pressure Phase Equilibria and Critical Curves of (Water + *n*-Butane) and (Water + *n*-Hexane) at Temperatures to 700 K and Pressures to 300 MPa"; *J. Chem. Thermodynam.* **1991**, *23*, 105.
30. Danneil, A.; Tödheide, K.; Franck, E.U.; "Verdampfungsgleichgewichte und kritische Kurven in den Systemen Äthan/Wasser und n-Butan/Wasser bei hohen Drücken"; *Chem.-Ing. Tech.* **1967**, *39*, 816.
31. Japas, M.L.; Franck, E.U.; "High Pressure Phase Equilibria and PVT-Data of the Water-Nitrogen System to 673K and 250 MPa"; *Ber. Bunsenges. Phys. Chem.* **1985**, *89*, 793.
32. Japas, M.L.; Franck, E.U.; "High Pressure Phase Equilibria and PVT-Data of the Water-Oxygen System Including Water-Air to 673K and 250 MPa"; *Ber. Bunsenges Phys. Chem.* **1985**, *89*, 1268.
33. Seward, T.N.; Franck, E.U.; "The System Hydrogen-Water up to 440°C and 2500 bar Pressure"; *Ber. Bunsenges. Phys. Chem.* **1981**, *85*, 2.
34. Tödheide, K.; Franck, E.U.; "Das Zweiphasengebiet und die kritische Kurve im System Kohlendioxide-Wasser bis zu Drucken von 3500 bar"; *Z. Phys. Chem. N.F.* **1963**, *37S*, 387.
35. Mather, A.E.; Franck, E.U.; "Phase Equilibria in the System Carbon Dioxide-Water at Elevated Pressures"; *J. Phys. Chem.* **1992**, *96*(1), 6.
36. Krader, T.; Franck, E.U.; "The Ternary Systems H_2O-CH_4-NaCl and H_2O-CH_4-$CaCl_2$ to 800 K and 250 MPa"; *Ber. Bunsenges. Phys. Chem.* **1987**, *91*, 627.
37. Alwani, Z.; Schneider, G.; "Phasengleichgewichte, kritische Erscheinungen und PVT-Daten in binären Mischungen von Wasser mit aromatischen Kohlenwasserstoffen bis 420°C und 2200 bar"; *Ber. Bunsenges. Phys. Chem.* **1969**, *73*(3), 294.

38. Gehrig, M.; Lentz, H.; Franck, E.U.; "The System Water-Carbon Dioxide-Sodium Chloride to 773 K and 300 MPa"; *Ber. Bunsenges. Phys. Chem.* **1986**, *90*, 525.

39. Michelberger, T.; Franck, E.U.; "Ternary Systems Water-Alkane-Sodium Chloride and Methanol-Methane-Sodium Bromide to High Pressures and Temperatures"; *Ber. Bunsenges. Phys. Chem.* **1990**, *94*, 1134.

40. Jasmund, K.; "Solubility of KCl in the Gas Phase of Supercritical Water"; *Heidelberger Beitr. Mineral. u. Petrogr.* **1952**, *3*, 380.

41. Martynova, O.I.; "Solubility of Inorganic Compounds in Subcritical and Supercritical Water"; In *High Temperature, High Pressure Electrochemistry in Aqueous Solutions*; Jones, D. de G., and Staehle, R.W., Eds.; National Association of Corrosion Engineers: Houston, TX, **1976**, pp. 131-138.

42. Ravich, M.I.; Borovaya, F.E.; "Phase Equilibria in the Sodium Sulphate-Water System at High Temperatures and Pressures"; *Russ. J. Inorg. Chem.* **1964**, *9*(4), 520.

43. Bischoff, J.L.; Pitzer, K.S.; "Liquid-Vapor Relations of the System NaCl-H$_2$O: Summary of the P-T-x Surface from 300 to 500°C"; *Amer. J. Sci.* **1989**, *289*, 217.

44. Pitzer, K.S.; Pabalan, R.T.; "Thermodynamics of NaCl in Steam"; *Geochim. Cosmochim. Acta* **1986**, *50*, 1445.

45. Linke, W.F.; *Solubilities—Inorganic and Metal-Organic Compounds, 4th edition*; D. Van Nostrand Co., Inc.: New York, **1958**.

46. Tödheide, K.; "Water at High Temperatures and Pressures"; In *Water: A Comprehensive Treatise*, Franks, F., Ed.; Plenum Press, Inc.: New York, **1972**, pp. 463-514.

47. Lamb, W.J.; Hoffman, G.A.; Jonas, J.; "Self-Diffusion in Compressed Supercritical Water"; *J. Chem. Phys.* **1981**, *74*(12), 6875.

48. Dudziak, K.H.; Franck, E.U.; "Messungen der Viskosität des Wassers bis 560°C und 3500 bar"; *Ber. Bunsenges. Phys. Chem.* **1966**, *70*(9-10), 1120.

49. Killilea, W.R.; Swallow, K.C.; Hong, G.T.; "The Fate of Nitrogen in Supercritical Water Oxidation"; Presented at the Second International Conference on Supercritical Fluids, Boston, MA, May 20-22, **1991**, and submitted for publication in *J. Supercritical Fluids*.

50. Titmas, J.A.; "Method and Apparatus for Conducting Chemical Reactions at Supercritical Conditions"; U.S. Patent 4,594,164, June 10, **1986**.

51. Titmas, J.A.; "Method and Apparatus for Enhancing Chemical Reactions at Supercritical Conditions"; U.S. Patent 4,792,408, Dec. 20, **1988**.

52. Swallow, K.C.; Killilea, W.R.; Hong, G.T.; Lee, H.W.; "Behavior of Metal Compounds in the Supercritical Water Oxidation Process"; Presented at the 20th Intersociety Conference on Environmental Systems, Williamsburg, VA, July 9-12, **1990**; SAE Technical Paper Series No. 901314.

53. Gulf Coast Waste Disposal Authority; "An Evaluation of Supercritical Deepwell Wet Oxidation of Sludge"; Houston, TX, November **1986**.

54. Stanford, C.C.; Gloyna, E.F.; "Energy Consumption in the Destruction of Wastewaters and Sludges by Supercritical Water Oxidation Deep-Shaft Reactors"; Technical Report No. CRWR230; University of Texas at Austin, **1991**.

55. Smith, J.M.; Raptis, T.J.; "Supercritical Deep Well Wet Oxidation of Liquid Organic Wastes"; Presented at the International Symposium on Subsurface Injection of Liquid Wastes, March **1986**.

56. Smith, J.M.; "Deep-Shaft Wet-Air Oxidation"; In *Standard Handbook for Hazardous Waste Treatment and Disposal*; Freeman, H.M., Ed.; McGraw-Hill: New York, **1987**, pp. 8.137-8.151.

57. Smith, J.M.; Hartman, G.L.; Raptis, T.J.; "Supercritical Deep Well Oxidation: A Low Cost Final Solution"; Presented at the APCA Specialty Conference, New Orleans, December **1986**.

58. Momont, J.A.; Berrigan, J.K. Jr.; Forbess, R.G.; Willett, M.R.; "Wet Air Oxidation for Detoxification of RCRA 'P' and 'U' Wastewaters"; Presented at the AIChE National Meeting, Los Angeles, CA, November 17-22, **1991**.

59. Dickinson, N.L; "Pollutant-Free Low Temperature Combustion Process Utilizing the Supercritical State"; U.S. Patent 4,292,953, October 6, **1981**.

60. Dickinson, N.L.; "Pollution-Free Pressurized Fluidized Bed Combustion Utilizing a High Concentration of Water Vapor"; U.S. Patent 4,377,066, March 22, **1983**.

61. Dickinson, N.L.; "Pollutant-Free Low Temperature Combustion Process Utilizing the Supercritical State"; U.S. Patent 4,380,960, April 26, **1983**.

62. Burleson, J.C.; "Method and Apparatus for Disposal of a Broad Spectrum of Waste Featuring Oxidation of Waste"; U.S. Patent 4,564,458, January 14, **1986**.

63. Dickinson, N.L.; "Combination of Supercritical Wet Combustion and Compressed Air Energy Storage"; U.S. Patent 4,593,202, June 3, **1986**.

64. Zeigler, J.E.; Peterscheck, H.W.; "Method and Apparatus for the Production and Liquefaction of Gases"; U.S. Patent 4,822,394, April 18, **1989**.

65. Hong, G.T.; Killilea, W.R.; Thomason, T.B.; "Method for Solids Separation in a Wet Oxidation Type Process"; U.S. Patent 4,822,497, April 18, **1989**.

66. Welch, J.F.; Siegwarth, J.D.; "Method for the Processing of Organic Compounds"; U.S. Patent 4,861,497, August 29, **1989**.

67. Zeigler, J.E.; Peterscheck, H.W.; "Method for Wet Oxidation Treatment"; U.S. Patent 4,891,139, January 2, **1990**.

68. Hossain, S.U.; Blaney, C.A.; "Method for Removing Polychlorinated Dibenzodioxins and Polychlorinated Dibenzofurans from Paper Mill Sludge"; U.S. Patent 5,075,017, December 24, **1991**.

69. Barton, D.M.; Schoeffel, E.W.; Zimmerman, F.J.; "Process and Apparatus for Complete Liquid-Vapor Phase Oxidation and High Enthalpy Vapor Production"; U.S. Patent 2,944,396, July 12, **1960**.

70. Thiel, R.; Dietz, K.-H.; Kerres, H.; Rosenbaum, H.J.; Steiner, S.; "Process for Wet Oxidation of Organic Substances"; U.S. Patent 4,141,829, February 27, **1979**.

71. Modell, M.; Reid, R.C.; Amin, S.I.; "Gasification Process"; U.S. Patent 4,113,446, September 12, **1978**.

72. Modell, M.; "Gasification and Liquefaction of Forest Products in Supercritical Water"; In *Fundamentals of Thermochemical Biomass Conversion*, Overend, R.P., Milne, T.A., and Mudge, L.K., Eds.; Elsevier Science Publishers: Amsterdam, **1985**, Chapter 6.

73. Li, L.; Eller, J.M.; Gloyna, E.F.; "Pilot-Scale Reactor Systems for Supercritical Water Oxidation of Wastewater and Sludge"; Presented at AIChE Summer National Meeting, San Diego, CA, August **1990**.

74. Tongdhamachart, C.; Gloyna, E.F.; "Supercritical Water Oxidation of Anaerobically Digested Municipal Sludge"; Technical Report No. CRWR229; University of Texas at Austin, February **1991**.

75. Wilmanns, E.G.; Gloyna, E.F.; "Supercritical Water Oxidation of Volatile Acids"; Technical Report No. CRWR218; University of Texas at Austin, November **1990**.

76. Takahashi, Y.; Wydeven, T.; Koo, C.; "Subcritical and Supercritical Water Oxidation of CELSS Model Wastes"; *Adv. Space Res.* **1989**, *9*(8), (8)99.

77. Shanableh, A.; Gloyna, E.F.; "Supercritical Water Oxidation—Wastewaters and Sludges"; *Water Sci. Technol.* **1991**, *23*(1-3), 389.

78. Lee, D.S.; Gloyna, E.F.; Li, L.; "Efficiency of H_2O_2 and O_2 in Supercritical Water Oxidation of 2,4-Dichlorophenol and Acetic Acid"; *J. Supercritical Fluids* **1990**, *3*(4), 249.

79. Buelow, S.T.; Dyer, R.B.; Atenolo, J.H.; Wander, J.D.; "Advanced Techniques for Soil Remediation of Propellant Components in Supercritical Water"; Report No. LA-UR-90-1338; Los Alamos National Laboratory, Los Alamos, NM, **1990**.

80. Bramlette, T.T.; Mills, B.E.; Hencken, K.R.; Brynildson, M.E.; Johnston, S.C.; Hruby, J.M.; Feemster, H.C.; Odegard, B.C.; Modell, M.; "Destruction of DOE/DP Surrogate Wastes with Supercritical Water Oxidation Technology"; Report No. SAND90-8229; Sandia National Laboratories, Livermore, CA, **1990**.

81. Huang, S.; Daehling, K.; Carleson, T.E.; Taylor, P.; Wai, C.; Propp, A.; "Thermodynamic Analysis of Corrosion of Iron Alloys in Supercritical Water"; In *Supercritical Fluid Science and Technology*; Johnston, K.P., and Penninger, J.M.L., Eds.; ACS Symposium Series 406; American Chemical Society: Washington, DC, **1989**, pp. 276-286.

82. Berry, W.E.; "The Corrosion Behavior of Fe-Cr-Ni Alloys in High-Temperature Water"; In *High Temperature, High Pressure Electrochemistry in Aqueous Solutions*; Jones, D. de G., and Staehle, R.W., Eds.; National Association of Corrosion Engineers: Houston, TX, **1976**, pp. 48-66.

83. Mills, B.E.; Kao, C.; Ottesen, D.K.; "Analysis of Oxidation and Corrosion Products in a Supercritical Water Oxidation Reactor Processing Simulated Mixed Waste"; Presented at the Third Annual Symposium on Emerging Technologies for Hazardous Waste Management, Atlanta, GA, October 1-3, **1991**.

84. MODAR, Inc.; "Detoxification and Disposal of Hazardous Organic Chemicals by Processing in Supercritical Water"; Final Report, U.S. Army Contract No. DAMD 17-80-C-0078, **1986**.

85. Thomas, A.J. III; Gloyna, E.F.; "Corrosion Behavior of High-Grade Alloys in the Supercritical Water Oxidation of Sludges"; Technical Report No. CRWR229; University of Texas at Austin, February **1991**.

86. Dickinson, N.L.; Welch, C.P.; "Heat Transfer to Supercritical Water"; *Transactions of the ASME* **1958**, April, 746.

87. Swenson, H.S.; Carver, J.R.; Kakarala, C.R.; "Heat Transfer to Supercritical Water in Smooth-Bore Tubes"; *J. Heat Transfer* **1965**, *87*(11), 477.

88. Yamagata, K.; Nishikawa, K.; Hasegawa, S.; Fujii, T.; Yoshida, S.; "Forced Convective Heat Transfer to Supercritical Water Flowing in Tubes"; *Int. J. Heat Mass Transfer* **1972**, *15*, 2575.

89. Michna, R.J.; Gloyna, E.F.; "Heat Transfer to Water in Countercurrent Flow Within a Vertical, Concentric-Tube Supercritical Water Oxidation Reactor"; Technical Report No. CRWR195; University of Texas at Austin, July **1990**.

90. Holgate, H.R.; Webley, P.A.; Helling, R.K.; Tester, J.W.; "Carbon Monoxide Oxidation in Supercritical Water: Effects of Heat Transfer and the Water-Gas Shift Reaction on Observed Kinetics"; Presented at the AIChE National Meeting, Los Angeles, CA, November 17-22, **1991**, and submitted for publication in *Energy & Fuels*.

91. Webley, P.A.; "Fundamental Oxidation Kinetics of Simple Compounds in Supercritical Water"; Doctoral Thesis in the Department of Chemical Engineering, Massachusetts Institute of Technology, Cambridge, MA, **1989**.

92. Paulaitis, M.E.; Penninger, J.M.L.; Gray, R.D. Jr.; Davidson, P.; *Chemical Engineering at Supercritical Fluid Conditions*; Ann Arbor Science: Ann Arbor, MI, **1983**.

93. McHugh, M.A.; Krukonis, V.J.; *Supercritical Fluid Extraction: Principles and Practice*; Butterworths: Boston, **1986**, pp. 199-215.

94. Johnston, K.P.; Haynes, C.; "Extreme Solvent Effects on Reaction Rate Constants at Supercritical Fluid Conditions"; *AIChE J.* **1987**, *33*(12), 2017.

95. Subramaniam, B.; McHugh, M.A.; "Reactions in Supercritical Fluids—A Review"; *Ind. Eng. Chem. Proc. Des. Dev.* **1986**, *25*, 1.
96. Staszak, C.N.; Malinowski, K.C.; Killilea, W.R.; "The Pilot-Scale Demonstration of the MODAR Oxidation Process for the Destruction of Hazardous Organic Waste Materials"; *Environ. Prog.* **1987**, *6*(1), 39.
97. Wightman, T.J.; "Studies in Supercritical Wet Air Oxidation"; M.S. Thesis in the Department of Chemical Engineering, University of California, Berkeley, CA, **1981**.
98. Helling, R.K.; Tester, J.W.; "Oxidation Kinetics of Simple Compounds in Supercritical Water"; Presented at the AIChE Summer National Meeting, Boston, MA, August **1986**.
99. Helling, R.K.; Tester, J.W.; "Oxidation Kinetics of Carbon Monoxide in Supercritical Water"; *Energy and Fuels* **1987**, *1*, 417.
100. Helling, R.K.; Tester, J.W.; "Oxidation of Simple Compounds and Mixtures in Supercritical Water: Carbon Monoxide, Ammonia and Ethanol"; *Environ. Sci. Technol.* **1988**, *22*(11), 1319.
101. Webley, P.A.; Tester, J.W.; "Fundamental Kinetics and Mechanistic Pathways for Oxidation Reactions in Supercritical Water"; Presented at the 18th Intersociety Conference on Environmental Systems, San Francisco, CA, July 11-13, **1988**; SAE Technical Paper Series No. 881039.
102. Webley, P.A.; Tester, J.W.; "Oxidation of Methane in Supercritical Water"; *Energy & Fuels* **1991**, *5*, 411.
103. Yang, H.H.; Eckert, C.A.; "Homogeneous Catalysis in the Oxidation of p-Chlorophenol in Supercritical Water"; *Ind. Eng. Chem. Res.* **1988**, *27*, 2009.
104. Webley, P.A.; Holgate, H.R.; Stevenson, D.M.; Tester, J.W.; "Oxidation Kinetics of Model Compounds of Human Metabolic Waste in Supercritical Water"; Presented at the 20th Intersociety Conference on Environmental Systems, Williamsburg, VA, July 9-12, **1990**; SAE Technical Paper Series No. 901333.
105. Webley, P.A.; Tester, J.W.; Holgate, H.R.; "Oxidation Kinetics of Ammonia and Ammonia-Methanol Mixtures in Supercritical Water in the Temperature Range 530-700°C at 246 bar"; *Ind. Eng. Chem. Res.* **1991**, *30*(8), 1745.
106. Webley, P.A.; Tester, J.W.; "Fundamental Kinetics of Methanol Oxidation in Supercritical Water"; In *Supercritical Fluid Science and Technology*; Johnston, K.P., and Penninger, J.M.L., Eds.; ACS Symposium Series 406; American Chemical Society: Washington, DC, **1989**, pp. 259-275.
107. Lee, D.S.; Gloyna, E.F.; "Supercritical Water Oxidation of Acetamide and Acetic Acid"; Technical Report No. CRWR209; University of Texas at Austin, September **1990**.
108. Thornton, T.D.; Savage, P.E.; "Phenol Oxidation in Supercritical Water"; *J. Supercritical Fluids* **1990**, *3*(4), 240.
109. Thornton, T.D.; LaDue, D.E. III; Savage, P.E.; "Phenol Oxidation in Supercritical Water: Formation of Dibenzofuran, Dibenzo-p-dioxin, and Related Compounds"; *Environ. Sci. Technol.* **1991**, *25*, 1507.
110. Holgate, H.R.; Tester, J.W.; "Fundamental Kinetics and Mechanisms of Hydrogen Oxidation in Supercritical Water"; Presented at the Second International Conference on Supercritical Fluids, Boston, MA, May 20-22, **1991**, and submitted for publication in *Combust. Sci. Technol.*
111. Amin, S.; Reid, R.C.; Modell, M.; "Reforming and Decomposition of Glucose in an Aqueous Phase"; Presented at the Intersociety Conference on Environmental Systems, San Francisco, CA, July 21-24, **1975**; ASME Paper No. 75-ENAs-21.
112. Antal, M.J. Jr.; Mok, W.S.-L.; Richards, G.N.; "Four-Carbon Model Compounds for the Reactions of Sugar in Water at High Temperatures"; *Carbohydr. Res.* **1990**, *199*, 111.

113. Antal, M.J. Jr.; Mok, W.S.L.; Richards, G.N.; "Mechanism of Formation of 5–(Hydroxymethyl)-2-Furaldehyde from D-Fructose and Sucrose"; *Carbohydr. Res.* **1990**, *199*, 91.

114. Antal, M.J. Jr.; Leesomboon, T.; Mok, W.S.-L.; Richards, G.N.; "Mechanism of Formation of 2-Furaldehyde from D-Xylose"; *Carbohydr. Res.* **1991**, *217*, 71.

115. Mok, W.S.-L.; Antal, M.J. Jr.; Varhegyi, G.; "Productive and Parasitic Pathways in Dilute Acid-Catalyzed Hydrolysis of Cellulose"; *Ind. Eng. Chem. Res.* **1992**, *31*(1), 94.

116. Ramayya, S.; Brittain, A.; DeAlmeida, C.; Mok, W.S.-L.; Antal, M.J. Jr.; "Acid-Catalysed Dehydration of Alcohols in Supercritical Water"; *Fuel* **1987**, *66*, 1364.

117. Narayan, R.; Antal, M.J. Jr.; "Kinetic Elucidation of the Acid Catalyzed Mechanism of 1-Propanol Dehydration in Supercritical Water"; In *Supercritical Fluid Science and Technology*; Johnston, K.P., and Penninger, J.M.L., Eds.; ACS Symposium Series 406; American Chemical Society: Washington, DC, **1989**, pp. 226-241.

118. Narayan, R.; Antal, M.J. Jr.; "Influence of Pressure on Acid-Catalyzed Rate Constant for 1-Propanol Dehydration in Supercritical Water"; *J. Am. Chem. Soc.* **1990**, *112*(5), 1927.

119. Xu, X.; DeAlmeida, C.; Antal, M.J. Jr.; "Mechanism and Kinetics of the Acid-Catalyzed Dehydration of Ethanol in Supercritical Water"; *J. Supercritical Fluids* **1990**, *3*(4), 228.

120. Xu, X.; DeAlmeida, C.; Antal, M.J. Jr.; "Mechanism and Kinetics of the Acid-Catalyzed Formation of Ethene and Diethyl Ether from Ethanol in Supercritical Water"; *Ind. Eng. Chem. Res.* **1991**, *30*(7), 1478.

121. Whitlock, D.R.; "Organic Reactions in Supercritical Water"; M.S. Thesis in the Department of Chemical Engineering, Massachusetts Institute of Technology, Cambridge, MA, **1978**.

122. Lawson, J.R.; Klein, M.T.; "Influence of Water on Guaiacol Pyrolysis"; *Ind. Eng. Chem. Fundam.* **1985**, *24*, 203.

123. Abraham, M.A.; Klein. M.T.; "Pyrolysis of Benzyl Phenylamine Neat and with Tetralin, Methanol and Water Solvents"; *Ind. Eng. Chem. Prod. Res. Dev.* **1985**, *24*, 300.

124. Townsend, S.H.; Abraham, M.A.; Huppert, G.L.; Klein, M.T.; Paspek, S.C.; "Solvent Effects During Reactions in Supercritical Water"; *Ind. Eng. Chem. Res.* **1988**, *27*, 143.

125. Huppert, G.L.; Wu, B.C.; Townsend, S.H.; Klein, M.T.; Paspek, S.C.; "Hydrolysis in Supercritical Water: Identification and Implications of a Polar Transition State"; *Ind. Eng. Chem. Res.* **1989**, *28*, 161.

126. Klein, M.T.; Torry, L.A.; Wu, B.C.; Townsend, S.H.; Paspek, S.C.; "Hydrolysis in Supercritical Water: Solvent Effects as a Probe of the Reaction Mechanism"; *J. Supercritical Fluids* **1990**, *3*(4), 228.

127. Klein, M.T.; Mentha, Y.G.; Torry, L.A.; "Decoupling Substituent and Solvent Effects During Hydrolysis of Substituted Anisoles in Supercritical Water"; *Ind. Eng. Chem. Res.* **1992**, *31*(1), 182.

128. Wu, B.C.; Klein, M.T.; Sandler, S.I.; "Solvent Effects on Reactions in Supercritical Fluids"; *Ind. Eng. Chem. Res.* **1991**, *30*(5), 822.

129. Antal, M.J. Jr.; Brittain, A.; DeAlmeida, C.; Ramayya, S.; Roy, J.C.; "Heterolysis and Homolysis in Supercritical Water"; In *Supercritical Fluids: Chemical and Engineering Principles and Applications*; Squires, T.G., and Paulaitis, M.E.; Eds.; ACS Symposium Series 329; American Chemical Society: Washington, DC, **1987**, pp. 77-86.

130. Mok, W.S.-L.; Antal, M.J. Jr.; Jones, M. Jr.; "Formation of Acrylic Acid from Lactic Acid in Supercritical Water"; *J. Org. Chem.* **1989**, *54*, 4596.

131. Deshpande, G.V.; Holder, G.D.; Bishop, A.A.; Gopal, J.; "Extraction of Coal Using Supercritical Water"; *Fuel* **1984**, *63*, 956.
132. Tiffany, D.M.; Houser, T.J.; McCarville, M.E.; Houghton, M.E.; "Reactivity of Some Nitrogen-Containing Compounds at Supercritical Water Conditions"; *Prepr. Pap. Am. Chem. Soc. Div. Fuel Chem.* **1984**, *29*(5), 56.
133. Houser, T.J.; Tiffany,D.M.; Li, Z.; McCarville, E.; Houghton, M.E.; "Reactivity of Some Organic Compounds with Supercritical Water"; *Fuel* **1986**, *65*(6), 827.
134. Houser, T.J.; Dyla, J.E.; Van Atten, M.K.; Li, Z.; McCarville, M.E.; "Continued Study of the Reactivity of Organic Compounds with Supercritical Water"; In *International Conference on Coal Science*; Moulijn, J.A., Ed.; Elsevier Science Publishers: Amsterdam, **1987**, pp. 773-776.
135. Houser, T.J.; Tsao, C.-C.; Dyla, J.E.; Van Atten, M.K.; McCarville, M.E.; "The Reactivity of Tetrahydroquinoline, Benzylamine and Bibenzyl with Supercritical Water"; *Fuel* **1989**, *68*, 323.
136. Amestica, L.A.; Wolf, E.E.; "Catalytic Liquefaction of Coal with Supercritical Water/CO/Solvent Media"; *Fuel* **1986**, *65*, 1226.
137. Mikita, M.A.; Fish, H.T.; "Reactions of Coal and Coal Model Compounds with Supercritical Water"; *Prepr. Pap. Am. Chem. Soc. Div. Fuel Chem.* **1986**, *31*(4), 56.
138. Horiuchi, A.K.; Fish, H.T.; Mikita, M.A.; "Amphoteric Reactions of Supercritical Water with Coal Models"; *Prepr. Pap. Am. Chem. Soc. Div. Fuel Chem.* **1988**, *33*(3), 292.
139. Greenkorn, R.A.; Chao, K.C.; "Supercritical Fluid Extraction of Coal"; Final Report, DOE Contract No. FG 22-83PC60036, **1987**.
140. Ross, D.S.; Green, T.K.; Mansani, R.; Hum, G.P.; "Coal Conversion in CO/Water. 1. Conversion Mechanism"; *Energy & Fuels* **1987**, *1*(3), 287.
141. Ross, D.S.; Green, T.K.; Mansani, R.; Hum, G.P.; "Coal Conversion in CO/Water. 2. Oxygen Loss and the Conversion Mechanism"; *Energy & Fuels* **1987**, *1*(3), 292.
142. Ross, D.S.; Hum, G.P.; Miin, T.-C.; Green, T.K.; Mansani, R.; "Isotope Effects in Supercritical Water: Kinetic Studies of Coal Liquefaction"; In *Supercritical Fluids: Chemical and Engineering Principles and Applications*; Squires, T.G., and Paulaitis, M.E., Eds.; ACS Symposium Series 329; American Chemical Society: Washington, DC, **1987**, pp. 242-250.
143. Penninger, J.M.L.; "Reactions of Di-n-butylphthalate in Water at Near-Critical Temperature and Pressure"; *Fuel* **1988**, *67*, 490.
144. Penninger, J.M.L.; Kolmschate, J.M.M.; "Chemistry of Methoxynaphthalene in Supercritical Water"; In *Supercritical Fluid Science and Technology*, Johnston, K.P., and Penninger, J.M.L., Eds.; ACS Symposium Series 406; American Chemical Society: Washington, DC, **1989**, pp. 242-258.
145. Abraham, M.A.; Klein, M.T.; "Solvent Effects During the Reaction of Coal Model Compounds"; In *Supercritical Fluids: Chemical and Engineering Principles and Applications*; Squires, T.G., and Paulaitis, M.E., Eds.; ACS Symposium Series 329; American Chemical Society: Washington, DC, **1987**, pp. 67-76.
146. Flarsheim, W.M.; Tsou, Y.M.; Trachtenberg, I.; Johnston, K.P.; Bard, A.J.; "Electrochemistry in Near-Critical and Supercritical Fluids 3. Studies of Br⁻, I⁻, and Hydroquinone in Aqueous Solutions"; *J. Phys. Chem.* **1986**, *90*, 3857.
147. Flarsheim, W.M.; Bard, A.J.; Johnston, K.P.; "Pronounced Pressure Effects on Reversible Electrode Reactions in Supercritical Water"; *J. Phys. Chem.* **1989**, *93*(10), 4234.
148. Schilling, W.; Franck, E.U.; "Combustion and Diffusion Flames at High Pressures to 2000 bar"; *Ber. Bunsenges. Phys. Chem.* **1988**, *92*, 631.

149. Franck, E.U.; "High Pressure Combustion and Flames in Supercritical Water"; Proceedings of the Second International Symposium on Supercritical Fluids, Boston, MA, May 20-22, **1991**, pp. 91-96.

150. Armellini, F.J.; Tester, J.W.; "Salt Separation During Supercritical Water Oxidation of Human Metabolic Waste: Fundamental Studies of Salt Nucleation and Growth"; Presented at the 20th Intersociety Conference on Environmental Systems, Williamsburg, VA, July 9-12, **1990**; SAE Technical Paper Series No. 901313.

151. Armellini, F.J.; Tester, J.W.; "Experimental Methods for Studying Salt Nucleation and Growth from Supercritical Water"; *J. Supercritical Fluids* **1991**, *4*(4), 254.

152. Killilea, W.R.; Hong, G.T.; Swallow, K.C.; Thomason, T.B.; "Supercritical Water Oxidation: Microgravity Solids Separation"; Presented at the 18th Intersociety Conference on Environmental Systems, San Francisco, CA, July 11-13, **1988**; SAE Technical Paper Series No. 881038.

153. Dell'Orco, P.C.; Gloyna, E.F.; "The Separation of Solids from Supercritical Water Oxidation Processes"; Presented at the AIChE National Meeting, Los Angeles, CA, November 17-22, **1991**.

154. Jin, L.; Shah, Y.T.; Abraham, M.A.; "The Effect of Supercritical Water on the Catalytic Oxidation of 1,4-Dichlorobenzene"; *J. Supercritical Fluids* **1990**, *3*(4), 233.

155. Rofer, C.K.; Streit, G.E.; "Kinetics and Mechanism of Methane Oxidation in Supercritical Water"; Report No. LA-11439-MS; Los Alamos National Laboratory, Los Alamos, NM, **1988**.

156. Rofer, C.K.; Streit, G.E.; "Phase II Final Report: Oxidation of Hydrocarbons and Oxygenates in Supercritical Water"; Report No. LA-11700-MS; Los Alamos National Laboratory, Los Alamos, NM, **1989**.

157. Helling, R.K.; "Oxidation Kinetics of Simple Compounds in Supercritical Water: Carbon Monoxide, Ammonia and Ethanol"; Doctoral Thesis in the Department of Chemical Engineering, Masssachusetts Institute of Technology, Cambridge, MA, **1986**.

158. Li, L.; Chen, P.; Gloyna, E.F.; "Generalized Kinetic Model for Wet Oxidation of Organic Compounds"; *AIChE J.* **1991**, *37*(11), 1687.

159. Melius, C.F.; Bergan, N.E.; Shepherd, J.E.; "Effects of Water on Combustion Kinetics at High Pressure"; *Symp. (Int.) Combust. (Proc.)* **1990**, *23rd*, 217.

160. Butler, P.B.; Bergan, N.E.; Bramlette, T.T.; Pitz, W.J.; Westbrook, C.K.; "Oxidation of Hazardous Waste in Supercritical Water: A Comparison of Modeling and Experimental Results for Methanol Destruction"; Presented at the Spring Meeting of the Western States Section of the Combustion Institute, Boulder, CO, March 17-19, **1991**.

161. Schmitt, R.G.; Butler, P.B.; Bergan, N.E.; Pitz, W.J.; Westbrook, C.K.; "Destruction of Hazardous Wastes in Supercritical Water. Part II: A Study of High-Pressure Methanol Oxidation Kinetics"; Presented at the Fall Meeting of the Western States Section of the Combustion Institute, University of California at Los Angeles, October 13-15, **1991**.

162. Margolis, S.B.; Johnston, S.C.; "Nonadiabaticity, Stoichiometry and Mass Diffusion Effects on Supercritical Combustion in a Tubular Reactor"; *Symp. (Int.) Combust. (Proc.)* **1990**, *23rd*, 533.

163. Sourirajan, S.; Kennedy, G.C.; "The System H_2O-NaCl at Elevated Temperatures and Pressures"; *Amer. J. Sci.* **1962**, *260*, 115.

164. Bischoff, J.L.; Rosenbauer, R.J.; Pitzer, K.S.; "The System NaCl-H_2O: Relations of Vapor-Liquid Near the Critical Temperature of Water and of Vapor-Liquid-Halite from 300 to 500°C"; *Geochim. Cosmochim. Acta* **1986**, *50*, 1437.

165. Parisod, C.J.; Plattner, E.; "Vapor-Liquid Equilibria of the NaCl-H_2O System in the Temperature Range 300-400°C"; *J. Chem. Eng. Data* **1981**, *26*, 16.

166. Ölander, A.; Liander, H.; "The Phase Diagram of Sodium Chloride and Steam Above the Critical Point"; *Acta. Chim. Scand.* **1950**, *4*, 1437.
167. Khaibullin, I.Kh.; Borisov, N.M.; "Experimental Investigation of the Thermal Properties of Aqueous and Vapor Solutions of Sodium and Potassium Chlorides at Phase Equilibrium"; *High Temperature* **1966**, *4*, 489.
168. Armellini, F.J.; Tester, J.W.; "Solubilities of Sodium Chloride and Sodium Sulfate in Sub- and Supercritical Water Vapor"; Presented at the AIChE National Meeting, Los Angeles, CA, November 17-22, **1991**.
169. Pitzer, K.S.; Tanger, J.C. IV; "Near-Critical NaCl-H$_2$O: An Equation of State and Discussion of Anomalous Properties"; *Int. J. Thermophys.* **1989**, *9*, 635.

RECEIVED September 3, 1992

Chapter 4

Photo-Assisted Mineralization of Herbicide Wastes by Ferric Ion Catalyzed Hydrogen Peroxide

Joseph J. Pignatello and Yunfu Sun

Connecticut Agricultural Experiment Station, 123 Huntington Street, P.O. Box 1106, New Haven, CT 06504

Photolysis of hydrogen peroxide and ferric ion in dilute aqueous acid completely mineralized the chlorophenoxyalkanoic acid herbicides 2,4-D and 2,4,5-T to chloride and CO_2 in less than 2 h. These results could be achieved using either a "black lamp", which emits in the near UV, or a conventional fluorescent lamp, which emits mostly in the visible region along with a small amount in the UV. Complete mineralization required as little as 5 moles of peroxide per mole of (0.1 mM) herbicide with either light source. The thermal reaction (Fe^{3+}/H_2O_2) gave only partial conversion to CO_2. Initial rates of transformation by the full system ($Fe^{3+}/H_2O_2/h\nu$) was greater than the sum of the rates of the component systems (Fe^{3+}/H_2O_2 and the photolyzed single-reactant systems, $Fe^{3+}/h\nu$ and $H_2O_2/h\nu$). Direct photolysis of H_2O_2 was negligible, but photolysis of Fe(III) species was inferred to play a major role in both transformation and mineralization of the herbicides.

Hydrogen peroxide has gained interest as a low temperature oxidant for treatment of organic chemical wastes (1-6). When combined with certain transition metal ions (7) or photolyzed with UV light (8), it can be a source of the hydroxyl radical (OH·), a potent oxidant of organic compounds (9). Technologies for treating waste water using direct UV photolysis of hydrogen peroxide are commercially available (e.g., 1,2). Hydrogen peroxide also augments ozonolysis (10) and heterogeneous TiO_2 photocatalytic oxidation (11) of organics.

We are investigating iron-catalyzed hydrogen peroxide for treating pesticide wastes (6,12-13). Fe(II) reacts with H_2O_2 to generate OH· in the Fenton reaction (equations 1-2) (7). Fe(III) catalyzes the decomposition of H_2O_2 to water and molecular oxygen; according to the classical "radical chain" mechanism (7,14), OH· is generated via transient Fe^{2+} (equations 3-6).

0097–6156/93/0518–0077$06.00/0

Fenton reaction:

$$Fe^{2+} + H_2O_2 \longrightarrow Fe^{3+} + OH\cdot + HO^- \qquad (1)$$

$$OH\cdot + Fe^{2+} \longrightarrow HO^- + Fe^{3+} \qquad (2)$$

Fe(III)-catalyzed reaction:

$$Fe^{3+} + H_2O_2 \rightleftharpoons Fe\text{-}O_2H^{2+} + H^+ \qquad (3)$$

$$Fe\text{-}O_2H^{2+} \longrightarrow HO_2\cdot + Fe^{2+} \qquad (4)$$

$$Fe^{3+} + HO_2\cdot \longrightarrow Fe^{2+} + O_2 + H^+ \qquad (5)$$

$$OH\cdot + H_2O_2 \longrightarrow H_2O + HO_2\cdot \qquad (6)$$

Reactive high valent iron-oxo complexes (e.g., ferryl, $Fe(IV)=O$) have also been implicated in reactions of both $Fe(II)$ (15) and $Fe(III)$ (16) complexes with hydrogen peroxide.

Fenton's reagent has been investigated for treatment of contaminants in waste-water and soil (3,5, citations in 6). Much less attention has been paid to the Fe(III)-catalyzed reaction for oxidizing organic wastes. Murphy et al. (4) found that Fe^{3+}/H_2O_2 was effective in mineralizing formaldehyde wastes. Recently we demonstrated that Fe^{3+}/H_2O_2 rapidly oxidizes chlorophenoxyalkanoic acid herbicides 2,4-D and 2,4,5-T in weakly acidic solution (6). Dechlorination was quantitative but mineralization of ring- and carboxy-^{14}C-labelled compounds to CO_2 was incomplete. Mineralization to CO_2 is greatly enhanced, however, by irradiating solutions with near-UV light (300-400 nm), or with visible light containing a near-UV component. Some of these results are presented here. Photolysis of H_2O_2 itself is shown to be unimportant in these systems. So far, little attention has been given to photolysis in combination with Fe^{3+}/H_2O_2 for destruction of wastes.

Experimental

Reaction mixtures (100 ml) containing $Fe(ClO_4)_3$, $HClO_4$, and pure herbicide were prepared in 250 ml Pyrex Erlenmeyer flasks, as previously described (6). Degradation was initiated by adding a small volume of concentrated H_2O_2 to vigorously stirred solutions. Oxygen in solution and in the headspace was in large excess with respect to mineralization of carbon. Irradiation was carried out with either four 15 W "black lamps" (F15T8/BLB, General Electric), which emit light in the near UV, or four 200 W "cool white" fluorescent lamps (F96T12/CW/HO, Westinghouse), which emit mostly in the visible region but contain a small UV component. Emission spectra of the lamps are given in Figure 1. Irradiation intensities are given in the respective figure legends. Photon flux was determined by chemical actinometry using ferrioxalate (17). Analyses were performed as follows: herbicide by HPLC, chloride potentiometrically using a halide-selective electrode, ^{14}C by liquid scintillation counting, and H_2O_2 by iodometric titration with thiosulfate in the presence of fluoride to complex Fe(III) (6).

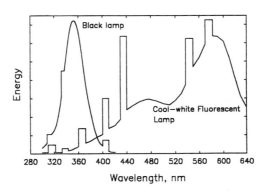

Figure 1. Emission spectra of black and visible fluorescent lamps supplied by the manufacturer.

Results and Discussion

Thermal Reaction. In the dark, dilute solutions of Fe^{3+} and H_2O_2 at pH 2.8 transform 2,4-D and 2,4,5-T in minutes. Table I shows that chloride ion was evolved stoichiometrically, but mineralization to CO_2 was incomplete, even at 500 mM H_2O_2 (1.7% by weight). Both ring- and carboxy-^{14}C-labelled 2,4-D were partially converted to [14]CO_2. In all cases, the remainder of the ^{14}C added was found in the reaction solution.

Table I. Oxidation of Chlorophenoxyacetate Herbicides by Fe^{3+}/H_2O_2 in the Dark; 1 mM Fe^{3+}, pH 2.8, 21° C

			% theoretical yield	
labelled cpd., mM		[H_2O_2], mM	Cl^- [a]	[14]CO_2[b]
ring-2,4-D	0.20	10	100	40
"	0.10	0	0	0
"	0.10	10	na[c]	45
"	0.10	100	na	56
"	0.10	500	na	69
carboxy-2,4-D	0.10	10	na	37
ring-2,4,5-T	0.175	10	103	41

[a]After 30-40 min. [b]After 2-3 h. Balance of radioactivity found in solution. [c]Not analyzed.

Other features of the thermal reaction are summarized below; they have been discussed in detail elsewhere (6).

1. With excess peroxide, the rate and extent of mineralization is independent of whether Fe^{2+} or Fe^{3+} is employed.

2. The reaction rate drops steeply on either side of the optimum pH 2.7-2.8. Hydrogen peroxide decomposition in the absence of herbicide gives the same pH profile. Above the optimum pH, reactivity decreases because Fe^{3+} precipitates as ferric oxyhydroxide; while below this optimum, the formation of $Fe-O_2H^{2+}$ (the initial step leading to generation of active oxidants, (equation 3) is inhibited.

3. The corresponding di- or trichlorophenol is a transient, low yield (~ 10%) intermediate. Chloride evolution tracks closely with loss of herbicide + the phenol, indicating no major concentrations of other chlorinated intermediates can be expected. Organic ^{14}C-containing products remaining after dechlorination of ring-labelled 2,4-D are not extractable by ethyl acetate, and therefore are highly hydrophilic.

4. The reaction is inhibited by sulfate or chloride ions. Sulfate reduces the rate by complexing with Fe^{3+}. Chloride intercepts $OH \cdot$ to give $Cl_2^{-\cdot}$, and results in a change in mechanism from preferential attack on the aromatic ring to attack on the side chain.

Photo-Assisted Reaction. Rapid and complete mineralization of both herbicides could be achieved by irradiating solutions. Figure 2 shows degradation of ring-labelled 2,4-D in reactions kept in the dark or irradiated with cool-white fluorescent light.

As Figure 2A indicates, transformation of 2,4-D was faster in the photo-assisted reaction ($Fe^{3+}/H_2O_2/h\nu$) than in the thermal reaction (Fe^{3+}/H_2O_2), and much faster than in the photolyzed single-reactant systems ($Fe^{3+}/h\nu$ and $H_2O_2/h\nu$). Figure 2B shows that $Fe^{3+}/H_2O_2/h\nu$ gave quantitative loss of solution-^{14}C in less than 2 h, whereas Fe^{3+}/H_2O_2 gave only about 40% loss of solution-^{14}C, with no further changes after 2 h. $Fe^{3+}/h\nu$ gave slow and incomplete mineralization. $H_2O_2/h\nu$ and the dark Fe^{3+}-only control showed no reaction. Results similar to these were obtained for 2,4,5-T (6).

Table II shows initial reaction rates for 2,4-D transformation under various conditions, using the black light for photolyzed reactions. Again, $H_2O_2/h\nu$ reactions were negligible. By inspection it can be seen that the rates in the full system, $Fe^{3+}/H_2O_2/h\nu$, are greater than the sum of the rates in the component systems, Fe^{3+}/H_2O_2, $Fe^{3+}/h\nu$, and $H_2O_2/h\nu$. This conclusion also holds for both 2,4-D and 2,4,5-T under visible light at a single H_2O_2 concentration (Table III).

On irradiation with the black light, complete mineralization of 2,4-D could be achieved in about 1 h using a molar ratio of H_2O_2 to 2,4-D as low as 5:1 (Figure 3). Even a 2:1 molar ratio gave substantial mineralization in 2 h. Ratios higher than 5:1 did not increase the mineralization rate. Under visible fluorescent light, a 5:1 molar ratio also proved effective, except that mineralization was somewhat slower; hence, in that case the solution ^{14}C curve at a 5:1 molar ratio was almost superimposable on the curve shown in Figure 2 where the ratio was 100:1 (6). Since the stoichiometry for oxidation of 2,4-D solely by peroxide requires a molar ratio of 15:1, these results indicate that $Fe^{3+}/H_2O_2/h\nu$ utilizes molecular oxygen, in part, for oxidizing equivalents.

Table II. Initial rates of 2,4-D disappearance (10^{-6} M/min) under UV light (black lamp) [a]

System	0.02 mM H_2O_2	1.0 mM H_2O_2	50 mM H_2O_2
$Fe^{3+}/H_2O_2/h\nu$	18	57	60
Fe^{3+}/H_2O_2	0.17	2.4	19
$Fe^{3+}/h\nu$	6.9[b]	6.9[b]	6.9[b]
$H_2O_2/h\nu$	0.1	0.2	0.3

[a]Common conditions: [2,4-D] = 0.1 mM; [Fe] = 1.0 mM; pH 2.8.   [b]No peroxide.

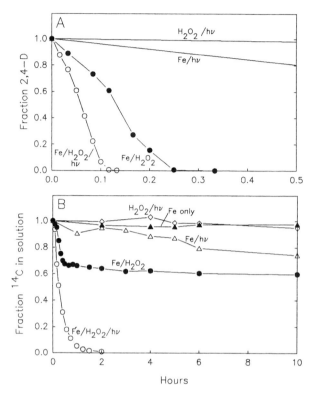

Figure 2. Oxidation of 2,4-D under various conditions. **A**, disappearance of parent compound; **B**, volatization of carbon starting with ring-labelled 2,4-D. The light source was the visible fluorescent lamp: 0.035 mW/cm² (290-385 nm) and 1.25 mW/cm² (400-700 nm). Conditions: [2,4-D] = 0.1, [Fe³⁺] = 1.0, [H₂O₂] = 10 mM; pH 2.8; ionic strength = 0.2 M (NaClO₄).

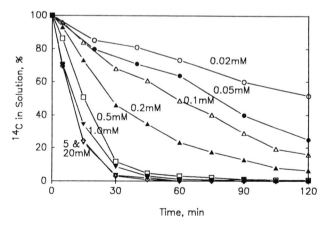

Figure 3. Mineralization of ring-¹⁴C-2,4-D under the black lamp at different hydrogen peroxide concentrations. Conditions: 2.5 mW/cm² (290-385 nm), 9 x 10¹⁷ quanta s⁻¹L⁻¹; [2,4-D] = 0.1, [Fe³⁺] = 0.2 mM; pH 2.8.

Table III. Initial rates of 2,4-D and 2,4,5-T disappearance (10^{-6} M/min) under visible light [a]

system	2,4-D	2,4,5-T
$Fe^{3+}/H_2O_2/hv$	13.2	8.5
Fe^{3+}/H_2O_2	5.4	4.7
Fe^{3+}/hv	0.7[b]	0.5[b]
H_2O_2/hv	~0.0	~0.0

[a]Common conditions: [2,4-D] = 0.1 mM; [Fe^{3+}] = 1.0 mM; [H_2O_2] = 10 mM; pH 2.8.   [b]No peroxide.

Mechanism. The thermal reaction leading to OH· presumably involves the chain outlined in equations 1-6. Photo-assistance is due potentially to photolysis of H_2O_2, Fe(III) species, or direct photolysis of the herbicide itself or its products.

Hydrogen peroxide absorbs only weakly above 300 nm, and the reaction vessel screened out wavelengths less than that. Photolysis generates two OH· :

$$H_2O_2 \xrightarrow{hv} 2\ OH· \tag{7}$$

Inspection of Figure 2 and Tables II and III reveals that both transformation and mineralization of 2,4-D and 2,4,5-T due to photolysis of H_2O_2 was insignificant. Moreover, the slow rates by H_2O_2/hv implies that direct photolysis of the herbicide was also insignificant.

This leaves photolysis of Fe(III) species. Above 300 nm, the monohydroxy pentaaquo complex $FeOH^{2+}$ is predicted to be the most important photo-labile species in solutions of ferric ion at pH 2.8 in the absence of non-complexing ligands (*18*). Of minor importance are the hexaaquo ion Fe^{3+}(aq) and the dimer $Fe_2(OH)_2^{4+}$. These complexes photolyze to give OH· by dissociation of the ligand-to-metal charge transfer excited state; e.g.,

$$FeOH^{2+} \xrightarrow{hv} Fe^{2+} + OH· \tag{8}$$

$FeOH^{2+}$ comprises about half of total dissolved Fe(III) at pH 2.75, and has a quantum yield for OH· production of 0.14 at 313 nm and 0.017 at 360 nm (*18*).

Since Fe^{2+} is also a product, photolysis of $FeOH^{2+}$ in the presence of hydrogen peroxide may generate a second OH· by the Fenton reaction (equation 1). This ancillary production of hydroxyl can therefore as much as double the contribution of Fe^{3+}/hv to degradation by $Fe^{3+}/H_2O_2/hv$, depending on [H_2O_2] (i.e., depending on the rate of reaction 1). However, Tables II and III show that the initial rate of 2,4-D disappearance in the full system exceeds the sum of the component systems even when the rate due to Fe^{3+}/hv is doubled.

Other reactions that potentially could contribute to degradation include photolysis of $Fe-O_2H^{2+}$, which exists in solution in small concentrations (*19*), and photolysis of Fe(III)-complexes of herbicide degradation intermediates bearing carboxylate groups. These complexes may photolyze via ligand-to-metal charge transfer excitation in the near UV in a manner analogous to $FeOH^{2+}$ (equation 8). An example of photolysis of an Fe-carboxylate is ferrioxalate (*17*). These reactions can result in direct degradation, by giving decarboxylation and an organoradical that can be further oxidized by O_2, and/or indirect degradation via production of Fe^{2+}, which can give more OH· by the Fenton reaction. Further study is needed to determine the photo-active species involved.

Acknowledgments. We are grateful for the help of Marta Day and grants from the USDA National Pesticide Impact Assessment and Water Quality Special Grants Programs.

Literature Cited

1. Sundstrom, D.W.; Weir, B.A.; Redig, K.A. In: D.W. Tedder and F.G. Pohland (eds.), *Emerging Technologies in Hazardous Waste Management*, ACS Symposium Series 422, American Chemical Society, Washington, D.C., **1990**, pp. 67-76.
2. Hager, D.G. In: H.M. Freeman (ed.) *Innovative Hazardous Waste Treatment Technology Series. Vol. 2. Physical/Chemical Processes*, **1990**, Technomic Publ. Co., Lancaster, PA. pp. 143-153.
3. Barbeni, M.; Minero, C.; Pelizzetti, E.; *Chemosphere*, **1987**, *16*, 2225-2237.
4. Murphy, A. P.; Boegli, W.J.; Price, M.K.; Moody, C.D.; *Environ. Sci. Technol.*, **1989**, *23*, 166-169.
5. Watts, R.J.; Udell, M.D.; Rauch, P.A. *Haz. Waste Haz. Mat.*, **1990**, 7, 335-345.
6. Pignatello, J.J. *Environ. Sci. Technol.*, in press.
7. Walling, C.H. *Acc. Chem. Res.*, **1975**, *8*, 125-131.
8. Vaghjiani, G.L.; Ravishankara, A.R.; *J. Chem. Phys.*, **1990**, 92,996-1003.
9. Dorfman, L.M., and G.E. Adams. **1973**. Reactivity of the hydroxyl radical in aqueous solutions; NSRDS-46; National Bureau of Standards (U.S.); U.S. Government Printing Office, Washington, D.C.
10. Peyton, G.R. In: D.W. Tedder and F.G. Pohland (eds.), *Emerging Technologies in Hazardous Waste Management*, ACS Symposium Series 422, American Chemical Society, Washington, D.C., **1990**, pp. 100-118.
11. Ollis, D.F.; Pelizzetti, E.; Serpone, N. *Environ. Sci. Technol.*, **1991**, *25*, 1522-1529.
12. Sun, Y.; Pignatello, J.J. *J. Agric. Food Chem.*, **1992**, *40*, 322-327.
13. Sun, Y.; Pignatello, J.J. *J. Agric. Food Chem.*, under review.
14. Barb, W.G.; Baxendale, J.H.; George, P.; Hargrave, K.R.; *Trans. Faraday Soc.*, **1951**, *47*, 591-616.
15. Rahhal, S.; Richter, H.W.; *J. Am. Chem. Soc.*, **1988**, 110, 3126-3133.
16. Kremer, M.L.; *Int. J. Chem. Kinetics*, **1985**, *17*, 1299-1314.
17. Hatchard, C.A.; Parker, A.C. *Proc. Roy. Soc. (London)*, **1956**, A235, 518.
18. Faust, B.C.; Hoigne, J. *Atmos. Environ.*, **1990**, 24A, 79-89.
19. Evans, M.G.; George, P.; Uri, N. *Trans. Faraday Soc.*, **1948**, *44*, 230-239.

RECEIVED October 1, 1992

Chapter 5

Toxic Pollutant Destruction

Comparison of the Oxidants Potassium Permanganate, Fenton's Reagent, and Chlorine Dioxide on the Toxicity of Substituted Phenols

Philip A. Vella and Joseph A. Munder

Carus Chemical Company, Ottawa, IL 61350

In recent years, there have been many advances in the detection of toxic phenolics in the environment. Following this trend, our government has been adding to the list of controlled waste substances and lowering their permitted discharge limits. The overall result is that the chemical industry must upgrade current treatment methods in order to comply with the new mandates.

One of the techniques used to meet toxic phenolic restrictions is chemical oxidation. The oxidants currently used for treating phenolic water include potassium permanganate, hydrogen peroxide, chlorine dioxide, and ozone.

This study was divided into three parts. Part one investigated the effects of pH and chemical inhibitors (phosphate, methanol) on the oxidation of simple phenol with potassium permanganate, chlorine dioxide, and Fenton's reagent (H_2O_2/Fe). Due to the limited use of chlorine and ozone, these oxidants were not evaluated. In the second portion, 9 of the 11 EPA-regulated phenols were oxidized in a pure water matrix. These compounds included 2-chlorophenol, 2-nitrophenol, phenol, 2,4-dimethylphenol, 2,4-dichlorophenol, 2,4,6-trichlorophenol, 4-chloro-m-cresol, 2,4-dinitrophenol, and 4,6-dinitro-o-cresol. The third part of the study compared these oxidants in an actual wastewater matrix. In addition to the 9 compounds used in the previous study, the matrix included 4-nitrophenol, two substituted anilines and numerous inorganic compounds. Further, toxicity was measured in synthetic and actual wastewater solutions before and after oxidation.

Background

Chemical Oxidation of Phenols. Literature and practice support a number of methods for the treatment of phenolic wastewaters. A common treatment is chemical oxidation. Listed in Table I is a comparison of some commonly used oxidants.

0097–6156/93/0518–0085$06.25/0

Table I. Chemical Oxidant Options

Oxidant	Complete Oxidation	Oxidant:Phenol Ratio w/w	pH	Catalyst Required	Treatment Form
O_3[1,2]	yes	2-6:1	7-12	no	Dissolved gas*
Cl_2[3]	yes	100:1	7-8	no	Liquid or gas**
ClO_2[1,4]	yes	1.5-3.5:1	2-10	no	Dissolved gas*
H_2O_2[1,5]	yes	2-5:1	4-6	yes	Liquid
$KMnO_4$[1,6]	yes	16:1	7-10	no	Crystal

*Generated On-Site
**Hypochlorite may be applied
SOURCE: Reproduced with permission from ref. 35. Copyright 1992.

Lanouette (*1*) reports that ozone will oxidize phenol (1.5-2.5 w/w ratio) over a broad pH range but is phenol selective at 11.5 to 11.8. At low phenol levels, the quantity of ozone required to oxidize phenol is greater than 2.5:1. Ozone requires on site generation and has a high capital cost.

Throop and Boyle (*7*) demonstrated a 100:1 chlorine to phenol ratio for complete oxidation, and a recommended pH range of 7-8.3 to minimize chlorophenol formation. Chlorine and hypochlorite are inexpensive but present control problems in the avoidance of chlorinated phenols formation (*1-7*).

Potassium permanganate is a strong chemical oxidant and has been widely used for the oxidation of phenol (*6-10*). The Water Pollution Control Federation lists potassium permanganate as an oxidation pretreatment chemical for phenol (*11*). The oxidation of phenol by potassium permanganate to CO_2 is given by the following: (*6*)

$$3C_6H_5OH + 28KMnO_4 + 5H_2O \longrightarrow 18CO_2 + 28KOH + 28MnO_2$$

The step-wise oxidation of phenol has been described as proceeding through oxalic acid, tartaric acid, formic acid and carbon dioxide. The actual oxidation pathway is matrix dependent. The weight ratio of $KMnO_4$ to phenol can be varied depending on the extent of the oxidation desired. Complete oxidation requires 15.7 parts of $KMnO_4$ for each part of phenol, but reports indicate that 6-7 parts of $KMnO_4$ will afford ring cleavage (*12*). The presence of other reducing compounds may increase the amount of $KMnO_4$ required.

Chlorine dioxide is a green-yellow gas that must be generated at the application site. At a pH below 10, it will oxidize phenol to 1,4 benzoquinone. The ClO_2 to phenol weight ratio under these conditions is 1.5 to 1. At pH values above 10, a weight ratio of 3.5 is required with the reaction by-products being maleic, fumaric and oxalic acids. One of the claimed advantages of chlorine dioxide, relative to chlorine, is that it does not form taste and odor compounds (chlorophenols)(*13-16*). However, reports have indicated that chlorination does occur depending on reaction conditions (*17-20*). These applications have by and large involved only phenol, not substituted phenols. Chlorine dioxide also reacts with some substituted phenols, but the reactions and their by products are not as well documented (*18, 20-21*).

Although there is evidence that phenolic compounds react with hydrogen peroxide alone (Greenberg), the reactions are extremely slow and severely limited. In order to increase the reactivity and rate, peroxide is usually catalyzed by a metal ion. The most common catalyst is ferrous sulfate, $FeSO_4$. The combination of iron and peroxide is commonly referred to as Fenton's reagent or Fenton's system. Several studies have been performed using Fenton's reagent for the oxidation of phenolic compounds (*22-29*). From these reports, conflicting information can be found. For instance, it was stated that "When all available positions were blocked (for example, pentachlorophenol) no reaction occurred (*22*)." Another paper (*23*) reported that the percent removal of pentachlorophenol was 100%. These inconsistencies seem to indicate that the reaction is variable and that environment is a major factor in the oxidation. The reactions of Fenton's reagent with the substituted phenols are just not well explored.

Alternative Treatment. Other phenol treatment options including physical and biological processes may be applied to the reduction of phenol concentrations in wastewater streams. As early as 1975 (*30*) studies demonstrated that activated carbon removes phenols down to several parts per billion in the effluent. Biological removal of phenols has also been recommended. (*31*) The effluent limitations for direct dischargers, using end-of-pipe biological treatment, are more restrictive (26 µg/L daily maximum and 15 µg/L maximum monthly average) (*31*) than those applied to non-biological treatment systems.

Experimental

Reagents. The reagents used were of technical grade or better. The hydrogen peroxide was 30% laboratory grade. The water used in all tests was treated using a softener, a reverse osmosis unit, and a demineralizer to produce a water with a resistivity of approximately 15 megohm/cm. Stock solutions of ClO_2, prepared as given below, were stored in amber bottles at 4° C.

Preparation of Chlorine Dioxide Solution. Chlorine dioxide was generated by reacting a 25% solution of H_2SO_4 with a 25% solution of $NaClO_2$. Nitrogen was bubbled through the mixture to strip off the chlorine dioxide. The gas was then bubbled through two consecutive saturated solutions of $NaClO_2$ which purified the chlorine dioxide by removing any Cl_2 which may have been present. The final step was bubbling the gas through ice cold water forming a ClO_2 solution. In each step, aerators were used to increase gas-liquid contact and reactivity. The concentration of the ClO_2 stock solution was determined spectrophotometrically at 445 nm in a 1 inch cell using a HACH DR-2000 spectrophotometer. The total chlorine content was determined by purging an aliquot of the ClO_2 solution and determining total Cl_2 using the HACH ampoule method. Chlorine residuals were determined to be less than 1 mg/L.

General Test Conditions. Except where noted, the reaction times were 15 minutes. Oxidant quenching was done with sodium thiosulfate at a minimum 3:1 ratio of $Na_2S_2O_3$ to original oxidant concentration. In the case of Fenton's reagent, excess peroxide was removed by adjusting the pH to 10 with lime. This was most important in the toxicity experiments where the possible effects of dithionate had to be minimized. All ratios are reported as w/w unless noted. Samples reacted with either $KMnO_4$ or ClO_2 were adjusted to pH 7.0 ± 0.5 prior to oxidant addition. Reactions involving Fenton's Reagent were adjusted to pH 4.0 prior to reaction. Iron was added as $FeSO_4$ at a ratio of 1:10 to peroxide. The oxidant to phenolic dose ratio was based upon the assumption that the phenols in the mix were each at a 50 mg/L concentration. Phenol residuals were calculated based on individual peak areas before and after oxidation. All results were corrected for dilution.

Phenol Analysis by Gas Chromatography. This procedure is a modified version of EPA Method 604. A 25 mL sample was placed into a 40 mL vial and adjusted to pH less than 2 with concentrated sulfuric acid. Methylene chloride (5mLs) was added to the vial which was shaken for 1 minute. The aqueous and organic layers were allowed to separate and a 3 μL sample of the organic layer was injected into the GC. The column used was a 6 ft x 1/4 inch OD x 2 mm ID glass column packed with 1% SP-1240 DA on 100/120 Supelcoport. The gas chromatograph was a Hewlett Packard 5880. The injection port was heated to 225° C, and the Flame Ionization Detector (FID) to 220° C. The oven was programmed to hold at 70° C for 4 minutes following injection and then increased to 180° C at a rate of 8°/min. The oven was maintained at 180° C for 15 minutes. The carrier gas was helium at a flow rate of 30 mL/min. A standard curve was generated from 10 to 500 mg/L phenol. Periodic standards were run to verify the curve. The Detection Limit (DL) of this procedure was 5 mg/L phenol. Samples which were oxidized to below this concentration were considered to be 100% oxidized.

MICROTOX Toxicity Test. Microtox toxicity is a measure of the quantity of aqueous toxicant required to reduce the light emission of a bioluminescent (phosphorescent) bacteria (*photobacterium phosphoreum*) by a specified amount. The amount of light emitted by the bacteria is proportional to their active concentration.

The bacteria are salt water organisms and, therefore, non-saline samples require the addition of a 2% sodium chloride solution MOAS (Microtox Osmotic Adjusting Solution). Generally, the salinity is maintained at approximately 2% in samples and blank.

The toxicity is expressed as the percent of the undiluted sample necessary to reduce the light emission of the test bacteria by 50% at a specified time, generally 5 and 15 minutes. Results are reported as EC-50 (Effective Concentration) values at the corresponding times.

Samples are prepared by serial dilution with Microtox® Diluent (a 2% saline solution)and are compared to a blank containing no sample. Samples and blank are prepared in cuvettes which are placed in an incubator (Row A) maintained at 15° C. Bacteria are prepared by reconstitution from freeze-dried samples, and are held in a separate incubator at 5° C. Ten μL of solution containing the bacteria are pipetted into cuvettes in row B which contain 500 μL of diluent and allowed to equilibrate for 15 minutes. After 15 minutes, the light emissions of the bacteria/diluent solutions from row B are recorded as initial values (I_0). Immediately, 500 μL of previously diluted sample from row A are transferred to the corresponding cuvettes in row B which contain bacteria. At 5 and 15 minutes after sample addition, the light emissions of these cuvettes are again recorded (I_5 and I_{15}). The blank solution is measured at 5 and 15 minutes and the R_5 and R_{15} values are calculated from these readings. The EC-50 is calculated from the ratio of the light lost to light remaining. Lower EC-50 percentages correspond to higher toxicity levels. For example, a sample with an EC-50 of 15% is much more toxic than one with an EC-50 of 60%.

For ease in reporting and clarity, the EC-50 percentages were transformed to give 'Toxicity Units'. These are simply the inverse of the decimal equivalent of the EC-50 percentage. For EC-50 values of 15 and 60%, the Toxicity Unit equivalent would be

6.67 and 1.67 respectively. Therefore, higher toxicity units correspond to higher toxicity. For work done here, samples with values above 10 are considered toxic and above 100 are extremely toxic.

Preparation of the Phenolic Solutions

Phenolics in a Pure Water Matrix. The stock solutions for the phenol study were prepared in DI water. In general, the stock solutions for the substituted phenols were made in methanol. However, methanol will dramatically interfere with the free radical generated in Fenton's system. Therefore, all phenol mixtures that were reacted with Fenton's reagent were prepared in DI water.

The concentrated stock solution (5,000 mg/L) of the substituted phenols were prepared by dissolving each phenol into methanol. From these standards, a mixture of the phenols (50 mg/L for each phenol) was prepared in DI water. Due to the insolubility of two of the 11 phenols, the solution was filtered through 0.05 μm membrane filters prior to any testing. Nine of the 11 phenols remained: 2-chlorophenol, 2-nitrophenol, phenol, 2,4-dimethylphenol, 2,4-dichlorophenol, 2,4,6-trichlorophenol, 4-chloro-m-cresol, 2,4-dinitrophenol, and 4,6-dinitro-o-cresol.

Phenolics in a Wastewater Matrix. A wastewater sample was obtained from an organics manufacturer. Analysis of the sample indicated the following organic compounds: 4-nitroaniline (392 mg/L), 2-nitroaniline (739 mg/L), 4-nitrophenol (1828 mg/L) and 2-nitrophenol (14 mg/L). Other organics typically present in the sample but not analyzed include xylenes, aniline, formanilide, 4-nitrodiphenylamine and n-methylaniline. The concentrations of these components range from 32 to 144 mg/L. In addition to the organics compounds, the sample usually contains the following inorganic materials in concentrations of 1 to 6%: ammonia, potassium formate, sodium chloride, potassium chloride, and potassium bicarbonate.

The test solution was prepared by fist making a 1:20 dilution of the wastewater. To this solution an aliquot of the nine standard phenol mixture was added to give a final concentration of 50 mg/L.

Methanol Oxidant Demand. There was an initial concern that the methanol used to make the test solutions would create a high oxidant demand. Subsequent testing indicated that ClO_2 did react slightly with the methanol (10%) during the time frame investigated. All data using ClO_2 was therefore corrected. Potassium permanganate did not react significantly during the 15 minute reaction period.

Results and Discussion

Oxidation of Phenol

Reaction with Potassium Permanganate. The ratios of permanganate to phenol examined were 1:1, 4:1, 7:1, and 8:1. Figures 1 and 2 give the results of these tests.

The reaction between potassium permanganate and 10 mg/L phenol was most effective at an 8:1 ratio (Fig 2). At this ratio, oxidation was 97-99% complete with in 15 minutes. With this same $KMnO_4$:phenol ratio and 1 mg/L phenol, oxidation was only 66% complete (Fig 1). It has been shown in a number of applications when low levels of phenol are to be oxidized, more than the theoretical amount of $KMnO_4$ is required.(10, 32)

At a concentration of 10 mg/L phenol, no pH effect was observed. At 1 mg/L, a pH of 9 proved to be the most effective.

Effect of Phosphate. The effect of complexing agents on the oxidation of phenol

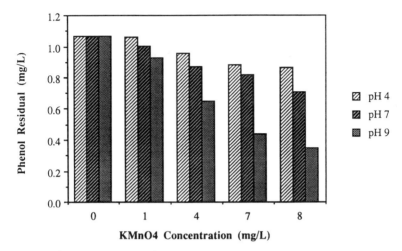

Figure 1   Effects of pH and Oxidant Dose: Oxidation of 1 mg/L Phenol with
KMnO$_4$ (Reproduced with permission from Reference 35. Copyright
1992 Technomic Publishing)

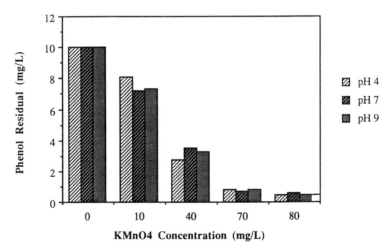

Figure 2   Effects of pH and Oxidant Dose: Oxidation of 10 mg/L Phenol with
KMnO$_4$ (Reproduced with permission from Reference 35. Copyright
1992 Technomic Publishing)

with KMnO$_4$ was examined. The KMnO$_4$ to phenol ratio was 4:1. Phosphate, in the form of NaH$_2$PO$_4$, was added in concentrations of 50, 100, and 200 mg /L. Adjustment to pH 9 was made after phosphate was added.

Figures 3 and 4 depict the effect of phosphate on the permanganate system. From Figure 3 there was a small improvement in the oxidation of 1 mg/L phenol at pHs of 4 and 7 (8 and 3% respectively) with a phosphate concentration of 200 mg/L. Figure 4 shows that addition of phosphate slightly hinders the oxidation of 10 mg/L phenol at pH 7 and 9 with a greater effect observed at a pH of 4.

Reaction of Phenol with Catalyzed Peroxide (Fenton's System). Tests were carried out at 1 and 10 mg/L phenol levels. All pH adjustments were made with dilute H$_2$SO$_4$ and NaOH. Iron catalyst was added as FeSO$_4$ in a 1:10 ratio to peroxide. During testing, the iron was added first and allowed to mix well before peroxide was added. Peroxide concentrations were at ratios of 1:1, 4:1, 6:1, and 8:1 w/w to phenol. Due to the pH dependency of this system only pH 4 was investigated.

Figures 5-7 show the results of the catalyzed peroxide testing. In a 1 mg/L phenol solution, oxidation was 98% complete with a peroxide dose of 6 mg/L after 15 minutes. After 30 minutes, the phenol was oxidized to below the detection limit (0.005 mg/L). The 10 mg/L phenol solution required only a 40 mg/L peroxide dose to oxidize the phenol to below detection limits in 15 minutes. Due to rapid oxidation, no graphs were generated at this phenol level.

Using this pH and dose, the effects of phosphate and calcium were evaluated. In addition, an experiment was performed to test if permanganate could be substituted for the iron catalyst. The results of this test were negative.

Addition of NaH$_2$PO$_4$ (Fig. 6 and 7) to the Fenton system had a negative effect. At 1 mg/L phenol the oxidation dropped from nearly 100% to near 30%. When added to the 10 mg/L phenol system, the oxidation was reduced to 15%. This result is likely due to the chelation of the iron by the phosphate ion rather than directly scavenging of the OH radical. With the iron bound, the generation of free radicals is limited.

Reaction of Phenol with Chlorine Dioxide

Chlorine dioxide was dosed at 0.5, 1, 2, and 4 times the phenol concentration. The pH values investigated were 4, 7 and 10. Due to the instability of the stock solution, the chlorine dioxide concentration was determined prior to each test. As with previous oxidants, the effects of phosphate was examined.

Figures 8 and 9 present the results from the chlorine dioxide experiments. In a 1 mg/L phenol solution, pH 10 showed an increase in oxidation at low ClO$_2$ dosages. At this pH, the minimum concentration of chlorine dioxide resulting in 100% oxidation was 2 mg/L. Similar results were also observed at pH 7.

In a 10 mg/L phenol solution there was very little pH effect. In this case, a 4:1 dose of chlorine dioxide to phenol exhibited the best oxidation (91%).

Chlorine dioxide proved to be a very effective oxidant, even in the presence of the additives. Addition of NaH$_2$PO$_4$ had no adverse effect on the oxidation of the phenol at either 1 mg/L or 10 mg/L levels.

Oxidation of Substituted Phenols.

Reaction with Potassium Permanganate.

Pure water matrix. From Figure 10, potassium permanganate was very effective in oxidizing all of the phenolic compounds. There appears to be a concentration effect that relates directly to the selectivity of permanganate oxidation. At low KMnO$_4$ concentrations (500 mg/L) the nitro-substituted phenols were slightly oxidized, while

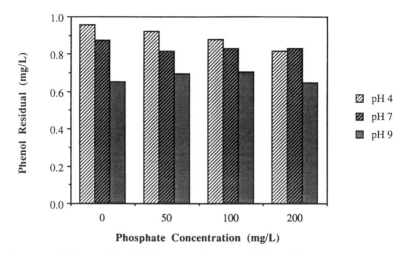

Figure 3   Effect of pH and Phosphate Dose: Oxidation of 1 mg/L Phenol
with 4 mg/L KMnO$_4$ (Reproduced with permission from
Reference 35. Copyright 1992 Technomic Publishing)

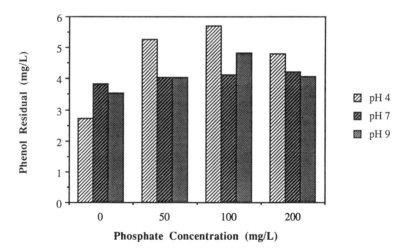

Figure 4   Effect of pH and Phosphate Dose: Oxidation of 10 mg/L Phenol
with 40 mg/L KMnO$_4$ (Reproduced with permission from
Reference 35. Copyright 1992 Technomic Publishing)

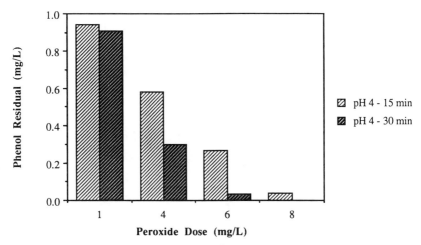

Figure 5   Effects of Fenton's Dose and Reaction Time: Oxidation of 1 mg/L
Phenol at pH 4 (Reproduced with permission from Reference 35.
Copyright 1992 Technomic Publishing)

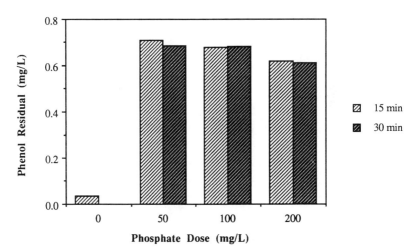

Figure 6   Effects of Phosphate and Reaction Time: Oxidation of 1 mg/L Phenol
with 4 mg/L H_2O_2 at pH 4 (Reproduced with permission from
Reference 35. Copyright 1992 Technomic Publishing)

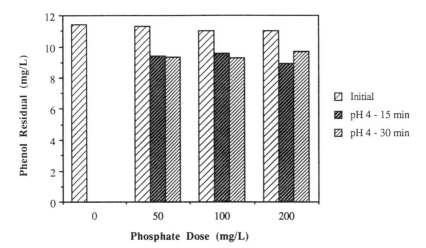

Figure 7   Effects of Phosphate Dose and Reaction Time Oxidation of 10 mg/L
Phenol with 40 mg/L H_2O_2 at pH 4 (Reproduced with permission
from Reference 35. Copyright 1992 Technomic Publishing)

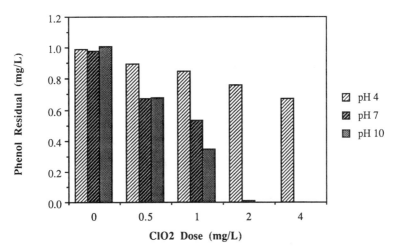

Figure 8   Effects of Chlorine Dioxide Dose and pH: Oxidation of 1 mg/L Phenol
(Reproduced with permission from Reference 35. Copyright 1992
Technomic Publishing)

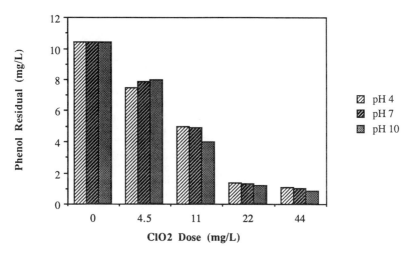

Figure 9   Effects of Chlorine Dioxide Dose and pH: Oxidation of 10 mg/L Phenol
(Reproduced with permission from Reference 35. Copyright 1992
Technomic Publishing)

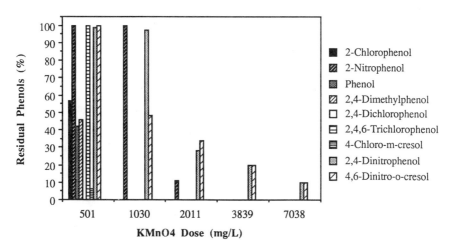

Figure 10.   KMnO4 Oxidation of Substituted Phenols in a Clean Matrix.

the 2,4-dichlorophenol and the 4-chloro-m-cresol were easily oxidized (100% and 93.6% respectively). The remaining phenols were removed between 40-50%. Increasing the permanganate concentration to 1029 mg/L markedly improved oxidation. At this concentration, six of the nine phenols were >99.9% oxidized. Again, the nitro-substituted phenols showed a resistance to oxidation. Of these, 2-nitrophenol remained stable and the disubstituted nitrophenols were oxidized by 3 and 51.7% respectively. At a permanganate level of 2011 mg/L oxidation of the nitrophenols begins to be significant. However, at the highest $KMnO_4$ concentration studied, neither 2,4-dinitrophenol nor 4,6-dinitro-o-cresol were > 99% removed. Paranitrophenol was fully oxidized at 3838 mg/L $KMnO_4$.

The observed results can generally be explained by electrophilic aromatic substitution. Reactivity in electrophilic aromatic substitution depends upon the tendency of a substituent group to release or withdraw electrons. A group that releases electrons activates the ring (faster reaction) and a group that withdraws electrons deactivates the ring (slower reaction). Based on this, phenols that have electron withdrawing groups such as halogen or nitro should be harder to oxidize. Phenols that contain electron donating groups such as methyl (cresol) should be more readily oxidized (33). Examination of the data indicates that this trend generally holds. The only exceptions are 2,4-dimethylphenol and 2,4-dichlorophenol. The methyl groups should have increased the reactivity of the phenol but showed no dramatic change. Chlorine substitution should have hindered oxidation instead of aiding it. These anomalies may be due to steric or matrix effects. Further work may be needed to resolve this issue.

Wastewater Matrix. As seen in Figure 11, the results in a wastewater matrix closely follow that observed in a pure water matrix. The additional amount of oxidant needed to achieve complete oxidation is due to the increase in total organics present and not due to the inorganic compounds in the sample. As seen previously, 2,4-dinitrophenol and 4,6-dinitro-o-cresol are least effected and as expected, 4-nitrophenol proved to be more difficult to oxidize.

Reaction with Fenton's Reagent

Pure Water Matrix. In free radical attack, there should be no great difference in aromatic reactivities having electron withdrawing or donating groups. However, it is generally accepted that high electron density on the ring will inhibit free radical oxidation. Therefore, phenols having electron withdrawing groups such as nitro or halogen, should be more reactive than phenol. The reactivity of monosubstituted phenol has been shown to be similar to phenol with the effect becoming more important with multisubstitution (22-24).

Reactions involving Fenton's reagent gave varying results and were repeated to insure that protocols and results were reproducible. These data are presented in Figures 12-13. Overall, Fenton's reagent is a very effective given sufficient reaction time. It reacts well toward all the substituted phenols studied with 4,6-dinitrophenol being the least reactive. These results agree with those reported elsewhere (23-26). The chlorinated phenols were completely oxidized with 2,4-dinitrophenol 30% oxidized. These data show complete reaction of the nitrophenol and less reactivity of the cresol. Also, the reactivity of 2-chloro and 2-nitrophenol is enhanced. These results agree with the electronegativity of the specific group attached to the ring.
Wastewater Matrix. The evaluation of Fenton's reagent in a wastewater environment is presented in Figures 14-15. As seen, the results are similar to those obtained in the pure water environment. The reactions are dependent upon time and substitutent groups.

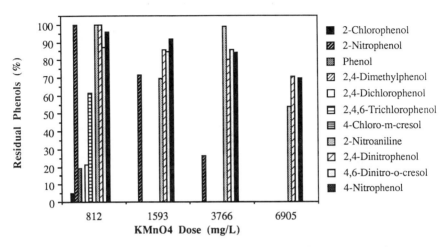

Figure 11.   KMnO$_4$ Oxidation of Substituted Phenols in a Waste Matrix.

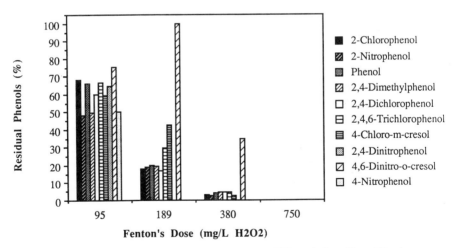

Figure 12.   Fenton's Oxidation of Substituted Phenols in a Clean Matrix: 15 Minutes Reaction Time.

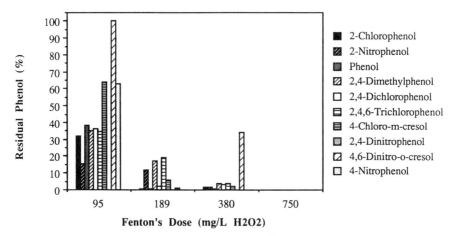

Figure 13.   Fenton's Oxidation of Substituted Phenols in a Clean Matrix: 4 Hour Reaction Time.

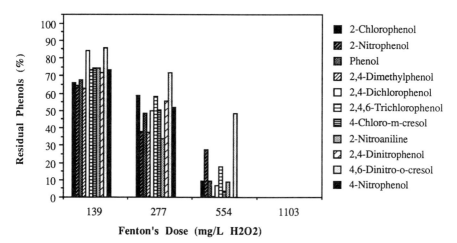

Figure 14.   Fenton's Oxidation of Substituted Phenols in a Waste Matrix: 15 Minute Reaction Time.

Reaction with Chlorine Dioxide.

Pure Water Matrix. Results from chlorine dioxide reactions (Figure 16) closely followed those seen with permanganate. The selectivity of ClO_2 compared to $KMnO_4$ is quite different. While $KMnO_4$ was effective on 2,4-dichlorophenol at lower concentration, ClO_2 was ineffective. Comparing 2-nitrophenol and 2,4,6-trichlorophenol, ClO_2 proved to be more selective. In the case of 2,4-dinitrophenol, ClO_2 was unable to oxidize the substrate at all concentrations studied. In an overall comparison with permanganate, a lower w/w dose is needed to achieve an equivalent level of oxidation.

Chlorine dioxide can react in a free radical mechanism similar to Fenton's reagent. Phenols containing electron-withdrawing groups ortho and para to the phenoxy group should be more reactive than phenol itself (*34*). As observed with the Fenton's system data, a number of substituted phenols do not fit this assertion. All but one of the phenols, 2,4-Dinitrophenol, reacts completely with ClO_2. A reasonable explanation of this result and the observed trend is lacking.

As seen in work done with phenol alone (*32*), new unidentified peaks were observed when substituted phenols were reacted with ClO_2. These observations were not made when $KMnO_4$ or Fenton's are used. The identification of these compounds is scheduled.

Wastewater Matrix. The evaluation of chlorine dioxide in a wastewater environment (Figure 17) follow that observed in a pure water matrix. Again, the additional amount of oxidant needed to achieve nearly complete removal is due to the increase in total organics present. There were no observable effects of the inorganic compounds present in the sample.

Toxicity Reduction Evlauation

Pure Water Matrix. To determine the effects of oxidation on toxicity, reactions were run to produce "complete" and incomplete oxidation. The term "complete oxidation" refers to the point where no substituted phenols could be detected. This is in contrast to total oxidation which would result in the ultimate formation of CO_2 and water. The concentrations of oxidants used and the resulting toxicities are given in Table II and Figure 18.

The results are generally consistent with other toxicity evaluations (*26,32*). Permanganate treated solutions were found to have the lowest toxicity. It achieved good oxidation and lowered the toxicity to 6.6 T.U. which translates to a 95.3% reduction in toxicity. The toxicity was dramatically lowered independent of the degree of oxidation.

Fenton's reagent also lowered the toxicity of the solution. In this case, the level of oxidation did have an effect on toxicity. The more complete the oxidation, the lower the toxicity. With complete oxidation, the toxicity was lowered from 140.9 to 33.1 TU. These results are in agreement with a similar toxicity evaluation (*26*).

Chlorine dioxide at 125 mg/L reduced the toxicity of the sample to 107.5 T.U. but when the concentration was increased to 1000 mg/L, the toxicity increased to 500 T.U. This increase may be due to various reaction by-products including quinones, chlorinated organics, or chlorite. A by-product from ClO_2 oxidation is chlorite which is known to be toxic to these bacteria. Although sulfite was added to quench the reaction and does react with chlorite. Residual chlorite was not determined.

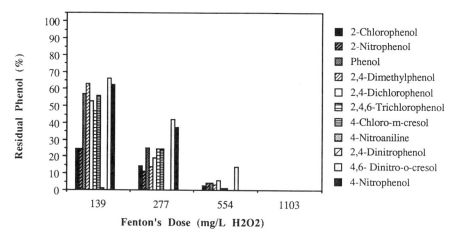

Figure 15.   Fenton's Oxidation of Substituted Phenols in a Waste Matrix:
4 Hour Reaction Time.

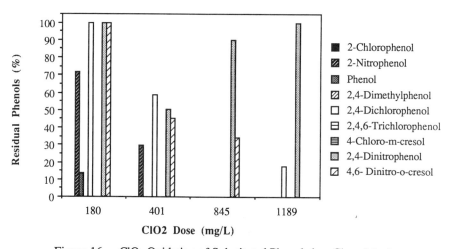

Figure 16.   ClO₂ Oxidation of Substituted Phenols in a Clean Matrix.

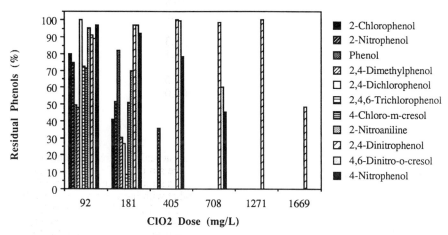

Figure 17.   ClO$_2$ Oxidation of Substituted Phenols in a Waste Matrix.

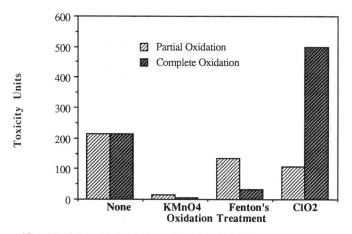

Figure 18.   Toxicity of Mixed Phenolics after Oxidation: Clean Water Matrix.

Table II. Toxicity Results after Oxidation in a Clean Water Matrix

Treatment	Dose (mg/L)	Toxicity (T.U.)
None	----	140.9
$KMnO_4$	1038.0	14.8
$KMnO_4$	2011.0	6.6
Fenton's	94.5	133.3
Fenton's	754.7	33.1
ClO_2	125.0	107.5
ClO_2	1000.0	500.0

Wastewater Matrix. The results obtained in an actual wastewater matrix are presented in Table III and Figure 19. The findings are similar to those obtained in a pure water matrix.   Both permanganate and Fenton's reduced the toxicity of the sample, with $KMnO_4$ having the larger effect. Note that with increasing oxidation, the toxicity using Fenton's increases. This could be due to by-products formed in the wastewater matrix (additional organics, inorganics) that were not present in the clean matrix. Specific causes for this increase can only be theorized. Oxidation by ClO_2 increased the toxicity in both runs with a minor effect on toxicity the degree of oxidation. As in the clean matrix, chlorite and unknown by-products are probable reasons for the increase.

Table III . Toxicity Results after Oxidation

Treatment	Dose (mg/L)	Toxicity (T.U.)
None	----	238.1
$KMnO_4$	2245.5	34.2
$KMnO_4$	4421.4	22.8
Fenton's	276.8	73.0
Fenton's	1103.9	137.9
ClO_2	289.0	1250.0
ClO_2	1000.0	1428.6

Conclusions

1) In the oxidation of unsubstituted phenol, chlorine dioxide was the most flexible oxidant evaluated. It was less susceptible to interferences and more effective over a wide pH range. The ClO_2 dose for phenol oxidation was the lowest of the oxidants tested. However, no comparisons were made to determine cost effectiveness.

2) The flexibility of potassium permanganate closely followed that of chlorine

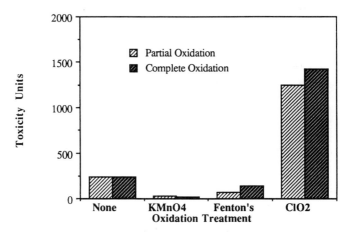

Figure 19.   Toxicity of Mixed Phenolics after Oxidation:Wastewater Matrix.

dioxide. At low phenol concentrations, a pH dependency was observed with a pH of 9 being optimum. Permanganate was only marginally effected by interferences.

3) Fenton's system performed well with in a narrowly defined pH range. It is also very susceptible to inhibitors and hydroxyl radical scavengers (phosphate and carbonate).

4) With substituted phenols, the above trends generally hold true. Fenton's system required the lowest dose on a w/w basis but remains pH dependent. Also, it may require longer reaction times to achieve the same degree of oxidation as potassium permanganate or chlorine dioxide. Although requiring higher dosages, both $KMnO_4$ and ClO_2 were quite effective.

5) The general effect on toxicity of the oxidants studied was $KMnO_4 <$ Fenton's$<<<ClO_2$. The degree of oxidation with $KMnO_4$ had little effect on the resulting toxicity. In the case of Fenton' the resulting toxicity could be higher or lower depending on the extend of oxidation. With ClO_2, the toxicity was always found to be greater after oxidation.

Literature Cited

1. Lanouette, K.H., "Treatment of Phenolic Wastes", Chemical Engineering Deskbook Issue, pp. 99-106, October 17, **1977**.
2. Conway, R.A. and Ross, R.D., "Handbook of Industrial Waste Disposal", pp. 224- 225, Van Nostrand Reinhold Company, New York, **1980**.
3. Throop, W.M., "Alternative Methods of Phenol Wastewater Control", J. of Hazr. Mater., pp. 319-329, 1(4), **1977**.
4. Masschelein, W. J., "Chlorine Dioxide Chemistry and Environmental Impact of Oxychlorine Compounds", Ann Arbor Science, Ann Arbor MI 1979.
5 Kibbble W.H., Jr., "Peroxide Treatment for Industrial Waste Problems", Industrial Water Engineering, pp. 6-11, August/September, **1976**.
6. Bobkov, V. N., Tr. VNII Vodosnabzh., Kanaliz, Gidrotekhn. Sooruzh. 1 Inzh. Gidrogeol. pp. 48-50, (50), (Russ.), **1975**.
7. Throop, W. M. and Boyle, W. C., "Perplexing Phenols Alternative Methods for Removal", Proc. Annu. Pollut. Control Conf Wastewater Equip. Manuf. Assoc., 3rd **1975**, pp. 115-143.
8. Kroop, R. H., "Treatment of Phenolic Aircraft Paint Stripping Wastewater", Engineering Bulletin of Purdue University, Engineering Extension Series, No. 142, pt. 2, pp. 1071-1087, **1973**.
9. Rosfjord, R. E., "A Water Pollution Control Assessment", Water and Sewage Works, pp. 96-99, March, **1976**.
10. Vella, P. A., Deshinsky, G., Boll, J. E., Munder, J., Joyce, W. M., "Treatment of Low Level Phenols with Potassium Permanganate", J. Water Poll. Control Fed., 62, 907-914, **1990**.
11. Pretreatment of Industrial Wastes, Manual of Practice No. FD-3 Facilities Development, p.116, Water Pollution Control Federation, Washington, D.C., 1981.
12. Humphrey, S. B., and Eikleberry, M. A., Unpublished Internal Report, Carus Chemical Company, Ottawa, Il., 1962.
13. Wallwork, J.F., Morris, J.C., Symonds, D.C. Water Treatment Exam., 18, 203-210, **1969**.
14. Theilemann, H. Gesund.-Ing. 92, 295-299, **1972**.
15. Stevens, A.A., "Reaction Products of ClO_2" Env. Health Perspect. 46,101-110, **1982**.

16. Wajon, J. E., Rosenblatt, D. H., Burrows, E. P. "Oxidation of Phenol and Hydroxylquinone by Chlorine Dioxide", Envir. Sci. and Technol. 16, 396, **1982**.
17. Paluch, K., Rocz. Chem., 38, 35-42, **1964**.
18. Masschelein, W. J., *Chlorine Dioxide,* Ann Arbor Science: Ann Arbor, MI. 1979.
19. Grimley, E.; Gordon, G. J. Inorg. Nucl. Chem. 35, 2383-2392, **1973**.
20. Gordon, G., Kieffer, R. G.; Rosenblatt, D. H. *Prog. Inorg. Chem.* 15, 201-286, 1972.
21. Amor, B. H., De Laut, J., Dore, M., Water Res. 18,12,1545-1560, **1984**.
22. Eisenhauer, H. R. "Oxidation of Phenolic Wastes". Journal of the Water Pollution Control Federation V 36, #9, 1116-1128, **1964**.
23. Greenberg, E. S., FMC Corporation
24. Keating, E. J., Brown, R. A. Greenberg, E. S. "Phenolic Problems Solved With Hydrogen Peroxide Oxidation" Proceedings of the 33rd Annual Purdue Industrial Waste Conference, Purdue University, Lafayette, IN 1978.
25. Bowers, A. R., Gaddipati, P., Eckenfelder, W. W., Monsen, R. M. "Treatment of Toxic or Refractory Wastewaters with Hydrogen Peroxide". Water Sci. Technol., 21, 477-486, **1989**.
26. Bowers, A. R., Kong, W. S., Eckenfelder, W. W., Monsen, R. M. "Treatment of Aromatic Organic Compounds with Hydrogen Peroxide". Industrial Waste Symposium, 62nd Annual Conference of the Water Pollution Control Federation, San Francisco, **1989**.
27. Kibbel, W. H., Raleigh, C. W., Shepard, J. A., "Hydrogen Peroxide for Industrial Pollution Control", Proceedings of the 27th Annual Purdue Industrial Waste Conference, Purdue University, Lafayette, IN 1972.
28. "Phenols in Refinery Wastewater Can Be Oxidized with Hydrogen Peroxide", FMC Chemicals, J. Oil and Gas, **1975**.
29. Kibbel, W.H., Jr., "Peroxide Treatment for Industrial Waste Problems", Industrial Water Engineering, pp. 6-11, August/September, **1976**.
30. Patterson, J. W., *Wastewater Treatment Technology*, p. 207, Ann Arbor Science, Ann Arbor, MI, 1975.
31. Newton, J., OCPSF "Effluent Limits, Pretreatment, and NSPS", pp, 48-51.Pollution Engineering, March **1988**.
32. Carus Internal Reports.
33. Stewart, R., *Oxidation in Organic Chemistry*, Part A, Academic Press, New York, pg 59, 1965.
34. Strumilla, G. B., Rapson, W. H., Trans. Tech. Sect. (Can. Pulp Pap. Assoc.) 3, 119-126, **1977**.
35. Vella, P., Munder, J. in *Chemical Oxidation: Technology for the Nineties*; Eckenfelder, W. W., Bowers, A., Roth, J. ED; Technomic Publishing Co. Inc., Lancaster, PA, 1992.

RECEIVED November 9, 1992

Chapter 6

Ultrasonically Accelerated Photocatalytic Waste Treatment

A. J. Johnston and P. Hocking

Bioengineering Research Laboratory, SRI International, Menlo Park, CA 94025

Semiconductors such as TiO$_2$ have been widely investigated as materials for the catalytic photodegradation of aqueous environmental contaminants by way of light-induced redox reactions at the semiconductor/liquid interface. Complete mineralization of organics upon UV irradiation of aqueous organics in the presence of TiO$_2$ has been reported for aliphatics and aromatics such as polychlorinated biphenyls and dioxins. TiO$_2$-mediated photodegradation may provide a safe and efficient means to destroy a variety of organic pollutants in groundwater and wastewater. Power ultrasound has shown potential for improving this process—concurrent UV and ultrasonic irradiation of the reaction mixture significantly increases degradation rate and efficiency. Potential reasons for the observed reaction rate acceleration are discussed, including cavitational effects, bulk and localized mass transport effects, and sonochemical reactions.

Background

Metal chalcogenide semiconductors (TiO$_2$, ZnO, CdS, WO$_3$, SnO$_2$) have been widely investigated as photocatalysts for the degradation of aqueous organics by means of light-induced redox reactions (1–3). The initiating step in this photocatalytic process involves the generation of conduction band electrons (e$^-_{cb}$) and valence band holes (h$^+_{vb}$) by illumination of the semiconductor with light of energy higher than the band gap of the semiconductor (<380 nm for anatase TiO$_2$, the most effective and widely studied catalyst). The reaction of adsorbed water with these electrons and holes to yield hydroxyl radicals OH· and hydrogen radicals H· is the generally accepted initial event in the photoreaction of aqueous substrates (4). Complete mineralization of chlorinated organics (via reaction with OH·) upon UV irradiation of TiO$_2$ suspensions of the aqueous substrate has been reported for simple aliphatics (5,6) and for chlorinated aromatics such as mono-

and polychlorinated phenols *(7,8)*, chlorobenzenes *(9,10)*, and polychlorinated biphenyls and dioxins *(11,12)*.

Although commercial implementation of photocatalytic technology has yet to be realized, results indicate that TiO_2-mediated photodegradation may provide a safe and efficient means to destroy a variety of organic pollutants in groundwater and wastewater *(13)*. Other destructive treatment methods, such as direct photolysis by UV light, often lead only to dechlorination *(14)* and partial degradation *(15,16)*, and in some cases result in the formation of compounds more toxic than the original pollutant *(17)*. Only the simultaneous use of UV light and an oxidant (H_2O_2, ozone, or oxygen with a photocatalyst) has been shown to be effective for complete destruction.

The approach we are investigating involves the *in situ* application of high-intensity (>10 W/cm^2) ultrasonic irradiation in conjunction with a heterogeneous photocatalytic process using photoreactive semiconductors such as titanium dioxide (TiO_2). This system has potential to provide increased organic degradation and mineralization rates and efficiencies as compared to the photocatalytic process alone, by

- Improved rates and efficiencies (hence throughput) under optimized conditions.
- The ability to use impure or cheaper forms or lower levels of the photocatalyst.
- Significant increases (up to fivefold in preliminary experiments) in the degradation rate of aqueous organics without poisoning of the catalyst.
- Improvements in conventional suspended-catalyst separation methods, such as ultrafiltration.

For applications involving treatment of wastewater containing suspended solids, the use of sonication may also assist in the release of hydrophobic organics adsorbed on soil particles, for subsequent reaction with OH·.

Experimental

Initial experiments were designed and conducted to evaluate the scope of this technology for selected pollutants. To this end, the degradation of aqueous solutions of chlorinated phenols and biphenyls in the presence of the heterogeneous photocatalyst TiO_2 was investigated. The photocatalysts zinc oxide and cadmium sulfide were also examined to determine the effect of sonication on photoreaction with these catalysts.

General Procedure. The aqueous solution containing the substrate (sample size 25–30 ml) was placed in a 40-ml cylindrical jacketed glass cell (~ 3 cm diameter) that allowed continuous water cooling. The temperature of the test solution was controlled at 35± 2°C. For experiments with TiO_2, titanium dioxide (Degussa P25, anatase form, surface area = 55± 10 m^2/g) was added to the cell solution to the desired concentration (normally 0.2% w/w).

The cell was irradiated by means of a Blak-Ray B-100A long-wave ultraviolet light source (100-W mercury bulb with a nominal intensity of

7000 μW/cm^2 at 350 nm) positioned 12.7 cm from the cell outer jacket wall. The solution was sonicated using a 1.27-cm titanium horn immersed 1 cm into the cell solution and powered by a 475-W Heat Systems XL2020 ultrasonic processor. The amplitude of ultrasonic vibration (20 kHz) at the tip of the horn was set to the maximum value (120 μm); for the particular cell geometry and contents, this required approximately 130 W input power. After sonication or photolysis (or both) for the desired time, the sample was analyzed for chloride ion production using a chloride ion selective electrode. The disappearance of the compound was monitored by UV/visible spectrophotometric analysis after filtration through a 0.20-μm Teflon filter. Phenol levels were determined colorimetrically at 510 nm using an aminoantipyrine/ferrocyanide assay *(18)*. Ferrioxalate actinometry was used to evaluate light output at 360 nm. Recovery of substrate from TiO$_2$ suspensions that were stirred in the dark for >90 minutes were greater than 95%, indicating that surface adsorption of unreacted substrate was not significant in analyses to determine degradation.

Parameters that were varied included time of sonication, concentration of pollutant, and effect of presonication of the catalyst. Control experiments (sonication only, photolysis with high-speed magnetic stirring, photolysis or sonication without catalyst) were also performed.

Results. The use of sonication during photolysis had a significant effect on the rate and efficiency of organic destruction as compared with photolysis alone. Significant enhancements in degradation rates were noted with TiO$_2$ and chlorophenols.

Rate enhancements effects due to decreases in particle size (increases in total surface area) for 0.2% TiO$_2$ suspensions subjected to sonication were insignificant: photolysis using TiO$_2$ that had been presonicated for 10 minutes showed minor differences in initial rate of disappearance of the substrate (the actual change in total surface area was not established). UV irradiation of catalyst-free solutions with and without sonication did not result in significant degradation of the organic substrate, except for the combined UV irradiation and sonication of 4-chlorophenol solutions, indicating that the UV energy alone was insufficient for significant direct photolysis. Sonication of catalyst/substrate suspensions without simultaneous UV irradiation did not result in significant degradation except, again, for 4-chlorophenol.

Pentachlorophenol. The photocatalytic degradation of pentachlorophenol (PCP, 2.4x10^{-4} M solution) with and without sonication was investigated. The initial rate of appearance of chloride was about 2.7 times faster with sonication than without (Figure 1). Also, in the absence of sonication, the percent degradation as measured by release of chloride ion reached a value of about 40% after 50 minutes, and continued irradiation up to 200 minutes did not result in any significant increase in degradation. In contrast, combined sonication/photolysis resulted in a rapid initial degradation, with near quantitative release of chloride after 120 minutes. A possible explanation for this observation is the poisoning of the catalyst by some relatively inert intermediate formed during photodegradation;

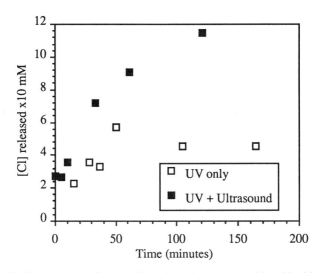

Figure 1. Degradation of pentachlorophenol as measured by chloride release

sonication serves to remove this intermediate from the active sites, allowing continued degradation. Alternatively, decreases in pH during photolysis may lead to a significant decrease in a photochemical degradation process.

3-Chlorobiphenyl. The sparingly soluble PCB isomer 3-chlorobiphenyl was added to water at a level of 75 ppm (4×10^{-4} M) and dispersed using an ultrasonic cleaning bath, resulting in a cloudy solution. TiO_2 was added to a level of 0.2% w/w, and 30-ml aliquots were subjected to UV irradiation and to combined UV/ultrasound irradiation. The chloride ion released was monitored over time with results as shown in Figure 2. The rate of appearance of chloride was fairly linear over time for both treatment conditions, but the rate using sonication was approximately three times greater with than without sonication. The highly hydrophobic nature of this solute, evidenced by its low solubility, may be a factor in the enhancement observed. That is, ultrasound may serve to deposit the substrate on the particle surface more effectively than stirring alone, analogous to ultrasonic phase transfer catalysis.

2,4-Dichlorophenol. Several reaction parameters were varied in the degradation experiments with 2,4-dichlorophenol (2,4-DCP). In addition to evaluating the rate increase upon sonication, we examined the effects of lower catalyst concentration (0.05% TiO_2), of using presonicated TiO_2 in UV-only exposures, and of varying the initial concentration of 2,4-DCP.

The use of sonication (1×10^{-3} M 2,4-DCP; 0.2% TiO_2) in photolysis resulted in enhancing the chloride release rate by a factor of four as compared with UV irradiation only (Figure 3). Experiments with the lower catalyst concentration yielded results similar to degradations using 0.2% TiO_2 (Figure 4; data from spectrophotometric measurement on phenol levels). This result suggests that the

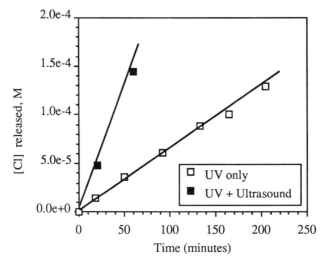

Figure 2. Degradation of 3-chlorobiphenyl as measured by chloride release

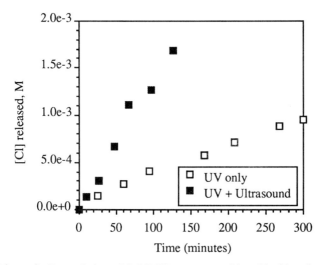

Figure 3. Degradation of 2,4-DCP as measured by chloride release

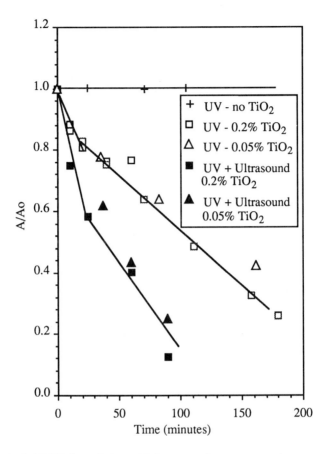

Figure 4. 2,4-DCP degradation with lower catalyst concentration, as measured by UV spectral analysis (510 nm).

incident light is completely absorbed by the lower level of TiO_2 in this experimental setup. The effect of presonicated TiO_2 (0.20%) in UV-only experiments was not significant, consistent with the conclusion that the amount of catalyst present is already sufficient to absorb all of the light. Additional experiments using presonicated TiO_2 at lower levels will have to be performed to determine if increases in particle surface area at low levels of TiO_2 will be a beneficial effect of using ultrasonic irradiation.

Decreasing the initial concentration of 2,4-DCP by one half resulted in little change in the initial rates of phenol disappearance for both sonication/photolysis and photolysis alone (Figure 5). This could indicate that we are near the plateau region of the initial-rate versus initial-concentration curve as described by the Langmuir-Hinshelwood mechanistic model. *(3)* Further experiments are required in order to more fully characterize the effect of initial substrate concentration on degradation rate and to evaluate the kinetic parameters according to the Langmuir surface-reaction model.

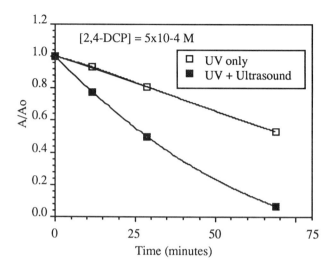

Figure 5. 2,4-DCP degradation with 1/2 initial concentration of 2,4-DCP
as measured by UV spectral analysis (510 nm).

4-Chlorophenol. 4-Chlorophenol (4-CP) degradation has been extensively
studied by other researchers as a substrate for TiO$_2$ heterogeneous
photodegradation *(7,19)*. For this substrate, combined UV irradiation/sonication of
catalyst-free solutions resulted in a significant decrease in the substrate
concentration over time, as compared with UV irradiation only, which had a
minimal effect. Sonication alone resulted in a similar decrease in 4-CP after 30
minutes, suggesting that homogeneous sonochemical reaction of the 4-CP is
occurring.

The higher vapor pressure and higher initial concentration (7×10^{-3}M) of
4-CP as compared with the other substrates studied may provide a rationale for the
sonochemical effects observed. If there is a significant percentage of 4-CP in the
vapor phase of the cavitational bubble upon collapse, the likelihood of cavitation-
related thermal decomposition increases.

However, sonication of UV-irradiated catalyst solutions resulted in similar
initial rates of disappearance of the phenol (Figure 6) and appearance of chloride
(Figure 7) as compared with UV irradiation alone. At longer treatment times, the
rate of disappearance of substrate was constant for sonication/photolysis but
started to level off in the photolysis-only treatment (Figure 6). The relatively high
concentration of 4-CP used may have some effect on the catalyst; after short
exposure times there was a definite change in the appearance of the TiO$_2$ to a
more flocculant yellow solid that settled out rapidly.

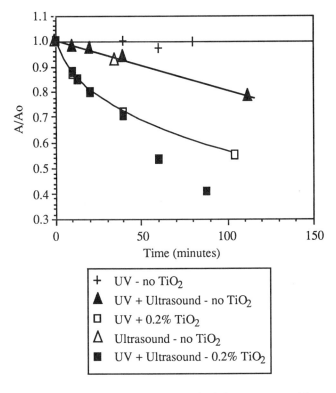

Figure 6. Disappearance of phenol in 4-CP as measured by
UV spectral analysis (510 nm)

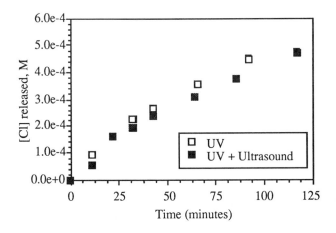

Figure 7. Appearance of chloride in 4-CP degradation

The possible mechanisms for these enhancements are discussed below. Depending upon the particular application (identity of substrate, level of destruction required, flow rates) and the relative costs associated with construction and operation of a sonication system, this technology has the potential to increase destruction efficiency and throughput in a cost-effective manner. For applications where TiO_2 photocatalysis has been identified as a viable or competitive water treatment option, power ultrasound may prove to be an important system addition to allow improved degradation rates, efficiencies, and scope for treatment of aqueous wastes.

Economic Analysis

Currently available data on the enhancement of degradation rate provided by a photolytic/sonication process do not allow a detailed analysis of the cost savings and cost effectiveness over those of conventional UV photocatalysis and other alternatives. However, we have made a rough estimate of the savings due to increased process throughput using data from Ollis *(19)* on costs of UV photocatalysis versus UV ozonation. Our estimate assumes that the increase in degradation rate and efficiency resulting from application of ultrasound to the photocatalytic process will translate into an increased throughput for a given plant, with certain additional construction and operation costs. The additional costs associated with construction can be broken down into these:

- Cost of ultrasonic transducer units and reaction chamber for simultaneous UV/ultrasonic irradiation.
- Potentially increased costs in postfiltration of fine suspensions (more significant if compared with conventional photolysis using immobilized catalyst).
- Increased costs associated with the increase in flow rates, made possible by more rapid mineralization with ultrasound.
- Costs associated with temperature and pressure control for process optimization, if such control is found to be beneficial.

These total construction costs have been estimated as contributing an additional 15% to the amortization costs. However, these are rough estimates only and a more detailed analysis is yet to be performed.

The adjustments to operating expenses include the cost of additional electrical energy required to run the ultrasound transducers, the cost of additional maintenance, and lower costs of catalyst, if smaller loadings and less expensive grades of catalyst (larger particle size, lower purity) can be substituted and if poisoning can be reduced using ultrasound.

The estimated increased construction and operating costs for implementation of ultrasonic irradiation in a photocatalytic process can be offset by the increased throughput made possible by the increased rate of degradation (increases of 50% or greater). Since preliminary data suggest that the rate enhancement for certain pollutants may be higher than 500%, we conclude that the process is worthy of further study to arrive at a detailed assessment of the potential cost effectiveness upon implementation.

In addition to the potential cost effectiveness of the technology as outlined above, advantages related to process efficiency may become important. These include reduction of poisoning effects of other solutes and intermediates produced during degradation, increases in active surface area and/or quantum yields allowing lower catalyst levels to be used, and assistance of ultrasound in increasing the flow rates during ultrafiltration.

Discussion

The use of ultrasonic irradiation in conjunction with photolysis of TiO_2 suspensions has been observed to increase the rate and efficiency of organic substrate mineralization. The following factors are potential reasons for the observed reaction enhancement:

- Cavitational effects lead to increases in temperature and pressure at the localized microvoid implosion sites.
- Cleaning or sweeping of the TiO_2 surface due to microstreaming allows more active sites.
- Mass transport of reactants and products is increased at the catalyst surface and in solution.
- Surface area is increased by fragmentation or pitting of the TiO_2 powder.
- Cavitation-induced radical intermediates participate in the destruction of the organic compounds.
- The organic substrate reacts directly with the photogenerated surface holes and electrons.

Most of these effects are likely to be occurring simultaneously, and their relative contributions to the ultrasonic enhancement of other heterogeneous process have not been definitely established *(20)*. The possible relative importance of each of these effects in ultrasonic irradiation of TiO_2 photocatalysis is discussed below.

Cavitational Effects. Since sonication of TiO_2 suspensions in the absence of UV irradiation did not result in appreciable degradation for most organic substrates, homogeneous cavitational collapse does not appear to be a major route for degradation of the substrate. Intermediates formed during photolysis, however, may react via nonphotolytic (thermal) pathways that are amenable to acceleration by ultrasound-induced cavitation. Hydrolysis reactions in particular may be accelerated using sonication *(19)*.

Microstreaming. When cavitational collapse occurs near a solid surface, the pressure distortion results in asymmetric implosion and generation of a jet of liquid directed at the surface. This jet can erode solid surfaces, remove nonreactive coatings, and fragment powders. Such collisions are so intense in metal powders that localized melting can occur at the point of impact *(21)*. Together with the high local temperatures and pressures generated near or on the surface, the surface of the catalyst is constantly swept clean of reactants, intermediates, products, and other species that may inhibit surface reactions by adsorption on active sites. For photodegradation where a less reactive intermediate, product, or inhibitor adsorbs in a competitive fashion on active sites, sonication may assist to remove those species, allowing access of the substrate (or water) to the active sites for photoreaction.

Increased Mass Transport. Both cavitational and noncavitational ultrasound will result in more effective mass transport at solid/liquid interfaces than mechanical stirring alone. If the rate-determining step in heterogeneous photocatalysis is transport of reactant to, or removal of product from, the surface sites, then sonication will accelerate reaction. The observation of zero-order kinetics for long periods in many of the degradation experiments suggests that sonication results in a significant change in the heterogeneous catalysis mechanism. If it is possible to maintain zero-order behavior during the course of the degradation, then significant reduction will be realized in the time required for destruction.

Increased Surface Area. Ultrasonic treatment of solid powders often results in fragmentation and pitting, increasing the overall active surface area. Photoreactions with presonicated TiO_2 did not show any appreciable effect on reaction rate, indicating that the rate increases observed in photoreactions performed using ultrasound are not due primarily to surface area increases, although experiments at low catalyst loadings must be performed.

Radical Formation. Most of the intermediates and products formed upon sonication (only) of aqueous organic substrates are due to reaction of the substrate with hydroxyl and hydrogen radicals produced from water sonolysis *(23)*. The high vapor pressure of water relative to other inorganics and organics ensures that sonochemical effects in aqueous solutions will result primarily from secondary reactions with these species. Spin trapping has provided evidence for the intermediacy of H· and OH· *(22)*, although other data have suggested that other high-energy species may be formed, including e^-_{aq} *(23)* and HO_2· *(24)*.

In a study on the trapping of OH· as a function of various organic solutes, Henglein *(25)* determined that the hydrophobicity of the organic solute was the important parameter in trapping hydroxyl radicals, suggesting that the reaction occurs at the interface between the cavitation bubble and the liquid where hydrophobic solutes will concentrate. This would be a major difference between free radical chemistry under sonolysis in water and that created by photochemical means.

The observation that sonication did not increase the initial rate of photodegradation of 4-CP where more hydrophobic substrates did show rate increases would be consistent with this argument. However, the rate of hydroxyl production by sonication is much lower than the rate via photolysis of TiO_2 suspensions, and the observed increases in destruction rate and efficiency cannot be explained solely by the additional hydroxyl radicals, although some contribution from ultrasound-generated hydroxyl radicals is possible. Sonication without UV irradiation did not result in appreciable degradation of most substrates over the same time period as did UV irradiation and sonication of the suspension.

Direct Reaction with Substrate. Acoustic and cavitational effects at the solid/liquid interface (mass transport, cavitational collapse leading to local pressure and temperature increases) may enhance pathways for the direct reaction of the organic substrate with photogenerated holes and electrons. For very insoluble substrates such as polychlorobiphenyls, direct deposition on the catalyst from a hexane solution followed by solvent evaporation is necessary *(12)*, suggesting that a direct attack of h^+ is possible. Although no firm evidence exists to support this rationale, it would be instructive to identify the key intermediates formed during combined sonication and photolysis and compare them with those obtained under photolysis only.

In conclusion, the process of *homogeneous* cavitational collapse does not contribute significantly to the enhanced degradation rates. Cavitation will improve mixing, but on a macroscopic scale it is probably no more effective than a high-speed stirrer. On a microscopic scale, however, mass transport is improved at solid surfaces in motion as a result of sound energy absorption, or acoustic streaming *(26)*. If surface adsorption of the substrate is an important first step in reaction with hydroxyl radicals (adsorbed or free), then ultrasonic irradiation may aid this process and reaction as a result of cavitational collapse proximal to the semiconductor particle. This would generate local high temperatures and pressures for the enhancement of reactions between the radical and substrate. Also, impingement of a jet of solution at the particle may result in a cleaning of the catalyst surface, allowing more effective adsorption and radical production as well as decreasing the thickness of the effective diffusion zone for reactants and products. The origin of the ultrasonic enhancement appears to be mainly acoustic or cavitational effects (or both) at the semiconductor/solution interface during UV irradiation.

Detailed studies on the effects of UV intensity, catalyst concentration, sonication intensity, pressure, temperature, substrate concentration and identity, mixtures of different substrates, and other factors on the rate and efficiency of substrate degradation are required in order to further elucidate this enhancement effect and provide guidelines for process optimization. For example, if cavitation is the major factor, then a gradual change in ultrasound intensity, from low intensities insufficient for cavitation to higher intensities, should result in a corresponding increase in the initial degradation rate as the cavitational threshold is reached.

Although the absolute contributions of the potential rate enhancement explanations will be difficult to establish, further experimentation will determine their relative importance in order to predict and optimize design and performance characteristics. Further work aimed at elucidation of the sonochemical contribution is under way. Parameters under investigation include catalyst type, loading, purity and source, particle size, reaction medium chemistry (pH, dissolved organics and inorganics, dissolved gases), suspended solids, temperature and pressure, and UV and ultrasound irradiation intensity and duration.

Literature Cited

1. Ollis, D. F., *Environ. Sci. Technol.* **1985**, *19*, pp. 480–484.
2. Ollis, D. F., in *Homogeneous and Heterogeneous Photocatalysis*, E. Pelizzetti and N. Serpone, Eds.; D. Reidel Publishing Co.: Norwell, MA, 1986, pp. 651–656.
3. Pelizzetti, E., in *Photocatalysis and Environment*, M. Schiavello, Ed.; Kluwer Academic Publishers: Amsterdam, 1988, pp. 469–497.
4. Turchi, C. S.; Ollis, D. F., *J. Catalysis*, **1990**, *122*, pp. 178–192.
5. Pruden, A. L.; Ollis, D. F., *Environ. Sci. Technol.*, **1983**, *17*, pp. 628–631.
6. Ahmed, S.; Ollis, D. F., *Solar Energy*, **1984**, *32*, pp. 597–601.
7. Barbeni, M.;Pramauro, E.; Pelizzetti, E.; Borgarello, E.; Gratzel, M.; Serpone, N.; *Nouv. Journal De Chimie*, **1989**, pp. 547–550.
8. Barbeni, M.; Pramauro, E; Pelizzetti, E; *Chemosphere*, **1985**, *14*, pp. 195–208.
9. Matthews, R. W; *J. Catalysis*, **1986**, *97*, pp. 565–568.
10. Tunesi, S.; Anderson, M. A.; Chemosphere, **1987**; *16*, pp. 1447–1456.
11. M. Barbeni, E. Premauro, E. Pelizzetti, E. Borgarello, N. Serpone, M.A. Jamieson, *Chemosphere*, **1986**, *15*, pp. 1913–1916.
12 Pellizzetti, E.; Borgarello, M.; Borgarello, E.; Serpone, N; *Chemosphere*, **1988**, *17*, pp. 499–510.
13. Peyton, G. R.; DeBerry, D. W; "Feasibility of Photocatalytic Oxidation for Wastewater Clean-Up and Reuse," SumX Corp., Office of Water Research and Technology Report OWRT/RU-81/1, 1981.
14. Carey, J. H.; Lawrence, J.; Tosine, H. M.; *Bull. Environ. Contam. Toxicol.*, **1976**, *16*, p. 697.
15. Wong, A. S.; Crosby, D. G.; *J. Agric. Food. Chem.*, **1981**, *19*, p. 125.
16. Cesareo, D.; di Domenico, A; Marchini, S.; Passerini, L.; Tosato, M. L.; in *Homogeneous and Heterogeneous Photocatalysis*, E. Pelizzetti and N. Serpone, Eds. D. Reidel: Norwell, MA, 1986, pp. 593–627.
17. Plimmer, J. R.; Klingbiel, U. I.; Crosby, D. G.; Wong,A. S.; *Adv. Chem. Ser.*, **1973**, *120,* p. 44.
18. Martin, R. W.; *Anal. Chem.*, **1949**, *21,* p. 1419.
19. Ollis, D. F.; in *Photocatalysis and Environment*, M Schiavello, Ed., Kluwer Academic Publishers: Amsterdam, 1988, pp. 663–667.
20. *Ultrasound: Its Chemical, Physical and Biological Effects*, K. S. Suslick, Ed., VCH Publishers, Inc.: New York, NY, 1988.
21. Doktycz, S. J.; Suslick, K. S.; *Science*, **1990**, *247*, p. 1066.
22. Riesz, P.; Berdahl, D; Christman, C. L.; *Environ. Health Persp.*, **1985**, *64*, p. 233.
23. Margulis, M.A.; Mal'tsev, A. N.; *Zh. Fiz. Khim.*, **1972**, *46*, p. 2970.
24. Lippitt, B.; McCord, J. M.; Fridovich, I.; *J. Biol. Chem.*, **1972**, *247*, p. 4688.
25. Henglein, A.; Kormann, C.; *Int. J. Radiat. Biol. Relat. Stud. Phys., Chem. Med.*, **1985**, *48,* p. 251.
26. Boudjouk, P.; in *High Energy Processes in Organometallic Chemistry*, K. S. Suslick, Ed., ACS Symposium Series 333, American Chemical Society, 1987, pp. 209–222.

RECEIVED September 3, 1992

Chapter 7

Adsorptive and Chemical Pretreatment of Reactive Dye Discharges

D. L. Michelsen[1], L. L. Fulk[1], R. M. Woodby[1], and G. D. Boardman[2]

[1]Department of Chemical Engineering and [2]Department
of Civil Engineering, Virginia Polytechnic Institute and State University,
Blacksburg, VA 24061-0211

The color from navy cotton reactive cold pad/batch and exhaust/jet dyeing processes using primarily C.I. Reactive Black 5 can be substantially lowered by the use of sorbents, flocculating polymers, reducing agents, and anaerobic biodegradation. Of the reducing agents sodium hydrosulfite, thiourea dioxide, sodium formaldehydesulfoxylate, and sodium borohydride, sodium hydrosulfite is the most economical, but the amount of thiourea dioxide needed to produce the same color change for both effluents is the least. Reduction via anaerobic digestion provides comparable color and increased TOC removal over chemical reduction of pad/batch rinse water. The amounts of sorbents and flocculants needed proved to be expensive and likely too massive for dye house effluents.

As the amount of cotton textiles dyed in this country has increased, the amount of color discharged into surface waters has increased. The cotton reactive dyes for the intense and bright colors favored today are difficult to remove from process water because of their high water solubility and chemical stability of the hydrolyzed form in wash waters (*1*). This leads to residual color in wastewater treatment plant effluents and colored surface waters, which are aesthetically unacceptable, if not toxic.

Our studies of color and total organic carbon (TOC) removal have been applied to a dark navy cotton reactive mixture of monoazo and disazo compounds used for dyeing 50/50 or 95/5 cotton/polyester fleece produced by a local textile manufacturer. This shade of navy is named Navy 106 by that plant. The dye mixture, consisting of black, red, and yellow dyes, is applied either by a cold pad /batch process or by an exhaust/jet process. The cold pad/batch dyed knit fabric is rinsed of excess dye and auxiliaries in a slack washer; this washer effluent was used for some of the treatment studies. The polyester in the 50/50 blends is first dyed in a jet machine, washed, and dried before padding. When the cotton of the 50/50 blend fleece is dyed in a jet machine, the polyester is first dyed and washed in the

0097–6156/93/0518–0119$06.00/0

same machine. The dye cycle effluent from the cotton dyeing in the jet machine was the other source of samples for treatment.

The Navy 106 mixture consists primarily of the bireactive C.I. Reactive Black 5, a disazo dye with two vinyl sulfone reactive groups. Much smaller quantities of a Remazol Red and a Remazol Golden Yellow, both azo dyes from Hoechst-Celanese, are used to obtain the dark navy shade. C.I. Reactive Black 5, one of the largest selling dyes in the world, is purchased in the salt form (1).

Upon mixing in a caustic medium the salt form is converted to the vinyl sulfone or reactive form (2). This vinyl sulfone form can either react with the nucleophilic cotton fiber hydroxyl groups by a Michael type 1-4 addition or hydrolyze in water to form the very water soluble, but colorful, chemically unreactive, hydroxyl form (3) that is quite stable (1). The pollution problem is further complicated by the frequent use of considerable sodium silicate and organic auxiliaries in pad dyeing, or sizeable amounts of sodium chloride, sodium carbonate, or sodium sulfate in jet dyeing to improve the mass transfer of the vinyl sulfone reactive dye molecules to the cellulose active sites.

Sorption Studies

Since the volume of process water used in textile dyeing operations is so large, one reasonable method of separating and reducing the volume of contaminated material would be sorption (adsorption and/or absorption) of the dye molecules onto a solid or into a liquid that is or can become insoluble. The segregated waste solid or liquid dye concentrate could then be treated by some form of chemical or aerobic oxidation, chemical or anaerobic reduction, and/or incineration to destroy the refractory species. Recycling of the adsorbent might also be possible. The effluent of vinyl sulfone reactive dye baths and washes contains the hydrolyzed form of this cotton reactive group with a terminal hydroxyl group that permits hydrogen bonding. In a strongly basic environment the hydroxyl group loses a hydrogen ion resulting in negative species, which are electrostatically attracted to cationic or basic functional groups on solids or liquids. This scenario is supported by our laboratory tests.

A series of room temperature laboratory tests at pH of 10.4 (the pH of the process water), 7.5, and 4.6 with 22 different solids were conducted in which 20,000 parts per million (ppm or mg/L) adsorbent was placed in the Navy 106 slack washer discharge. These were shaken at room temperature for a day. The ADMI color value and TOC were determined on the treated effluent after filtering through a 1.2 micron glass microfibre filter to obtain a clear liquid. Data from these adsorption studies are shown on Table I.

The ADMI color value units were used to evaluate the extent of color removed by the different treatments (2). This ADMI system is more accurate than other color systems because it is somewhat independent of hue (3), but most sensitive to red. To determine the ADMI color value of a liquid, a spectrophotometer is used to measure the percent transmittance of the liquid at 438, 540, and 590 nm. A series of equations using the Adams-Nickerson color difference equation are used to calculate the ADMI value from the transmittances. The treatment performance is

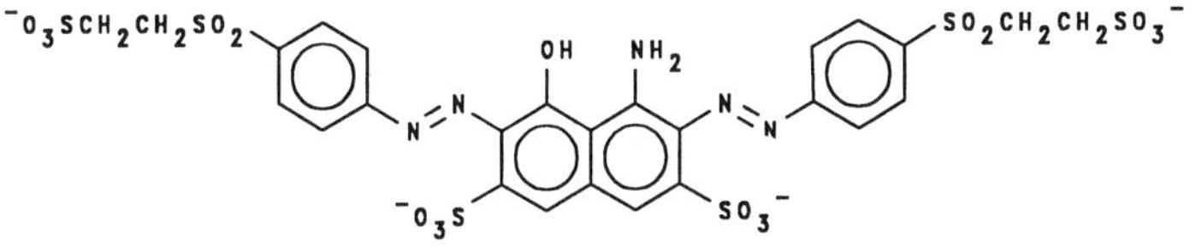

1

Structure of Parent Remazol Black B

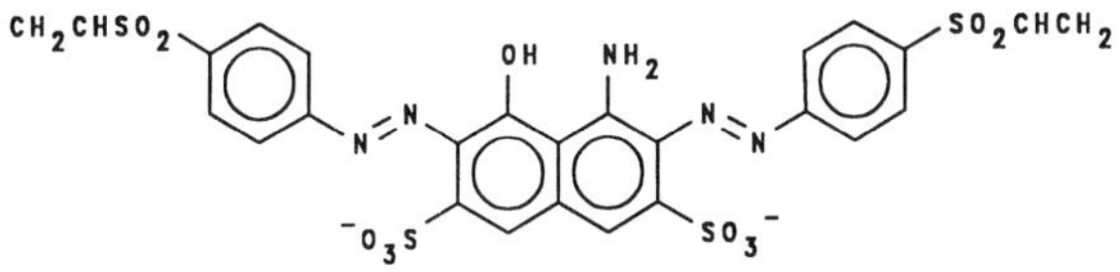

2

Structure of Vinyl Sulfone Remazol Black B

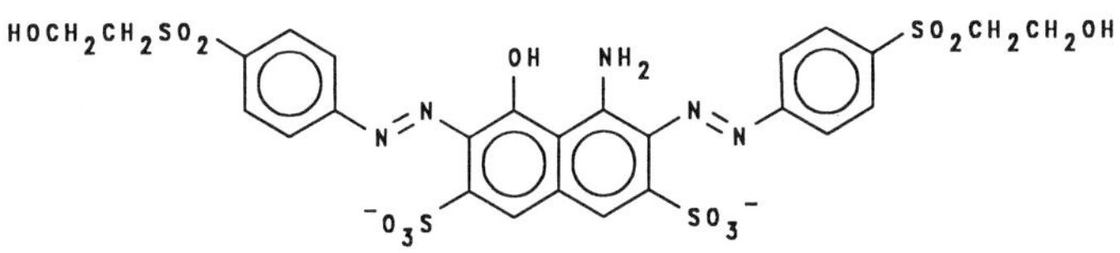

3

Structure of Hydrolyzed Remazol Black B

Table I. **Removal of Color and TOC from Navy 106 Slack Wash Water Discharge Using Solid Adsorbents, Room Temperature**

Description	pH = 10.4		pH = 7.5		pH = 4.6
	ADMI	TOC	ADMI	TOC	ADMI
As is	4165	274	4486	290	4481
TM-399	99		197		65
Rice germ crushed	1752	306	3060 753	323	3097
Wheat bran	2368		2534 1223		2228
Chitosan, sea cure 543	2250 3116	361	1298 1232	219	609
Chitin, VNS-379 (Vanson)	2142 2510	277	1650 1305	193	923
Crab meal	2331 2306	577	1657 1335	410	1937
Ground hydrolyzed pig hair	3902 2944	910	1358	322	3241
Crab meal, treated with HCl, and trypsin	2291 2160	273	1407	222	1063
Sludge, dewatered Blacksburg	2728	1266	1562	695	3986
Partially hydrolyzed human hair	1814				4033
Oat bran	2475	1030	3208 1641	385	3175
Crab meal (low HCl)	1785 (c)		1657		1044
Peanut shells	3712 3015	484	3208 1992	260	4285

Table I. Continued

Description	pH = 10.4		pH = 7.5		pH = 4.6
	ADMI	TOC	ADMI	TOC	ADMI
Chitin, Kodak	2418 3397	242	3208 2674	209	1361
Meat and bone meal		1789	2215	620	2777
Ducon chitosan	2536 (c)		2797		1337
Sea cure 140	2150 (c)		2579		611
Vermiculite	2545				4372
Organic brown rice	2394	874	3911	275	2726
White pine needles	3377		4441	445	4225
Millet	2345	406	4093 3983	229	3907
Oxycellulose	2987		3598		2611

then determined by changes in ADMI values before and after treatment. For all our studies, a Bausch & Lomb Spectronic 20 was used to measure the transmittances.

Since many of the solid adsorbents tended to disintegrate partially into solution, the net TOC was determined by subtracting the TOC of a filtered reagent blank from the final TOC of the sample. All TOC's were determined by a Shimadzu TOC-5000 total organic carbon analyzer. The tests at pH 7.5 and 4.6 were conducted with Navy 106 slack washer effluent after first decanting to remove the precipitated silicates.

In general, the color of the Navy 106 was removed more effectively by sorption as the pH decreased. The product which performed best was TM-399, a bentonite clay modified with a quaternary ammonium surfactant produced by Technical Minerals in Jackson, Mississippi. We were unable to regenerate the TM-399 by removing the sorbed dyes with hot water, basic solution, acidic solution, or combinations of these. The natural products which were most effective for color removal are protein based substances such as rice germ, wheat bran, chitosan, chitin, crab meal, hydrolyzed pig hair, and crab meal treated with acid and trypsin to isolate the chitin crudely. While color removal was effective, the dye solution often partially disintegrated the adsorbent so that the TOC after pretreatment was greater than the TOC of the untreated dye wastewater. These tests were conducted within closed containers under more or less anaerobic conditions. The most active anaerobic activity (judged by the foulness of the odors) and color removal occurred in solutions containing crab meal, meat and bone meal, dewatered sludge from the Blacksburg POTW, and pig hair. Earlier testing of color removal from a number of dye wastes showed pig hair at a concentration of 1 g in 60 mL of dye water readily went anaerobic. Color removal was essentially complete in two days by both anaerobic reduction and sorption.

Color removals from the slack wash water as a function of ppm sorbent are shown in Figure 1 for TM-399, activated carbon (BL Pulverized from Calgon Carbon Corp.), chitosan (Sea Cure 543), and pure chitin (VNS-379). A carbon dosage of about 2500 ppm changed the color from 4100 to 900 ADMI units. These products are all too expensive to use and are required in excessive quantities. Even at an estimated $0.10 per pound, impure chitin from crab meal represents an unappealing choice. A total of 12,000 ppm would be needed for 56% color removal, or 100,000 lb at a cost of $10,000 per million gallons of slack water treated.

A sample of navy jet dye bath effluent was diluted 1 to 20. The pH was lowered from 11.8 to 7.0 with hydrochloric acid. The resulting solution was agitated with varying amounts of activated alumina (Fisher), activated carbon (BL Pulverized from Calgon Carbon Corp.), TM-399, and chitosan (practical grade from crab shells, Sigma). The initial and final ADMI color values are plotted in Figure 2. Activated carbon performed best with color changes of 4000 to 1500 ADMI color units with a dosage of 150 ppm, and changes of 4000 to 5 ADMI with a dosage of 2500 ppm. Alumina and the TM-399 yielded similar results. In this test chitosan was the poorest performer. However, Smith and Hudson found that crab shell derivatives with up to 80% chitosan rapidly adsorbed all color from dye bathes of all classes except the basic ones (4).

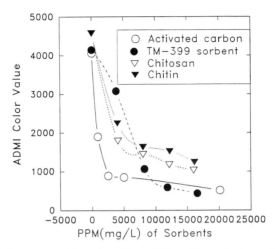

Figure 1. Color removal by adsorption of dyes from slack wash water from rinsing 50/50 cotton/polyester fleece dyed with Navy 106.

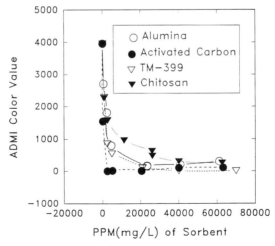

Figure 2. Color removal by adsorption of dyes from Navy 106 jet dye cycle effluent diluted 1 to 20 and adjusted to pH 7.0.

Flocculation Studies. Polyelectrolytes used as coagulants and flocculants can be thought of as sorbents for concentrating and separating pollutants from wastewater. These large molecules have various polar and ionic substituents that render them soluble in water. These reactive substituent groups are exploited for their affinity for reactive species in solution. Our laboratory tests of twelve Nalco polymers, Clarifloc L-311, and Riverclear were performed by weighing the polymers into 25-mL bottles, adding the Navy 106 slack washer water, agitating overnight, and determining the ADMI color value of the supernatant.

Table II shows the results of color removal from the Navy 106 wash water using various polymers. In general, with ADMI color starting at 3557 to 3595 in the pH range 4.3 to 10.8, very little color removal was observed. Only Nalco 7122, which is a strongly cationic, high molecular weight flocculant, yielded marked color removal, as seen in Table II. Unfortunately, Nalco 7122 was required at a concentration of 7632 mg/L to lower the color value of the dye wash from 3593 to 1728 ADMI color units at a pH of 4.3. In summary, with these highly soluble, hydrolyzed reactive dye species, flocculation does not look promising as a pretreatment for dye baths and washes.

Chemical Reduction

One way of destroying the chromophore system of an azo dye is to break the nitrogen-nitrogen double bond by chemical reduction. Since 50 to 60% of all dyes have azo linkages, reduction could be applied to dye types such as acid, direct, and disperse, in addition to the reactives in these studies. Strong reducing agents such as sodium hydrosulfite (also known as sodium dithionite), thiourea dioxide (also known as formamidine sulfinic acid, FAS), sodium formaldehydesulfoxylate, and sodium borohydride, produce significant color removal upon reaction with azo dyes. The reduction products of the hydrolyzed form of Reactive Black 5 are a substituted aniline (4) and a substituted naphthalene (5). These products of chemical or biological reduction are probably more toxic than the dye itself and exhibit more hydrophobic tendencies. Reife (5) has published color and TOC removal studies of an extensive group of hydrolyzed reactive dye species before and after reduction. In essentially all cases the products of dyes reduced with sodium hydrosulfite were adsorbed better onto activated carbon than the unreacted hydrolyzed soluble dyes.

Again, our reduction studies were carried out on discharge wash waters taken from slack washers after pad dyeing and dye bath effluent for exhaust dyeing in a jet machine. Laboratory reductive reactions were conducted at 60 to 80 $^{\circ}$C and pH 10.7 and 11.8 to match slack wash water and jet dye bath effluent conditions.

Experimental. Ten mL aliquots of either the slack wash water discharge or cotton reactive dye bath effluent from a jet machine dyeing 50/50 cotton/polyester knit with Navy 106 were placed in test tubes. These test tubes were heated in a water bath to 70 $^{\circ}$C. Aqueous solutions of sodium hydrosulfite, thiourea dioxide, and sodium formaldehydesulfoxylate were prepared and immediately added dropwise to the tubes. The contents of the test tubes were stirred with a Vortex mixer and returned to the water bath for 30 minutes. If a visible change in color occurred, the transmittances

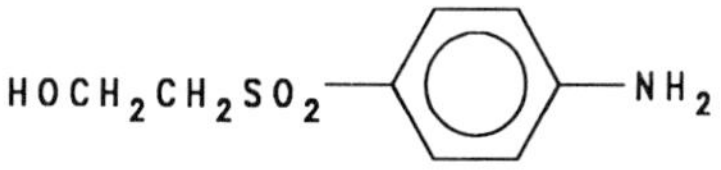
HOCH2CH2SO2
NH2

4

p-(2-hydroxyethylsulfone)-aniline

A reduction product of hydrolyzed Remazol Black B

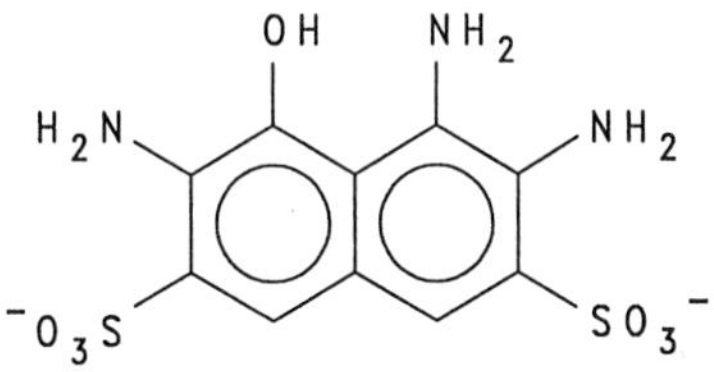
OH
NH2
H2N
NH2
-O3S
SO3-

5

H-acid Derivative

possible reduction product of

Remazol Black B

Table II.   Color removal from Navy 106 slack wash waters using various polymers

Nalco Polymers ID# and Description	pH	PPM of Polymer	ADMI
Navy 106 dye wash water as is	4.3	0	3595
	7.2	0	3658
	10.8	0	3557
high cationic coagulant, 7107	4.3	16200	2769
	7.2	11200	2291
	10.8	19200	2236
moderately cationic, 7123 *HMW flocculant	4.3	14056	too viscous
	7.2	29504	too viscous
	10.8	8292	too viscous
moderate cationic, 8103 coagulant	4.3	14896	3036
	7.2	9208	3051
	10.8	4592	2698
low cationic coagulant, 7134	4.3	9832	3056
	7.2	13776	2938
	10.8	14048	3097
high cationic coagulant, 8100	4.3	10984	3056
	7.2	4184	2972
	10.8	5552	2245
highly cationic, 7122 *HMW flocculant	4.3	7632	1728
	7.2	7132	3053
	10.8	3864	3025
moderately anionic, 7763 *HMW flocculant	4.3	6084	3025
	7.2	7340	3025
	10.8	4336	too viscous
slightly cationic, 7129 *HMW flocculant	4.3	4328	3053
	7.2	4676	3025
	10.8	5400	3055
nonionic *HMW flocculant, 7181	4.3	4108	3025
	7.2	4300	3214
	10.8	5796	3108
slightly anionic, 7182 *HMW flocculant	4.3	4068	3025
	7.2	4596	3025
	10.8	5944	3210

Table II. Continued

Nalco Polymers ID# and Description	pH	PPM of Polymer	ADMI
moderately anionic, 7766 *HMW flocculant	4.3	6668	3025
	7.2	6624	3025
	10.8	5004	3025
highly cationic, 7120 *HMW flocculant	4.3	3428	3218
	7.2	3788	3459
	10.8	3672	3793
Clarifloc L-311 cationic flocculant	10.80	8724	3551
	7.20	7984	3288
	4.30	7036	3091
Riverclear dewatering polymer	10.80	6128	3056
	7.20	5300	2632
	4.30	8964	3166

*HMW = high molecular weight

of the cooled solutions were measured directly for slack wash water, or after dilution of 1 to 25 for jet dye bath discharge. The ADMI color value was calculated for each solution and plotted versus the concentration of reducing chemical in parts ppm.

Figure 3 shows typical results for reducing the same Navy 106 wash water with an ADMI of 3395 at different pH using sodium hydrosulfite. With this series of tests at pH 4.0, the most color removal was achieved with the least amount of hydrosulfite. As seen in Figure 4, about half as much thiourea dioxide as sodium hydrosulfite produces the most color change at a pH of 10.7, but the reagent cost of the thiourea dioxide would be greater, assuming a cost of $0.62 per lb for sodium hydrosulfite and $2.00 per lb for thiourea dioxide. Thiourea dioxide was used only at high pH since the manufacturer's literature suggests that it be used in a caustic environment (6). Screening tests with thiourea dioxide verified that at low pH, color is removed less effectively. All color removal was complete in 10 to 15 minutes at elevated temperatures. A dose of 403 ppm sodium borohydride, needed to change the color value of a slack wash sample from 5700 to 1350 ADMI, makes it the most expensive of the four reducing agents tested.

A series of reductive studies were conducted on the very concentrated jet dye bath discharge from dyeing the cotton in 50/50 cotton/polyester blends with Navy 106 after the polyester has been dyed. The reduced dye solutions changed from opaque dark purple with an ADMI color value of about 100,000 to clear golden yellow with a color value of about 10,000, using a 1 to 25 dilution to measure the transmittances. The reactions with sodium hydrosulfite and thiourea dioxide were visible within three minutes, but the solutions with sodium formaldehydesulfoxylate did not show a visible change in color within three hours. However, after 12 hours, the color change was similar to that provided by the other two reagents. These color removal tests were conducted at 70 °C and a pH of 11.83 to match jet dye bath conditions. The results using freshly mixed solutions of the three reducing reagents are shown in Figure 5. Starting at 100,000 ADMI color units, final color values of 9,000 to 17,000 units were reached using approximately 400 ppm sodium hydrosulfite, 200 ppm thiourea dioxide, and 400 ppm sodium formaldehydesulfoxylate. The differential in price would again make the sodium hydrosulfite more economical. The effectiveness of sodium formaldehydesulfoxylate is diminished by a reaction time of half a day, which is too slow and would severely limit field pretreatment applications. Sodium borohydride is by far the most expensive treatment requiring 1450 ppm to change the jet color from 90,700 to 14,000 ADMI units.

The contribution and value of chemically reducing the Navy 106 slack wash water prior to biological treatment was the subject of our initial effort (7). Table III is an abbreviated summary of those results. Initially, the slack wash water had a 2650 ADMI color value. Mixing three parts of this wash water with one part sanitary wastewater resulted in a solution with a 2040 ADMI color value. A two-day residence time in a sequencing batch reactor resulted in a final color of 1370. With 100% slack wash water, 225 ppm sodium hydrosulfite reduced the color to 1000 ADMI units. Adding three parts of slack wash water, chemically reduced with 225 ppm sodium hydrosulfite and then oxidized with 150 ppm hydrogen peroxide, to one part sanitary wastewater produced a solution with an ADMI color value of 750. Two

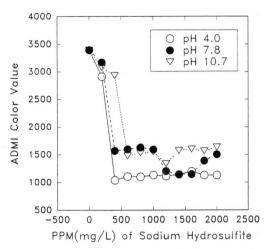

Figure 3. Color removal by reduction with sodium hydrosulfite of slack wash water from rinsing 50/50 cotton/polyester fleece dyed with Navy 106.

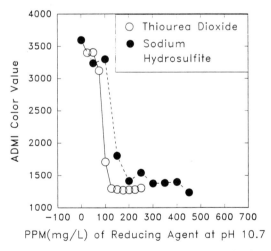

Figure 4. Color removal by reduction with thiourea dioxide and sodium hydrosulfite added to slack wash water at pH 10.7 from rinsing 50/50 cotton/polyester fleece dyed with Navy 106.

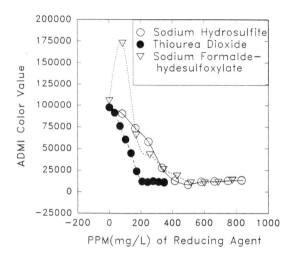

Figure 5.   Color removal by reduction with sodium hydrosulfite, thiourea dioxide, and sodium formaldehydesulfoxylate added to the dye cycle effluent from jet dyeing 50/50 cotton/polyester fleece with Navy 106.

Table III. Removal of Color and TOC Using Chemical Reduction and Oxidation Combined with Biodegradation for Treating Navy 106 Slack Wash Water Discharge from Dyeing Cotton in Polyester/Cotton Blend Knits. (Data from ref. 7.)

Treatment	No Chemical Reduction/Oxidation		Hydrosulfite Reduction/Oxidation	
	Color	TOC	Color	TOC
Initial Condition	2650	210	2650	211
After Reduction	NA	NA	1000	212
After H_2O_2 Addition and Dilution[*]	2040	188	750	189
After Aerobic Biodegradation	1370 [**]33% removal	130 [**]31% removal	615 [***]67% removal	82 [**]57% removal

[*]3 parts slack wash water to 1 part sanitary discharge
[**]Biological only
[***]Chemical reduction + biological ((2650-615)-(1000-750))/2650

days aerobic degradation of the solution reduced the color further to 615 ADMI units. The hydrogen peroxide was added to decrease the toxicity of the reduced solution in the aerobic reactor; however, simple aeration may suffice.

Another method of providing a reducing environment without the high operating cost of chemicals is anaerobic degradation prior to discharge to or during the processing at a POTW. Anaerobic digestion is only feasible for the rinse water after padding, because the exhaust/jet dye cycle effluent has too high a salt content and a high pH. A recent study by Loyd (8) using a concentrated slack wash water discharge with an ADMI of 9500 has shown that no color removal occurred with aerobic treatment, compared to McCurdy's 33% color removal with a more diluted slack washer discharge (2040 to 1370 ADMI, Table III). The TOC in Loyd's aerobic study dropped from 237 to 103 ppm, or 57% removal, versus 188 to 130 ppm, or 31% removal, in McCurdy's study. Both studies utilized a 3 to 1 dilution of dye slack wash water with sanitary discharge prior to aerobic treatment. These stated removals reflect changes due to aerobic biological treatment alone. As seen in Table IV, Loyd's color removal with combined aerobic/anaerobic treatment was 79% with 69% TOC removal, compared with McCurdy's 67% color removal and 57% TOC removal from reduction and aerobic treatment.

Table IV. Removal of Color and TOC Using Anaerobic and Aerobic Biodegradation for Treating Navy 106 Slack Wash Water Discharge from Dyeing Cotton in Polyester/Cotton Blend Knits. (Data from ref. 8.)

Treatment	ADMI Color	TOC (ppm)
Initial Condition	9630	326
After Dilution and Seeding	5220	228
After Anaerobic/Aerobic Biodegradation	1110 (79% removal)	71 (69% removal)

A more promising approach for chemical pretreatment may be reducing the Navy 106 jet dye bath discharge from cotton dyeing. Note on Figure 5 that approximately 400 ppm sodium hydrosulfite can reduce color in the jet bath discharge from 100,000 to 10,000, a 90,000 ADMI color change. On the other hand, 200 ppm sodium hydrosulfite can reduce color in the slack washer discharge from 3500 to 1400, only a 2100 ADMI color change (Figure 4). This is certainly bigger bang for the dollar, and testing is underway to determine the biological degradability of the chemically-reduced, jet bath concentrate. In one study navy jet discharge and navy slack wash water samples were diluted approximately 1 to 20 for the jet liquor and 3 to 5 for filtered slack washer water to obtain solutions with an ADMI color value of about 3500. These two diluted solutions were reduced with thiourea dioxide and measured for color. The results are plotted in Figure 6. The large spikes to the ADMI seen in Figures 5 and 6 occurred when the hue changed

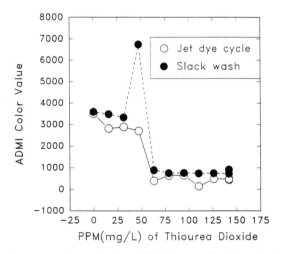

Figure 6. Color removal by reduction with thiourea dioxide added to Navy 106 slack wash water and jet dye bath effluents diluted to approximately 3500 ADMI color units.

from purple to red. The ADMI color measurement system is most sensitive to red. Thiourea dioxide at a concentration of 63 ppm effected dramatic drops in the color value of both jet and slack washer discharges. The TOC of the 3500 ADMI jet effluent was 19 ppm, while that for the 3500 ADMI slack wash was 168 ppm. The TOC from the dyes themselves in both 3500 ADMI pad and jet solutions is approximately 10 ppm. This means that about 50% of the jet solution TOC is from the dyes, while only approximately 6% of the slack solution TOC is attributed to the dyes. The additional organic compounds in the slack wash can be attributed to auxiliaries, such as surfactants used in padding, and possibly a small percentage from the degradation of the polyester in the caustic environment. Samples of the undyed 50/50 cotton/polyester knit were subjected to water and the pad dye liquor of auxiliary chemicals minus the dyes for 12 hours to mimic actual pad dyeing conditions. Analyses of the resulting solutions revealed that 22% of the total TOC in the solution of auxiliary chemicals contacted with the fabric could not be attributed to the auxiliaries or the tints, starches, and oils used in spinning and knitting. Unfortunately, none of the significant peaks could be identified as the breakdown products of polyester from analysis by high pressure liquid chromatography (HPLC) with a diode array detector. From the studies conducted by McCurdy (7) and Loyd (8), we see that after chemical or biological reduction followed by aerobic treatment there are TOC residues. At this point in our studies, the identity of all of the refractory compounds is unknown.

Conclusions

From our studies of the cotton reactive navy dye discharges consisting of primarily C.I. Reactive Black 5, sorbents and flocculants were found to be of questionable value for removing color and TOC because of the massive amounts needed and the costs of those that performed well. Sodium hydrosulfite, thiourea dioxide, sodium formaldehydesulfoxylate, and sodium borohydride all yielded good color removal for both pad washes and jet effluents. More color removal per gram of each reducing agent was observed for the more highly colored jet discharges. Thiourea dioxide performed the best by requiring the least amount of reagent for the greatest color change, but still is more expensive than the second best reagent, sodium hydrosulfite, which was most effective at a pH of 4.0. However, sodium hydrosulfite solids and aqueous solutions require more careful handling and safety precautions than the thiourea dioxide. Thus, thiourea dioxide may be preferable, if sodium hydrosulfite is not already in use at a site. Anaerobic followed by aerobic treatment gives good color and TOC removals with the advantages of low operating costs, but has the disadvantages of requiring relatively large amounts of space and yielding sludges. Toxicity before and after treatment is the next major problem to be addressed.

Acknowledgements

The authors would like to thank the U.S. EPA, Tultex Corp., the City of Martinsville Water Pollution Control Plant personnel, and ARCO for financial and technical support. Also, we thank Dr. Andrea Dietrich and Clifton Bell for the HPLC analyses and confirmation of the proposed C.I. Reactive Black 5 reduction products.

Literature Cited

1. Weber, E.J.; Sturrock, P.E.; Camp, S.R. "Reactive Dyes in the Aquatic Environment: A Case Study of Reactive Blue 19;" EPA/600/M-90/009, Aug., 1990.
2. American Public Health Association; American Water Works Association; Water Pollution Control Federation, *Standard Methods for the Examination of Water and Wastewater*, 17th ed.; American Public Health Assoc.: Washington, DC, 1989; pp. 2-9 to 2-11.
3. Horning, R.H. "Characterization and Treatment of Textile Dyeing Wastewaters;" *Text. Chem. & Colorist* **1977**, 9(3), 24-27.
4. Smith, B; Hudson, S. "A Novel Decolorization Method Using Crabshell Waste;" Preprinted and presented at Environmental Awareness: Targeting the Textile Industry Symposium; AATCC Meeting, Charleston, SC, March, 1992.
5. Reife, A. "Waste Treatment of Soluble Azo Acid, Direct, and Reactive Dyes Using a Sodium Hydrosulfite Reduction Treatment Followed by Carbon Adsorption; "*AATCC 1990 International Conference, Boston, October 1-3, Book of Papers*; American Association of Textile Chemists and Colorists: Research Triangle, NC, 1991; pp. 201-204.
6. Degussa Corp. "FAS (Formamidine sulfinic acid): Its use for reductive post-bleaching in the deinking process;" information pamphlet.
7. McCurdy, M.W.; Boardman, G.D.; Michelsen, D.L.; Woodby, R.M. "Chemical Reduction and Oxidation Combined with Biodegradation for the Treatment of a Textile Dye Wastewater," preprinted and presented at Purdue Industrial Waste Conference, to be published in proceedings, 1991.
8. Loyd, K. *Anaerobic/Aerobic Degradation of a Textile Dye Wastewater*; M.S. Thesis, Virginia Polytechnic Institute and State University, Mar. 1992.

RECEIVED November 10, 1992

BIOLOGICAL TREATMENT

Chapter 8

Applicability of Biological Processes for Treatment of Soils

W. Wesley Eckenfelder, Jr., and Robert D. Norris

Eckenfelder Inc., 227 French Landing Drive, Nashville, TN 37228

Biological processes are used to treat excavated soils, saturated and unsaturated soils in situ, and recovered groundwater. Application of these technologies utilizes the experience gained from waste water treatment and various soils and groundwater remediation methods. These processes can incorporate physical removal along with biooxidation. Site conditions and contaminant properties determine which, if any, biological treatment process is appropriate for a given site. This chapter is an introductory summary of these biological processes.

Biodegradation of organic wastes was used in other forms for many years before commercialization of what we refer to as bioremediation. Land treatment also referred to as "landfarming" has been practiced for several decades (1). A wide variety of food-processing wastewaters including meat, poultry, dairy, brewery, and winery wastewaters have been applied successfully to the land. Sludges and refinery wastes have also been treated by land application. In recent years the use of biological degradation to treat a wide variety of organic compounds has been extended to treat soils and groundwater impacted by accidental releases to the environment (2,3).

The microbiological processes for treating contaminated soils and groundwater are generally referred to as bioremediation. The development of bioremediation technology has benefited from the incorporation of the engineering, hydrogeology, and chemistry developed for other environmental processes such as groundwater production, pump and treat technology, and in situ vapor stripping. The relatively rapid development of bioremediation technology has been possible in part because of the extensive understanding of biodegradation that resulted from the long history of wastewater treatment (4) and interest by the academic community. For the most part, the development of commercial bioremediation technology has focused on the critical engineering and hydrogeological aspects of the processes while relying on established microbiological principles and data.

Years of biological wastewater treatment experience identified the critical parameters that effect biodegradation and determined the ease of biodegradation of most of the common environmental contaminants. Extensive information has been generated concerning both aerobic and anaerobic biodegradability of hundreds of compounds (5-9). Extensive data are available on degradation pathways, the effect of oxygen levels on aerobic and anaerobic biodegradation mechanisms, the dependency of aerobic degradation rates on oxygen concentrations, and nutrient requirements including nitrogen and phosphorus (which are commonly called macronutrients) and several other elements including calcium, sulfur, iron, potassium, sodium, magnesium, and manganese (sometimes referred to as micronutrients). The effects of other important factors such as pH and temperature have also been well defined.

The Bioremediation Processes

The goal of commercial bioremediation processes is to modify environmental conditions such that specific constituents are converted to species of lesser environmental concern at acceptable rates. The parameters that are important for bioremediation are basically those that have been defined from years of wastewater treatment experience.   These include temperature, pH, moisture, nutrient availability, oxygen levels, and absence of toxic or inhibitory species as well as the presence of a consortium of bacteria that is capable of effecting the desired conversions. In most cases pH, moisture, and temperature are within acceptable levels and an effective microbial consortium is present.   The focus of most bioremediation processes then becomes the provision of nutrients and oxygen.

As a result, bioremediation can be viewed as an engineered process for efficiently providing adequate sources and amounts of macronutrients (phosphorus and nitrogen sources) and oxygen. How this is accomplished depends on a number of factors including types of contaminants, soil properties, contaminant distribution, and hydrogeology. Nutrients are added either continuously or in batches. Oxygen is provided continuously in either a passive form or through active transport of a liquid phase (water) or a gas phase (air).

Various bioremediation processes incorporate pump and treat techniques to introduce oxygen and nutrients through movement of groundwater or in situ vapor stripping techniques to move air to introduce oxygen. The movement of fluids to transport an oxygen source and/or nutrients causes other changes as well as biodegradation to occur. The movement of air or water can result in the transport of volatile or water soluble species, thus removing some portion of these species from the contaminated zone. In some cases these effects will substantially reduce the time of remediation and the demands on biodegradation.

Effect Of Physical Properties

The physical properties of the contaminant(s) will affect the rate at which remedial goals are reached and the amount of groundwater or air treatment required. In each of the processes, the soils will be exposed to a moving fluid, either air or water. When air is the fluid, the ease of vaporization of the contaminant(s) and the air flow as well as many other factors will determine the extent to which the contaminant is transferred to the vapor phase. When the fluid is water, the ease of transport of the individual constituents and the rate of groundwater movement as well as many other factors will determine the extent to which the contamination is transported and partially removed in the dissolved phase.

Relevant physical constants are available for many compounds of environmental interest or can be estimated from properties of compounds with similar structures and molecular weight (*10-13*). For conceptual planning purposes, it is adequate to utilize available physical properties to estimate the relative importance of transport of dissolved or vapor phase constituents to the overall process, the extent to which nutrient and oxygen demands will be reduced, and the extent of treatment of recovered water or air that will be required. The same physical properties also provide good insight into which methods of air or water treatment are likely to be feasible. For instance, highly water soluble compounds are not effectively removed from the aqueous phase by either air stripping or carbon adsorption.

Table I lists readily available physical constants for several commonly encountered contaminants. These are: the solubility in water at 25°C which represents the maximum concentration of the substance that will be dissolved in water under equilibrium conditions; log of the octanol/water partition coefficient (log K_{ow}) which is defined as the ratio of the concentration of the compound in octanol divided by the concentration of the same compound in water under equilibrium conditions; vapor pressure which is a measure of the equilibrium between the liquid and vapor phases of a pure compound; and the Henry's Law Constant which is the partition coefficient for the equilibrium of a compound between the vapor and dissolved (in water) phases. These constants are useful for anticipating the relative ease of transport of specific compounds. Transport in the aqueous phase is favored by high solubility and low (or more negative) log K_{ow} values. Transport in the vapor phase is favored by high vapor pressures and Henry's Law Constants. For example, a comparison of the data in Table I shows that benzene is almost 20,000 times more soluble than pyrene and has a log K_{ow} value that is three units less than that of pyrene. In the absence of other effects, benzene will be transported in water much more rapidly than pyrene. Benzene also has a much higher vapor pressure and much larger Henry's Law constant than does pyrene and will be much more easily transferred to the vapor phase from either adsorbed (to soils) or dissolved phases.

While these constants serve to allow a general appreciation of the relative ease of transport and comparisons between specific compounds, many other factors affect the rate of transport of all compounds under specific site conditions. These factors include soil type and distribution, organic content (humates, etc.) of the

Table I. Physical Properties Important To Bioremediation Processes

Compound	Water-Solubility (mg/l)	Log K_{ow}	Vapor Pressure (mm Hg)	Henry's Law Constant (m^3/mole)
Benzene	1,791	2.13	95	5.4×10^{-3}
TCE	1,100	2.40	69	1.0×10^{-2}
Acetone	Misc.	-0.24	231	3.7×10^{-5}
Butanol	77,000	0.88	7	5.6×10^{-6}
Phenol	87,000	1.46	0.5	4×10^{-7}
Napthalene	31	3.28	0.2	4.6×10^{-4}
Pyrene	0.1	5.20	6.9×10^{-7}	1.1×10^{-5}
DEHP	0.3	5.11	5.0×10^{-6}	1.1×10^{-5}
Phenanthrene	0.9	4.46	6.8×10^{-4}	3.9×10^{-5}

Data obtained from:
Howard, Phillip H. et al., *Fate and Exposure Data*, Vol. I, 1989 and Vol. II, Lewis Publishers, New York, 1990.
Montgomery, John H. and Linda M. Welkins, *Groundwater Chemicals Desk Reference*, Vol. 1, Lewis Publishers, New York, 1990.

soils, presence of non-aqueous liquids, presence of solubilizing species (surfactants), temperature, etc. Typically, transport of specific compounds is calculated using the retardation coefficient which expresses the relative amount of time that it would take for the compound to move through the aquifer compared to the rate of movement of groundwater (14). The retardation factor (R) can be calculated from K_{OW} and the average percent carbon in the soils according to

$$R = 1 + (\rho/\eta) \, K_{OW}F_{OC}$$

where

ρ = bulk solids density (g/ml)
η = effective porosity of the solids and
F_{OC} = fractional soil organic content.

Transfer of a compound from the aqueous (dissolved) phase to the vapor phase is reasonably predictable from the Henry's Law Constant. Vaporization of compounds adsorbed to dry surfaces relates well to the vapor pressure. In unsaturated soils the rate of vaporization will depend on many factors including soil moisture, presence of other organic compounds, and whether the soil particles are organic wet or water wet. Some studies indicate that predictions of rate of vaporization from soils are best made from "lumped partition coefficients" which are determined from laboratory tests conducted using site soils (15). This coefficient is more appropriate than the Henry's Law coefficient or the vapor pressure but is site specific. In the absence of site specific data, use of the Henry's Law Coefficient provides a workable estimate. Actual rates of transport are also dependent on rates of diffusion and air flow patterns.

Each of the bioremediation processes described below incorporate the potential for physical removal of contaminants from the soils as well as the primary mechanism of biodegradation. Each process contains procedures to create conditions favorable to microbial growth and generally use the contaminant(s) as food and energy sources for the bacteria in order to promote conversion of the contaminant(s) to substances of lesser or no environmental concern. In some processes degradation of the contaminant occurs through co-oxidation as a result of exocellular enzyme activity in conjunction with another substance that serves as the metabolite. The specific procedures in each process are utilized to take advantage of a variety of potential site conditions and metabolic behavior.

In designing a bioremediation system, the impact of transport of contaminants needs to be understood. The proportions of the constituents of interest that will be addressed through biodegradation versus physical removal from the soils will affect the demand for nutrients and oxygen and for treatment of recovered air and/or water. Since treatment of water and air can represent a significant portion of total remediation costs, a properly designed system will balance the benefits and potential costs resulting from contaminant transport.

Treatment Of Excavated Soils

Land Treatment. Land treatment, sometimes referred to as land farming, incorporates batch addition of nutrients using physical mixing of the nutrients into the soils and continuous passive introduction of oxygen (air) (*16*).

Excavated soils are spread over a relatively large area (approximately 0.5 acres per 1,000 cubic yards of soil) to a depth of 6 to 18 inches depending on soil type, land availability, and time constraints. Nutrients are added in a dry form or as a dilute solution which also provides moisture. The soils are periodically tilled and/or plowed and nutrients added as required.

Nutrients consist of a phosphorus source that is most typically a salt of phosphoric acid and a nitrogen source that may be an ammonium salt, a nitrate salt, urea or a combination of sources. Nutrient requirements can be estimated from the contamination concentrations or laboratory treatability tests. Ten to fifty percent of the total anticipated nutrient requirement is added during the initial batch addition. Subsequent nutrient additions are made over the course of the project with the amounts estimated based on consumption during the process as determined from periodic analysis of soil samples.

High degrees of conversion and/or low residuals relative to remediation standards are possible with low molecular weight contaminants (and mixtures such as gasoline). However, the levels achievable with other contaminants such as polyaromatic hydrocarbons (PAH's) are unclear because their rate of solubilization may not support sufficient microbial growth (*17*).

For many compounds including most hydrocarbons, the naturally occurring bacteria are able to degrade the organic hydrocarbons converting them to innocuous materials such as carbon dioxide, water, cell material, and, possibly, high molecular weight byproducts provided that sufficient nutrients and oxygen are available and moisture and pH conditions are within acceptable limits. It is relatively easy to add allochthonous (non-indigenous) bacteria that are available from several commercial sources. To date, the benefits of adding bacteria have not been sufficiently documented to support recommendations for their use (*18*). Some studies have indicated that after initial addition of allochthonous bacteria, the preponderance of degradation can still be attributed to the indigenous bacteria. Other studies have shown that the allochthonous bacteria must be added every few days as their survival in the soils is limited. The need for regular addition of bacteria greatly adds to the cost. A number of firms claim success with added bacteria but do not provide good control data.

Prior to designing a treatment system, laboratory programs are frequently conducted to determine the suitability of the method and to establish engineering data (*19*). Microbial enumeration using plate counting techniques and soil pH data can provide a strong indication of any potential interferences with biodegradation. Such tests may be sufficient for small systems with contaminants that are known to be readily biodegradable. For larger systems or more recalcitrant contaminants, treatability studies are used to estimate degradation rates, extent of degradation, and nutrient requirements. Because land treatment does not require transport of nutrients and/or oxygen, these tests can simulate field conditions and predict extent and rate of degradation more closely than do tests for in situ methods. It is, however, necessary to be conservative in using the data to predict full scale results.

Design of a treatment system includes selection of a sufficiently large area for treatment of the soils in one or more batches, determination of liner specifications, when necessary, nutrient requirements, provisions for handling runoff, and arrangements for security.

In some instances depending on the contaminant, soil types, and risk to the underlying soils and the aquifer, steps need to be taken to prevent contamination of the underlying soils and groundwater that could result from leaching of soluble constituents following precipitation. This is accomplished by placement of a synthetic liner (some firms have used clay liners) over the treatment area prior to placement of the contaminated soils. If a synthetic liner is used, a cover of clean soils or sand must be put down to protect the liner from damage during tilling or plowing.

Unless the system is protected with a tent-like cover (20), the system may have to be designed to handle run-off resulting from precipitation. To control run-off, the treatment area is graded to permit the water to drain to a sump where the water can be collected and treated, sprayed back on the soils, or discharged to an existing on site treatment facility or sewer.

Monitoring includes sampling of the treatment soils prior to, during, and after treatment for contaminant parameters, pH, nutrients, soil moisture, and bacteria counts. Additionally, the soils beneath the treatment system are sampled and analyzed prior to and after remediation to document the lack of impact on the surficial soils.

The construction and operation of a land treatment system exposes the soils to large volumes of air (and thus oxygen) which allows volatile species to evaporate. Unless the system is contained within a tent-like structure, volatiles will not be contained and thus will be discharged to the atmosphere. Uncontrolled discharge of large amounts of volatile species is typically inappropriate and nonconforming with most state regulations. For soils contaminated with less volatile contaminants such as diesel fuel and especially the heavier heating oils, the volatile losses may be acceptable depending on the location relative to human activities and on state regulations. However, soils with significant levels of volatile components are more appropriately treated by other methods such as soil cell treatment.

Soil Cell Treatment. Excavated soils are mixed with nutrients in a single batch addition and placed on a liner (21). Air is provided continuously through a series of slotted PVC pipes that are manifolded to a blower (see Figure 1).

A treatment area of approximately 0.1 acres per 1,000 cubic yards is required. The area is prepared by removing debris and grading to provide a shallow slope to collect any water that may drain from the soils. The area is covered with a suitable synthetic liner which is then covered with approximately six inches of gravel or coarse sand. A geotechnical fabric is placed over the gravel or sand layer to prevent intrusion of the soils. Nutrients of a similar composition to those used for land treatment are mixed with the contaminated soils which are then placed over

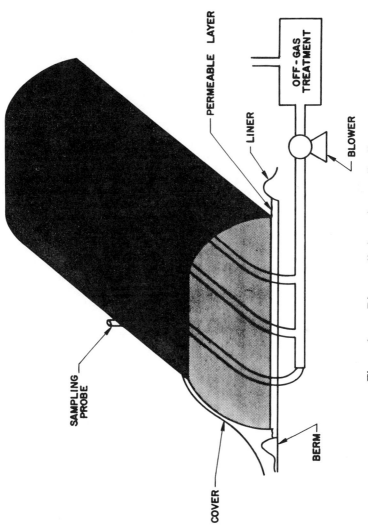

Figure 1.   Bioremediation using a soil cell.

the geotechnical fiber in two to three foot lifts. Slotted PVC pipes are placed at each lift. The PVC pipes are manifolded to a blower or vacuum pump. The soils are covered with an impervious material which extends over shallow berms located around the pile. The cover controls moisture in the soils, prevents runoff, and minimizes both dust and vapor emissions. The PVC pipe/blower system moves air through the soil pile to provide one to three pore volumes of air per day. The off-gas from the blower is treated if necessary.

Laboratory programs are similar to those conducted for land treatment. Treatability studies can mimic field conditions in terms of rate of provision of nutrients and oxygen. Differences in moisture distribution, blending of nutrients with the soils, and temperature variations will also impact the rate of extent of degradation attained during full scale application but it is difficult to design laboratory tests to approximate the range of these parameters that will be experienced in the field.

Once soils have been excavated or the volume of soils to be excavated has been estimated, a suitable area is selected. Laboratory and site data are used to design the cell with specifications for grading and surface preparation, liner composition and thickness, berm dimensions, nutrient requirements, PVC manifold design and blower specifications, off-gas treatment if needed, cover specifications, and operating and maintenance schedules. As with land treatment, surficial soils below the treatment area are sampled and analyzed prior to construction and after demolition.

In addition to providing oxygen, the movement of air through the soils will capture volatile species. The extent to which the process physically removes specific contaminants depends on the lumped partitioning coefficients and the air flow through the soils. Operating at a low flow rate favors degradation over volatilization. Flow rates that exchange two to three pore volumes of air per day will probably support the maximum degradation rates. If time is a factor or a cost efficient off gas treatment system is already available, the system can be designed to maximize physical removal of the volatile contaminants. In such systems, the flow rate of air through the soils should be designed based on the type of vapor treatment system and the mass of volatile species.

Operation is relatively simple. In addition to routine maintenance of the blower and off-gas treatment system, vapor samples from probes placed within the cell, the PVC manifold, and the blower off-gas are analyzed for temperature, carbon dioxide, oxygen and volatile contaminants. Once the carbon dioxide and oxygen data indicate that remediation is nearing completion as indicated by either material balance estimates based on initial contamination levels and the total carbon dioxide produced or by a reduced rate of oxygen consumption and carbon dioxide production, additional soil samples are collected and analyzed to determine the residual levels of contamination.

In Situ Treatment

In situ treatment methods offer the advantage of minimal disruption to the site. In some instances in situ treatment is the only approach that can result in site remediation without demolishing buildings or other structures. In situ treatment methods that destroy or convert the contaminants to innocuous materials mitigate long term liability and maintain property values. In situ bioremediation processes can be designed to treat the unsaturated and/or saturated zones (*22,23*).

In Situ Bioremediation Of Unsaturated Soils. In situ bioremediation of unsaturated soils utilizes systems for batch addition of aqueous solutions of nutrients and continuous addition of oxygen through induced air movement.

Nutrients, if required, are added by one of several types of surface or near surface percolation systems. If the surface soils are exposed or covered with gravel or sand or even vegetation, nutrients are added to the surface in a dry form or as a dilute solution. Water is sprayed on the surface to carry the nutrients through the soils. Soils that are located under pavement or buildings are more difficult to address. If the pavement or building floor is underlain with a gravel or sand layer, a nutrient solution is carefully forced into this layer and will spread horizontally before percolating into the soils. In other cases, closely spaced shallow wells are used as shown in Figure 2.

Nutrient selection is more important for in situ systems than for the ex situ methods. Selection of nutrients must take into account the degree of retention of the nutrient components by the soils. While some retention is beneficial, the degree of retention directly impacts the volume of water and time required to distribute the nutrients. Other considerations are also important. If sodium salts are used, clayey soils may swell and reduce permeability to both water and air. Use of nitrate as the nitrogen source may result in concern for increased nitrate levels in the groundwater as nitrate will be readily leached from the soils. Residual ammonium ion levels and potential conversion of ammonium ions to nitrate may also be of regulatory concern.

Oxygen is provided as air utilizing the methods of in situ vapor stripping (*24*). Wells screened within the unsaturated zone are manifolded to a blower or vacuum pump which creates a reduced pressure within the well bore. The partial vacuum causes air to enter the soils from the surface and sweep through the contaminated soils. Exchanging the air in the unsaturated zone from one to three times per day provides adequate oxygen for the bacteria. Air flow through the soils may be reduced as a result of the water added to introduce nutrients, particularly in finer grained soils. The design of the system must balance the need for both oxygen and nutrients.

The movement of air and water through the contaminated soils may mobilize some of the contaminants. Water soluble compounds may be carried downward with the percolating water. The extent of vertical migration will depend on the properties of the contaminant most appropriately represented by the octanol/water coefficient, the organic content of the soils, and the volume and percolation rate of the water as well as the rate of degradation under the established conditions. The degree to which volatile components are removed are, as for a soil cell, dependent upon the lumped partitioning coefficient, the rate of air flow through the contaminated soils, the rate of degradation, and to some extent, the

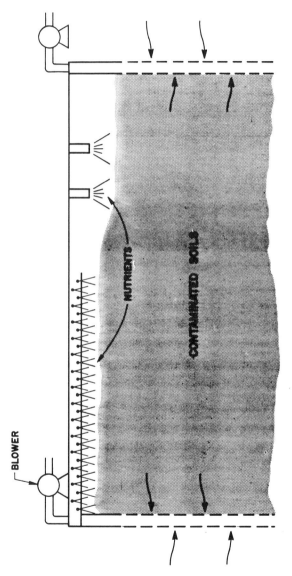

Figure 2.   In situ bioremediation of unsaturated soils.

solubility of the contaminant and the amount of water present. Volatilization and vertical migration can also be competitive.

While removal of volatiles eliminates the contaminants from the treatment zone, downward migration may result in the contaminants reaching the aquifer. If the aquifer is already contaminated and has a treatment system in place, the leachate compounds will be addressed by the aquifer treatment system. However, if the groundwater has not been previously impacted, careful consideration should be given before selecting this method. Groundwater monitoring should be implemented and possibly, groundwater recovery wells installed. Minimization of downward migration can be achieved by operating the air system for a period before nutrients are added even when few volatiles are present. Most soils contain sufficient nutrients for biodegradation to occur and in many in situ vapor stripping systems, biodegradation can account for a portion of the reduction in contaminant levels even when nutrients are not added. Measurements of carbon dioxide levels in the off gas can provide an indication of the rate of biodegradation, but may be misleading because of other carbon dioxide sources and the complex equilibrium between carbon dioxide vapors and the dissolved and precipitated forms of carbon dioxide. Carbon isotope ratios can provide a more accurate estimation of biodegradation of petroleum derived substances.

Although the removal of volatile contaminants contributes to the cleaning of the soils, it may require the inclusion of a treatment system to prevent discharge of these constituents to the atmosphere. Treatment of air can be accomplished with activated carbon, a catalytic converter, an incinerator, or a vapor phase bioreactor.

In addition to the type of laboratory test program discussed for the land treatment and soil cell methods, tests are conducted to estimate nutrient adsorption during percolation and the area of influence of air extraction wells. The latter is determined by conducting one to two day field tests in which air is extracted from a well and partial vacuum readings are made at monitoring points located at multiple distances from the extraction point. Design parameters include air recovery well screening depth and locations, air blower vacuum and flow rate, nutrient addition system, nutrient addition levels and schedule, maintenance, and monitoring.

Monitoring consists of measuring pressure readings at monitoring probes and sampling of air from the probes and the recovered air for analysis for oxygen, carbon dioxide, and specific volatile constituents. As for soil cells, once the carbon dioxide and oxygen levels approach ambient air levels, soils samples are obtained for analysis of the constituents of interest. Monitoring may also include groundwater sampling for contaminants and nutrient constituents.

In Situ Bioremediation of Saturated Soils. Groundwater is recovered from the aquifer, treated, and reinjected into the formation after amendment with nutrients and an oxygen source (*25*).

Groundwater recovery is designed to capture the contaminant plume and the injected nutrients and oxygen amended water. The recovered water is, in most cases, treated at the surface using activated carbon, an air stripper, or a fixed film bioreactor. A portion of the treated groundwater is amended with nutrients and an oxygen source and introduced into the formation through a series of injection wells or trenches (see Figure 3).

Nutrient selection is more critical for this type of bioremediation than for the other processes. Sodium salts can cause swelling of clays and should be avoided for soils with appreciable clay contents. Orthophosphates can cause precipitation of iron, calcium, and magnesium and thus should be used with great care. Tripolyphosphate salts solubilize iron, calcium, and magnesium and are much less likely to result in blockage of the formation. Nitrogen is usually provided as an ammonium salt. Nitrate has some advantages but is usually restricted by state regulations.

Oxygen can be provided by sparging air into the injection well, but this can provide only 8 ppm of dissolved oxygen (plus some entrained gas) unless injected at considerable depth below the water table. Because the remediation process cannot proceed any faster than oxygen is provided, other sources of oxygen are typically used. Sparging of pure oxygen gas instead of air into the injection water will result in a five fold increase in the rate of oxygen introduction into the aquifer. However, unless oxygen gas is already available, this method is usually not practical. Hydrogen peroxide, which undergoes conversion to oxygen and water, can be introduced at up to levels of 500 to 1,000 ppm corresponding to 250 to 500 ppm of oxygen. In practice only some of this oxygen is likely to be lost to the unsaturated zone due to too rapid abiotic or microbial induced decomposition (26). Hydrogen peroxide is more costly than air and should be used only where the increased rate of remediation more than offsets the added chemical costs. Recently, there has been a trend to use the direct injection of air into the aquifer. Air is forced into the formation several feet below the water table using sparging wells which are screened over a one or two foot interval. The air moves radially outward before reaching the unsaturated zone, thus providing dissolved oxygen for aerobic biodegradation.

In designing an in situ bioremediation system for saturated soils it is necessary to understand the distribution and total mass of the contamination so that the amount of nutrients and oxygen can be estimated. The practicality and time for remediation then depends on how long it will take to introduce sufficient nutrients and oxygen to degrade the constituents of concern.

Generally, it is not practical to implement in situ bioremediation if more than intermittent pockets or very thin films of free phase - lighter than water material is present. Free phase material should be removed using skimmers or dual phase pumps. Residual or thin layers of free phase volatile contaminants such as gasoline can be removed using in situ vapor stripping which will also serve to provide oxygen in the capillary zone and to soils exposed to air during periods of low groundwater levels.

Approximately three pounds of oxygen are needed for every pound of contaminant to be converted to carbon dioxide. This requirement is reduced by the extent to which the contaminant is converted to cell material. Nutrient requirements have been approximated by a carbon to nitrogen to phosphorus ratios ranging from 100:10:1 to 350:10:1. The amount required depends on the presence of nutrients already in the soils, the degree of recycling of nutrients by the bacteria, and the

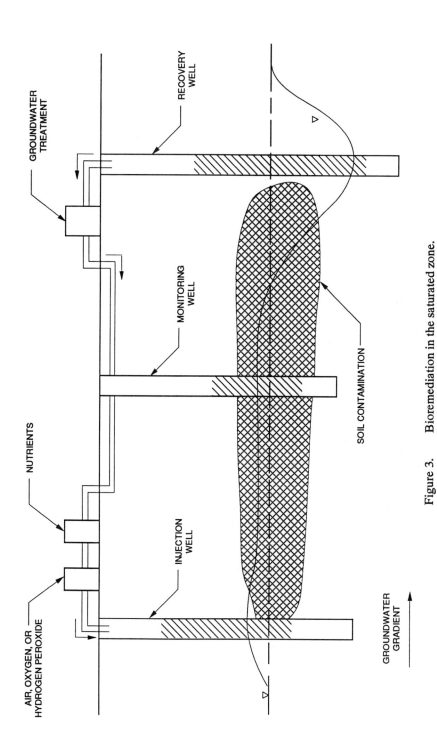

Figure 3.   Bioremediation in the saturated zone.

fraction of the contaminant converted to cell material. The length of time that it takes to provide the nutrient and oxygen requirements is dependent on the concentration of nutrients and oxygen that can be introduced to the formation, the rate of reinjection. The rate of distribution of nutrients is also dependent on the extent of adsorption of the nutrients by the soils and the efficiency of distribution.

The nutrient and oxygen requirements as well as the remediation time are further dependent on the extent to which the degradable species are removed from the aquifer. Soluble species are removed with the captured groundwater at rates which are a function of the octanol/water coefficient of the contaminants; the percent carbon in the soils; the rate of groundwater recovery; and the location of the recovery wells relative to the contaminant mass. If air sparging is used to provide oxygen, volatile constituents will be transferred to the unsaturated zone where they are captured by an in situ vapor stripping system. The extent to which they are removed will depend on the Henry's Law Constant of the volatile constituents and the air flow volumes and patterns.

For treatment of lighter petroleum hydrocarbons and many easily degraded compounds, the critical issues of feasibility are not rate or extent of biodegradation that might be indicated from laboratory experiments, but rate of injection of amended water and groundwater velocities under pumping and injection conditions. For many sites it is better to first determine the optimum pumping and injection scenario using simple modeling techniques and aquifer test data to determine if sufficient groundwater recirculation flow rates and pathways can be established (27). From these an estimate of the time of remediation can be made. If the aquifer data are already available, the modeling effort can be done much more quickly and at lower cost than can laboratory tests.

During operation of an in situ bioremediation system, groundwater is sampled from recovery wells and monitoring wells located within and outside of the treatment area. Groundwater is analyzed for nutrient constituents, bacteria populations, dissolved oxygen, pH, nitrate, calcium, iron, magnesium and contaminant levels. Groundwater flow rates and piezometric surfaces are determined and used to compare system operation to predictions of the model to design the system. When necessary, the model can be used to determine changes in flow rates needed to provide a more advantageous distribution of nutrients and/or oxygen.

Effluent Treatment

The same biological water treatment methods that have been developed for wastewater treatment can be utilized to treat recovered groundwater during in situ bioremediation or other aquifer remediation processes. The various biological water treatment processes in practice today are shown in Figure 4. Selection of the most appropriate process requires an understanding of the significant differences in the treatment of groundwaters compared to industrial wastewaters. In most cases, groundwater contamination levels are low compared to industrial wastewaters, e.g., 10 mg/L, and frequently the majority of the constituents are volatile. The

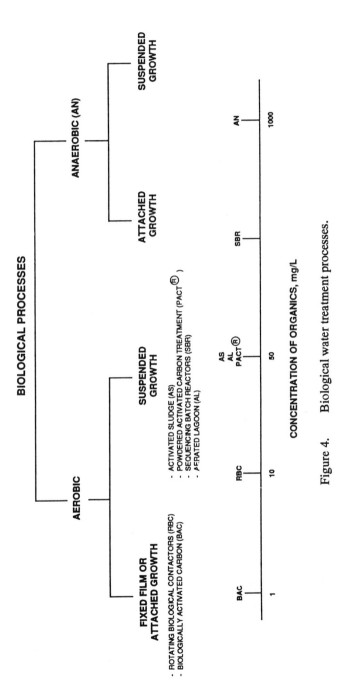

Figure 4.   Biological water treatment processes.

concentration and volatility of the contaminants frequently limits the choices of processes.

Biologically Activated Carbon Beds. These systems can be successfully employed for very low levels of organic constituents, i.e., less than 5 mg/L. In these systems, oxygen is added to the water either by aeration or as hydrogen peroxide in solution with nutrients to provide one milligram of oxygen for each milligram of carbon. Biological growth occurs on the carbon and organic removal results from both adsorption and biodegradation resulting in typical retention time requirements of 5 to 15 minutes.

Granular Activated Carbon Fluid Bed Reactors. This type of reactor has recently been developed for groundwater treatment. In this process high purity oxygen is dissolved in the water prior to passing through the reactor. The use of pure oxygen as the oxygen source eliminates any off-gas from the system. Organics resistant or refractory to biological treatment can be removed by adsorption on the carbon. Hydraulic retention times are on the order of five minutes with organic loadings (chemical oxygen demand) of up to three kg COD/cubic meter-day.

Fixed Film Processes. Waters with organic contents of 5-50 mg/L can be treated with fixed film reactors. Organic slime layers are grown on plastic surfaces. The organic constituents in the water diffuse into the slime layer where they under go oxidation. Traditionally, rotating biological contactors (RBC) consisting of plastic disks that rotate through the water exposing the microbial slime to both the organic containing water and air (as the oxygen source) have been widely used. Fixed film reactors consisting of a series of aerated flow-through chambers containing sheets of slime coated static plastic media have become common in the United States (*28*).

Activated Sludge Processes. Either conventional configurations or sequencing batch reactors (SBR), can be employed when the organic concentration exceeds 50 mg/L. Consideration needs to be given to the volatile emissions. If the groundwater contains high volatile carbon concentrations, either upstream air stripping or capture and treatment of the off-gas from the aeration basin may be required. If the groundwater contains non-degradable organics, powdered activated carbon (PAC) can be added to the activated sludge process for adsorption of the refractory organics.

If the organic concentration expressed as biological oxygen demand (BOD) is less than 50 mg/L, suspended growth processes such as the activated sludge process can not be used. This is because the net cellular growth is less than the normal suspended solids loss in the effluent and wash out can occur. In some instances, this can be compensated for by adding effluent filtration with the backwash returned to the aeration basin.

Anaerobic Reactors. High concentrations of some kinds of organics can be treated with anaerobic reactors. In order to meet effluent requirements, the effluent from the anaerobic reactor will require aerobic polishing.

Any of the various water treatment methods when used in conjunction with in situ bioremediation result in treated water containing low levels of nutrients and some levels of bacteria capable of degrading the constituents of interest being reinjected into the aquifer. The bacteria are likely to be filtered out by the soils in the vicinity of the injection well. Care needs to be taken to prevent loss of injectivity of the wells.

Remediation Times

Estimating remediation times can be difficult, especially for an in situ system where the heterogeneity of the soils and contaminant distribution will not be well defined. For in situ treatments of aquifers, treatment times are most appropriately estimated by calculating the time it will take to distribute sufficient nutrients and oxygen across the site. Typically, oxygen distribution is the limiting factor. In these cases, remediation times can be estimated from the total oxygen demand (three pounds of oxygen for each pound of biodegradable organic), the groundwater reinjection rate, and the concentration of oxygen in the reinjected groundwater. Making reasonable estimates of remediation times requires adequate modeling of the aquifer and a reasonable estimate of the contaminant mass. Appropriate and cost effective numerical models are available for this purpose. These models can also be used for developing conceptual and detailed designs and making adjustments to operating systems (27).

For ex situ treatment of soils, remediation times are generally not dependent on transport of nutrients or oxygen and rough estimates can be made from degradation rates determined from laboratory tests conducted with site soil samples. Estimations from literature degradation rates are useful but less accurate because they do not take into account soil specific effects, temperature, moisture content, and acclimation of the bacteria to the specific contaminant(s). However, such estimates may be adequate for preliminary evaluations.

Biodegradation kinetics can be approximated by first-order kinetics. This allows the concentration at any time to be predicted from the biodegradation reaction half-life. The half-life is the time that it takes for the concentration to decrease by one half. Thus if the half-life was ten days and the initial concentration was 1,000 ppm, the concentration would be 500 ppm after 10 days, 250 ppm after 20 days, and 125 ppm after 30 days. Provided that optimum conditions are maintained, the remediation time will depend on the half-life of degradation, the initial concentration of the most highly contaminated soils, and the final concentration that must be obtained. This approach is least reliable at very high and very low contaminant concentrations.

Extent Of Remediation

The above discussion on estimation of remediation times assumes that the remediation goals can be met. There has been considerable discussion about the extent of remediation that can be obtained. Unfortunately, there exists meager good quality case history data. In some instances, low ppm levels or levels below the detection limit have been attained. Typically laboratory treatability tests will yield lower levels than will full scale systems where control of critical parameters is more difficult.   For highly degradable constituents such as gasoline and many non-chlorinated solvents, low ppm or lower levels are reasonable to expect provided adequate control can be maintained over the system. For slower-to-degrade systems such as heavier petroleum hydrocarbon mixtures, it is best to consider the percent conversion required to attain the remediation goals. There are no hard and fast rules for estimating the extent of conversion that might occur.   The less readily degradable the target constituents, the greater care that needs to be exerted when high percent conversions are required.

Process Selection

Which bioremediation process, if any, is the most appropriate solution to environmental problems at a specific site depends on several factors. Selection of bioremediation as the sole method of remediation requires that all of the contamination of interest be biodegradable or capable of being physically removed by the specific process under consideration.   Otherwise bioremediation must be used in conjunction with other remediation technologies (29) or not used at all.

At many sites, excavation is not practical due to site activities or structures and thus in situ methods offer the most attractive approach to remediation. In situ methods may also provide the most cost effective solution.   Other advantages include avoidance of soils being classified as hazardous waste upon excavation, minimum disruption of the site, minimal public exposure during treatment, and, in some cases, provision of a ready remedy for potential future releases. On the other hand, in situ methods are applied under frequently highly heterogeneous conditions and provide the least control over the treatment conditions.

Ex situ methods require excavation of soils which is a significant cost factor and either require the excavation to be left open for a period of time or clean soils to be obtained to fill the excavation. However, ex situ treatment is generally fast compared to in situ methods, allowing the contaminated area to be returned to unimpaired or less restricted uses or to facilitate sale of the property. Ex situ methods also allow some level of blending of soils to reduce heterogeneity and simplify treatment and monitoring. Land treatment in particular provides a great deal of control over the soils. Nutrients can be readily added and the soils can be thoroughly mixed and worked. It is also relatively easy to monitor progress. Land treatment does have the disadvantages that a large area is required unless treatment is done sequentially, and volatile emissions can not be easily controlled. Soil cell treatment has the advantage that a smaller area is required, that emissions can be controlled, and that operations are relatively simple. Costs per unit soil volume, however, tend to be higher than for land treatment.

In selecting a biological treatment process for treating groundwater, it is necessary to take into account the concentration of the organic constituents, the presence of volatile compounds, and the changes in dissolved constituents that will occur over the lifetime of the project.   The change in constituent level and composition will occur much more rapidly for systems where active in situ treatment such as bioremediation is employed than when only pump and treat methods are used. The selection of a biological groundwater treatment system also needs to take into consideration the availability of trained personnel at the site. Many in situ remediation systems require only limited presence of trained personnel. In such cases, the surface treatment system should also require a low level of maintenance.

Literature Cited

1.   Overcash, M.R. and Dhiraj, P., *Design of Land Treatment Systems for Industrial Wastes*, Ann Arbor Science, Ann Arbor, MI, 1979.
2.   Ross, D. Application of Biological Processes to the Clean Up of Hazardous Wastes. Presented at The 17th Environmental Symposium: Environmental Compliance and Enforcement at DOD Installations in the 1990's, Atlanta, Georgia, 1990.
3.   U.S. Environmental Protection Agency. Technology Screening Guide for Treatment of CERCLA Soils and Sludges. EPA/540/2-88/004, 1988.
4.   Eckenfelder, W. Wesley, Jr., *Industrial Water Pollution Control*, McGraw-Hill Book Company, New York, 1967.
5.   Gibson, D.T., *Microbial Degradation of Organic Compounds*, Microbiology Series, Marcel Dekker, Inc., New York, 1984.
6.   Munnecke, D.M., Johnson, L.M., Talbot, H.W., and Barik, S. Microbial Metabolism and Enzymology of Selected Pesticides. In: Biodegradation and Detoxification of Environmental Pollutants. A.M. Chakrabarty, ed. CRC Press, Boca Ration, Florida, 1982.
7.   Pitter, P. and Chudoba, J., Biodegradability of Organic Substances in the Aquatic Environment. CRC Press, Boca Raton, Florida, 1990.
8.   Reineke, W. and Knackmuss, H.J., Microbial Degradation of Haloaromatics. *Ann. Rev. Microbial.* 42, pp. 263-287, 1988.
9.   U.S. Environmental Protection Agency. Microbiological Decomposition of Chlorinated Aromatic Compounds. EPA 600/2-86/090, 1986.
10. Montgomery, J.H. and Wilkins, L.M., *Groundwater Chemical Desk Reference*, Lewis Publishers: New York, Vol. I, 1990.
11. Howard, P.H., *Handbook of Environmental Fate and Exposure Data for Organic Chemicals: Volume I Large Production and Priority Pollutants*, Lewis Publishers, Chelsa, Michigan, 1989.
12. Howard, P.H., *Handbook of Environmental Fate And Exposure Data For Organic Chemicals: Volume II Solvents*, Lewis Publishers, Chelsa, Michigan, 1990.
13. Verschueren, K., *Handbook of Environmental Data on Organic Chemicals*, 2nd Edition, VanNostrand Reinhold, New York, 1983.
14. Devinny, J.S., Everett, L.G., Lee, J.C.S., and Stollar, R.L., *Subsurface Migrations of Hazardous Wastes*, Van Nostrand Reinhold, New York, 1990.
15. Wilson, D.J., Clarke, A.N., and Clarke, J.H., "Soil Clean Up by In Situ Aeration I, Mathematical Modeling," *Separations Science and Technology*, 23, 1988, pp. 991-1037.
16. Loehr, R.C., "Land Treatment as a Waste Management Technology: An Overview. Land Treatment: A Hazardous Waste Management Alternative." R.C. Loehr, et al., eds. Center for Research in Water Resources, The University of Texas at Austin, 1986, pp. 7-17.

17. Brubaker, G.R., "In Situ Bioremediation of PAH-Contaminated Aquifers," in the *Proceedings of the Petroleum Hydrocarbons and Organic Chemicals in Ground Water: Prevention, Detection, and Restoration.* Houston, 1991, pp. 377-390.

18. Atlas, R.M., "Bioremediation of Fossil Fuel Contaminated Soils," *In Situ Bioreclamation: Application and Investigations for Hydrocarbons and Contaminated Site Remediation.* Eds. Hinchee, R.E. and Olfenbuttel, R.F., Butterworth-Heinemann, 1991, pp. 14-32.

19. U.S. Environmental Protection Agency, "Guide for Conducting Treatability Studies Under CERCLA: Aerobic Biodegradation Remedy Screening. Interim Guidance," EPA/540/2-91/013A.

20. Yare, B., "A Comparison of Solid-Phase and Slurry-Phase Bioremediation of PNA-Containing Soils" Presented at *In Situ and On-Site Bioremediation - an International Symposium,* San Diego, CA, March 19-21, 1991.

21. Jacobson, J.E. and Hoehn, G.D., "Working with Developing Air Quality Regulations: A Case Study Incorporating On-Site Aeration Standards for Vadose Zone Remediation Alternative", in the Proceedings from *Petroleum Hydrocarbons and Organic Chemicals in Groundwater - Prevention, Detection and Restoration,* NWWA API Conference, Houston, TX, 1987.

22. Norris, R.D., Muniz, F.P., and Crosbie, J.R., "A Survey of Current Groundwater Remedial Methods", *Proceedings of the Construction and the Contaminated Site,* Connecticut Society of Civil Engineers and Connecticut Groundwater Association, November 2-3, 1989, Berlin, CT.

23. Downey, D.C., and Guest, P.R., "*Physical and Biological Treatment of Deep Diesel-Contaminated Soils,*" in the proceedings of the Petroleum Hydrocarbons and Organic Chemicals in Groundwater: Prevention, Detection and Restoration. Houston, 1991, pp. 361-376.

24   United States Environmental Protection Agency, *Soil Vapor Extraction Technology; Reference Handbook,* EPA/540/2-91/003.

25. Norris, R.D., and Brown, R.A., "In Situ Bioreclamation - A Complete On Site Solution", *Proceedings of the Hazardous Waste Management Conference - Hazmat West*" Long Beach, CA, December 1987.

26. Lowes, B.C., "Soil-Induced Decomposition of Hydrogen Peroxide," *In Situ Bioreclamation: Application and Investigation for Hydrocarbons and Contaminated Site Remediation,* Eds., Hinchee, R.E. and Olfenbuttel, R.F., Butterworth-Heinemann, 1991, pp. 143-156.

27. Falatico, R.J. and Norris, R.D., "The Necessity of Hydrogeological Analysis for Successful In Situ Bioremediation", *Proceedings of the Haztech International Pittsburgh Waste Conference.* Pittsburgh, PA, October 2-4, 1990.

28. Skladany, G.J., "Onsite Biological Treatment of an Industrial Landfill Leachate: Microbiological and Engineering Considerations," *Hazardous Waste and Hazardous Materials.* 6:213-222, 1989.

29. Norris, R.D., Sutherson, S.S., and Callmeyer, T.J., "Integrating Different Technologies to Accelerate Remediation of Multiphase Contamination", presented at *NWWA Focus Eastern Regional Groundwater Conference,* Springfield, MA, October 17-19, 1960.

RECEIVED November 12, 1992

Chapter 9

Development of Nonlinear Group Contribution Method for Prediction of Biodegradation Kinetics from Respirometrically Derived Kinetic Data

Henry H. Tabak[1] and Rakesh Govind[2]

[1]Risk Reduction Engineering Laboratory, U.S. Environmental Protection Agency, Cincinnati, OH 45268
[2]Department of Chemical Engineering, University of Cincinnati, Cincinnati, OH 45221

The fate of organic chemicals in the environment depends on their susceptibility to biodegradation. Recent studies have attempted to correlate the kinetics of biodegradation with the compound's molecular structure. This has led to the development of structure-biodegradation relationships (SBRs) using the group contribution approach. In this paper, a non-linear group contribution method has been developed using neural networks, which is trained using literature data on the first order biodegradation kinetic rate constant for a number of priority pollutants. The trained neural network is then used to predict the biodegradation kinetic constant for a new list of compounds. It has been shown that the non-linear group contribution method using neural networks is able to provide a superior fit to the training set data and produce a lower prediction error than the previous linear method.

The number and amount of synthetic organic chemicals produced commercially is large and increasing every year. The presence of many of these chemicals in the ecosystem is a serious public health problem. Biodegradation is an important mechanism for removing these chemicals from natural ecosystems (1). Biodegradation can eliminate hazardous chemicals by biotransforming them into innocuous forms, or completely degrading them by mineralization to carbon dioxide and water.

Kinetic data for calculating biodegradation rates in natural ecosystems are important for several reasons. Kinetic data is needed to estimate the role of biodegradation in the presenece of other competing

0097–6156/93/0518–0159$09.00/0

mechanisms, such as volatilization and adsorption, on the distribution and concentration of organic compounds in the environment. Information regarding the extent and the rate of biodegradation of organic chemicals is very important in evaluating relative persistence of the chemical in the environment, and for regulating their manufacture and use. Due to the large number of chemicals obtaining this information is labor intensive, time consuming and expensive. Thus, there is a need to develop correlations and predictive techniques to assess biodegradability (2).

Structure-activity relationships (SARs) are used to predict intrinsic properties of many chemicals and to estimate the kinetic constants for important transformation processes. SARs approach can be effectively used to shorten the list of thousands of chemicals to a few hundred key chemicals, for detailed laboratory and field testing. In a recent review of SARs (3) it was concluded that application of SARs has great potential in predicting the fate of organic chemicals and these techniques are being accepted to a greater extent by regulatory agencies in decision making and policy implementation.

In this paper, a quantitative structure-biodegradation relationship (SBR) has been developed using the group contribution approach. This method is widely used in chemical engineering thermodynamics to estimate pure compound properties such as liquid densities, heat capacities and critical constants.

The group contribution method is similar to the Free-Wilson model (4,5) widely used in pharmacology and medicinal chemistry. Using this method, a very large number of chemicals can be constituted from perhaps a few hundred functional groups. Using this method, the compound's property is predicted from its molecular structure, which is structurally decomposed into groups or fragments, each group or fragment having a unique contribution towards the specific value of the property.

Techniques for Measuring Biodegradation Kinetics

Techniques for evaluating biodegradation kinetics have been reviewed in the literature (6,7) and hence would not be repeated here. In the following section, the electrolytic respirometric method will be presented in some detail, since this method was used in obtaining experimental data on biodegradation kinetics utilized in the development of the non-linear model.

Electrolytic Respirometry Studies. This study was conducted using an automated continuous oxygen uptake and BOD measuring Voith Sapromat B-12 (12 unit system). The instrument consists of a temperature controlled waterbath, containing measuring units, an on-line microcomputer for data sampling, and a cooling unit for continuous recirculation of waterbath volume. Each measuring unit consists of a reaction vessel, containing the microbial inoculum and test compound, an oxygen generator, comprised of an electrolytic cell containing copper sulfate and sulfuric acid solution, and a pressure indicator which triggers oxygen generation. The carbon dioxide produced is absorbed by soda lime, contained in the reaction flask stopper. Atmospheric pressure fluctuations do not affect the results since the measuring unit forms an air sealed system. The uptake of oxygen by the microorganisms in the sample during biodegradation is compensated by the electrolytic generation of oxygen in the oxygen generator, connected to the reaction vessel. The amount of oxygen supplied by the electrolytic cell is proportional to its amperage requirements, which is continuously monitored by the microcomputer and the digital recorder.

Materials and Methods. The nutrient solution used in our studies was an OECD synthetic medium (8) consisting of measured amounts per liter of deionized distilled water of (1) mineral salts solution; (2) trace salts solution; and (3) a solution (150 mg/l) of yeast extract as a substitute for vitamin solution.

The microbial inoculum was an activated sludge from The Little Miami wastewater treatment plant in Cincinnati, Ohio, receiving municipal wastewater. The activated sludge sample was aerated for 24 hours before use to bring it to an endogenous phase. The sludge biomass was added to the medium at a concentration of 30 mg/l total solids. Total volume of the synthetic medium was 250 ml in the 500 ml capacity reaction vessels.

The test and control compound concentration in the media were 100 mg/l. Aniline was used as the biodegradable reference compound, at a concentration of 100 mg/l.

In a typical experimental run, duplicate flasks were used for the test compound, and reference compound, aniline, a single flask for toxicity control (test compound plus aniline at 100 mg/l each) and an inoculum control.

The reaction vessels were incubated in the dark at 25° C in the temperature controlled bath and stirred continuously throughout the run. The microbiota of the activated sludge were not pre-acclimated to the substrate. The incubation period of the experimental run was between 28-50 days. A more comprehensive description of the procedural steps involved in the respirometric tests has been presented elsewhere (9).

Evaluation of Biodegradation Kinetics

First Order Rate Constants. Biodegradation rates were quantified by measuring the ratio of the net biological oxygen demand (BOD) values in mg/l (oxygen uptake values of test compound minus endogenous oxygen uptake values [inoculum control]) to the theoretical oxygen demand (ThOD) of substrate i.e., the ratio BOD/ThOD. The values of theoretical oxygen demand (ThOD) were calculated by using the stoichiometric balanced oxidation equation.

The BOD/ThOD curves calculated from electrolytic respirometric data were characterized by four indices (10) (1) the lag time (t_1) which gives the adaptation time; (2) the rate constant (k); (3) the biodegradation time (t_d) before the endogenous respiration period; and (4) ratio of BOD/ThOD at time t_d. The values of k can be calculated from the slope of the straight line obtained by plotting log(BOD) vs time (t) for values of t such that $t_1 < t < t_d$. The appropriate equations for calculating the value of k are given as follows:

$$d(BOD)/dt = k'(BOD) \tag{1}$$
$$\log (BOD) = (k'/2.3)t = kt + constant \tag{2}$$
where $t_1 < t < t_d$

It should be noted that the above kinetic model for obtaining biodegradation kinetics differs from the traditional Monod equation which has been used extensively in the literature to analyze oxygen uptake data. However, the above model was selected for several reasons: (1) it has the ability to represent oxygen uptake data between time t_1 and t_d by a single parameter (k); (2) it follows the and Kato model (10), so that some of their kinetic constant values could be used in our training set; (3) the simple model provided an acceptable fit with the experimental data; and (4) the model results allowed the development of the prediction approach (linear and non-linear) for estimating the kinetic constant values for a variety of test compounds. This allowed us to compare the linear and non-linear approaches using the same data set.

The kinetic constant (k) values for the compounds were divided into two sets: (1) the training set; and (2) a testing set. This division of the compound list was based on the criterion that the chemical groups or fragments comprising the testing set compounds were all present in the training set and there were at least 5 compounds in the training set for each chemical group selected.

The main motivation for constructing the training and testing sets was to develop structure-biodegradation relationships using the training set data and then test the relationship using the testing set compounds.

The Monod constants. The Monod model provides an adequate description of microbial growth behavior. It states that the cell growth rate is first order with respect to biomass concentration (X) and mixed order with respect to substrate concentration (S).

$$dX/dt = (S\mu_m X)/(K_s + S) \tag{3}$$

Also cell growth is linearly related to substrate removal rate

$$dX/dt = - Y(dS/dt) \qquad (4)$$

The kinetic parameters of interest are maximum specific growth rate μ_m, half saturation constant K_s (it is the concentration of substrate when $\mu = 0.5\mu_m$) and the yield coefficient Y. Several authors, especially Schroeder (11) and Grady and Lim (12), have analyzed in depth the factors to be considered in biological kinetics and the pitfalls with indiscriminate use of the Monod equation.

Structure-Activity Relationships

Structure-activity relationships have been widely used in pharmacology and medicinal chemistry. In the field of biodegradation, interest in structure-activity relationships between the biodegradability of the chemical and its structure started many years ago (13). There are several studies which have attempted to correlate some physical, chemical or structural property of a chemical with its biodegradation. Literature reveals both qualitative and quantitative structure-biodegradability correlations. Lyman et al. (14) summarized rules of thumb which may be used to make qualitative predictions of biodegradability. These rules are based on degree of branching, chain length, oxidation and on number, type and position of substituents on simple organic molecules. Geating (15) developed a predictive algorithm based on the literature published between 1974 and 1981. Based on the type and the location of substituent groups, the model predicts biodegradability in qualitative terms. The algorithm was applied to group of compounds of known biodegradability and it predicted correctly for 93% of compounds, incorrectly for 2% of compounds and 5% could not be predicted at all. When applied to known nonbiodegradable compounds it was not as successful, predicting 70% correctly and of the remainder roughly half were predicted incorrectly while half could not be predicted at all. Others (16,17) have also investigated qualitative relationships for certain class of chemicals; however while these studies are useful, none provide the kind of prediction power needed for regulatory decision making, for which quantification is necessary.

Quantitative correlations relating either a biodegradation rate constant or 5-day biological oxygen demand (BOD) with different physicochemical properties or rate of other transformation processes have appeared in literature. Most of these correlations are linear single parameter models. Paris and co-workers (18,19,20,21,22) were among the first to investigate quantitative correlations using a microbial transformation rate coefficient. They used the Monod equation with the assumption that the substrate concentration is less than the half saturation constant so that the transformation rate becomes first order with respect to both substrate concentration as well as microbial concentration. They called this resulting rate coefficient a second order rate coefficient. They investigated several groups of chemicals : pesticides, substituted phthalates, mono-substituted phenols, carboxylic acid esters of 2,4-D, ethyl esters of chlorine substituted acetic acids and substituted anilines. These studies were conducted either with pure culture or mixed populations of organisms from natural environment. In almost all cases they obtained good correlations with a particular property of the chemicals. In case of substituted anilines and mono-substituted phenols, transformation rate was related with the van der Waal's radii of the substituent groups. The transformation rate in case of phthalates and pesticides was correlated with their alkaline hydrolysis rate constant. They attempted to correlate the rates of carboxylic acid esters of 2,4-D and ethyl esters of chlorine substituted acetic acids with the lipophilicity, specifically octanol-water partition coefficients. They were successful with the first group but not with the second.

Reinke and Knackmuss (23) studied di-oxygenation of substituted benzoic acids by two species of Pseudomonas and were able to obtain a good correlation of the rate coefficient with the Hammett constant for one species but not for the other. Pitter (24) obtained a linear relation between the logarithm of the biological degradation rate of substituted phenols and anilines and the Hammett constant of the substituent. Of all the substituents (OH, CH3, Cl, NO_2 and NH_2), only the amino group led to deviations from the linear correlations for mono-substituted phenols. He also attempted correlations using the steric and lipophilic constants but failed.

Vaishnav et al. (25) correlated biodegradation of 17 alcohols and 11 ketones as well as a series of alicyclic chemicals with octanol-water partition coefficient, log P. Alcohols revealed a biphasic relationship with an apparent change in slope at a log P of about 3. The relationship for ketones was parabolic or bilinear with a peak at a log P value of about 1. Statistically the difference between the parabolic and bilinear relationships was marginal, but the bilinear model gives a closer fit to experimental data. Degradability of the hydrophilic members of alicyclics was apparently not related to log P but degradability of the more hydrophobic members decreased with increasing lipophilicity. Banerjee et al. (26) studied biodegradation of phenol, resorcinol, p-cresol, benzoic acid and various chloro derivatives of phenol, resorcinol and anisole. The biodegradation rate was related to lipophilicity, where the rate increased with decreasing lipophilicity and then levelled off for chemicals with log P less than 2.

Deardan and Nicholson (27) studied aromatic and aliphatic amines, phenols, aromatic and aliphatic aldehydes, carboxylic acids, halogenated hydrocarbons and amino acids. They calculated different parameters for each compound ; molecular connectivities up to seventh order, log P values, molecular volume, accessible molecular surface area, Sterimol steric parameters and atomic charges. They correlated 5-day BOD of these compounds with the atomic charge difference across the bond(s) common to all compounds in the series. The regression coefficient and the constant term for each of the series of compounds were close enough to combine all the data into a single, all embracing equation covering amines, phenols, aldehydes, carboxylic acids, halogenated hydrocarbons and amino acids.

Another approach is to seek direct correlation between biodegradation and molecular structure of the chemical. The structural features of a molecule such as shape, size, branching and nature of atom-atom connections are expressed in terms of numerical descriptors called topological indexes. Many such indexes have been proposed, but the most successful of them in SAR are molecular connectivity indexes, which were introduced by Randic (28) and then developed extensively by Kier and Hall (29). Govind (30) has correlated first order biodegradation rate constants of priority pollutants with the first order molecular connectivity index. Boethling (31) has correlated log rate constant for 2,4-D alkyl esters, log percent degraded for carbamates, log percent theoretical oxygen demand for dialkyl ethers, rate constant for dialkyl phthalates and percent theoretical oxygen demand for aliphatic acids with molecular connectivity indexes. All these were single variable models. Two variable models substantially improved results for aliphatic alcohols and acids.

Most of the correlations that have appeared in literature are single parameter relations applicable to a particular class of compounds. This demonstrates that correlations are possible, but also that single parameter correlations are limited in their applicability. Babeu and Vaishnav (32) calculated 5-day BOD for 45 organic chemicals including alcohols, acids, esters, ketones and aromatics. The BOD data were correlated with water solubilities, log P, molar refractivities and volumes, melting and boiling points, number of carbon, hydrogen and oxygen atoms, molecular weight and theoretical BOD of chemicals. The experimental BOD values for 43 additional chemicals were compared with values predicted by the model and for 84%-88% of the test chemicals prediction was within 80% of the experimental values. Desai et al. (33) have predicted first order biodegradation rate constants within 20% of the experimental values using group contribution approach. The model based on group contributions was a first order linear model which neglected the interactions between groups. Table I summarizes the work done in SARs for predicting the biodegradability of chemicals.

Group Contribution Approach

Using a group contribution approach, a very large number of

Table I. Background Summary for SAR

Author(s)	Year	Parameters	Comments/Compounds studied
QUALITATIVE			
Lyman, et al	1982	-	summarized rules of thumb
Geating	1981	-	developed qualitative predictive algorithm
QUANTITATIVE			
Reinke and Knackmuss	1978	Hammett constant of substituent	substituted benzoic acids
Wolfe, et al	1980	Alkaline hydrolysis rate constant	phthalate esters and pesticides
Paris, et al	1982 1983	van der Waal's radii of substituent	substituted phenols
Banerjee, et al	1984	Hammett constant of substituent	phenols and its chloro derivatives
Paris, et al	1984	octanol-water partition coefficient	esters of chlorinated carboxylic acids
Pitter	1984	Hammett constant of substituent	substituted phenols and anilines
Boethling	1986	molecular connectivity	2,4-D alkyl esters, carbamates, aklyl ethers, dialkyl phthalates and aliphatic acids
Deardan and Nicholson	1986	atomic charge difference	amines, phenols, aldehydes, carboxylic acids, halogenated hydrocarbons and amino acids

Table I. Background Summary for SAR (continued)

Author(s)	Year	Parameters	Comments/Compounds studied
Babeu and Vaishnav	1987	theoretical BOD of compounds,melting point and number of carbon atoms	alcohols, acids, esters, ketones and aromatics. Multiparameter model
Govind	1987	molecular connectivity	priority pollutants
Paris and Wolfe	1987	van der Waal's radii of substituent	substituted anilines
Vaishnav, et al	1987	octanol-water partition coefficient	alcohols, ketones and alicyclics

chemicals of interest can be constituted from perhaps a few hundred functional groups. The prediction of the property is based on the structure of the compound. According to this method, the molecules of a compound are structurally decomposed into functional groups or their fragments, each having a unique contribution towards the compound property. The advantage of this approach is that the molecules of the compounds may be structurally dissected in any convenient manner and no independently measured group constants are required in the analysis.

The biodegradability rate constant, k, can be expressed as a function of contribution α, of each group or fragment of the compound

$$\ln(k) = f(\alpha_1, \alpha_2, \ldots \ldots \ldots \alpha_j) \qquad (5)$$

In general, the above functional relationship can be classified into two types: (1) linear function; and (2) non-linear function. The linear group contribution method is based on a linear function and the non-linear group contribution method results from using a non-linear function.

Linear Group Contribution Method. The above general function can be expanded in terms of taylor series. If the terms from second order onwards are neglected, a linear first order model for biodegradation rate constant k (1/hr) is obtained and this can be expressed as

$$\ln(k) = \sum_{j=1}^{L} N_j \alpha_j \qquad (6)$$

where N is the number of groups of type j in the compound, α_j is the contribution of group of type j and L is the total number of groups in the compound. For each compound a linear equation in α's is constructed. The linear equations are solved for α's using the method of least squares.

The above model, being first order approximation, will breakdown if interaction between groups become important. The interaction of different groups can be treated by considering second and higher order terms in the series.

Data generated by Urano and Kato (10) using electrolytic respirometry was used in applying the linear group contribution method. This ensured that the test conditions for obtaining the data were the same for all the compounds. The experimental conditions used by Urano and Kato (10) were: temperature 20° C, pH of solution 7, sludge concentration 30 mg/l and compound concentration 100 mg/l.

Non-linear Group Contribution or Neural Network Method. To incorporate the effects of interactions between the chemical groups used in the group contribution approach, it was necessary to develop a non-linear method. It was important that the non-linear method included not only all the interactions between the groups but also the algebraic form (type of non-linearity, such as square, cubic, etc.) of such interactions. Since this information was not known apriori, a neural network model was used to include all possible interactions between the groups and the algebraic form of these interactions was implicitly determined from the training set data using a large number of adaptable parameters or network weights.

Neural Network Models. There is extensive literature on mathematical models of artificial neural networks (34). Neural networks have found applications in image and speech recognition, on-line diagnosis of process faults, process control and in optimization of complex functions. Artificial neural network models consist of many nonlinear computational elements operating in parallel and arranged in patterns similar to biological neural nets. The nodes or computational elements are connected via weights that are typically adapted during use to improve performance. Superior performance is achieved via dense interconnection of simple computational elements.

Computational elements or nodes used in neural net models are nonlinear. In our model, each node has a large number of inputs and a single output. Each input value has an associated activation and weight. Each node or computational element applies an activation function to the sum of the products of the input activations and weights, and thereby generates the output value. The output of each computational element or node can be expressed as follows:

$$O_{pj} = 1/[1 + \exp(- \Sigma W_{ji}O_{pi} + O_j)] \qquad (7)$$

where O_{pj} = output value of node j
O_{pi} = output value of node i
W_{ji} = connection weight between the ith and jth nodes
O_j = bias of the jth node

Hence, each node or computational element forms a weighted sum of N inputs and passes the result through a nonlinearity, mathematically expressed by the above equation. More complex neural network models may include temporal integration or other types of time dependencies and more complex mathematical operations.

The neural network consists of interconnected nodes or computational elements. In our study, a three layer neural network was used, with eight input nodes, eight hidden or intermediate layer nodes, and one output node, as shown in Figure 1. It consists of a hidden or intermediate layer

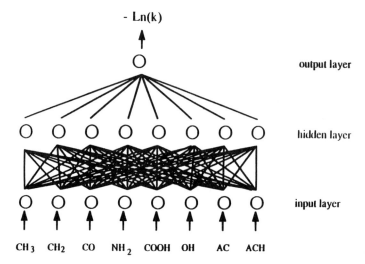

Figure 1. Neural Network architecture used to evaluate -ln(k).

between the input and output nodes. Multi-layer neural networks overcome some of the limitations of the single layer models. The capabilities of multi-layer neural networks stem from the nonlinearities used within nodes, which allow arbitrarily complex decision regions.

The neural network was trained by using a back-propagation algorithm, which used a gradient search technique to minimize a cost function equal to the mean square difference between the desired and the actual net outputs. The desired output of all nodes is typically "low" (0 or < 0.1) unless that node corresponds to the class the current input is from, in which case its output is "high" (1.0 or > 0.9). The net is trained by initially selecting small random weights and internal thresholds and then presenting the training data (number of each type of chemical groups present and the experimentally measured biodegradation kinetic rate constant) for each chemical repeatedly. Weights are adjusted after each trial, until the weights converge and the cost function is reduced to an acceptable value.

The output of the network depends on the weights assigned to each connection between the layers. Training of the network corresponds to the assignment of weights, which are determined in the back-propagation algorithm by minimizing the error function, E_p, written as follows

$$E_p = 1/2 \Sigma (t_{pj} - O_{pj})^2 \qquad (8)$$

where t_{pj} and O_{pj} are the desired and actual activation values of the output node j, due to an input pattern p. The generalized back propagation algorithm was used by Rumelhart et al. (35) to obtain the minimum value of E_p. In the first step an input pattern is presented and propagated forward through the network to compute the output value O_{pj} for each node. The output is compared with the desired value, resulting in an error d_{pj} for each output. For nodes in the output layer, the error signal is given by

$$d_{pj} = (t_{pj} - O_{pj})O_{pj}(1-O_{pj}) \qquad (9)$$

The error signal for a node in the intermediate layer is given by

$$d_{pj} = O_{pj}(1-O_{pj}) \Sigma d_{pk}W_{kj} \qquad (10)$$

In the second step, a backward pass is made through the network and the error signal is passed to each node in the network and the appropriate weight changes are made using the following equation

$$W_{ij}(t+1) = W_{ij}(t) + \beta(W_{ij}(t) - W_{ij}(t-1)) \qquad (11)$$

where β is the learning rate. In our study, a relatively fast learning rate (β) of 0.5 was chosen.

In our study, there were eight input nodes, each node corresponding to a specific chemical group, previously used in the linear group contribution analysis. The inputs to the neural network corresponds to the number of each type of chemical group present in the chemical structure. For example, in the case of ethyl alcohol, there is one methyl group, one methylene group, and one hydroxy group. The number of each group is entered corresponding to the node representing the group. The output is the biodegradation kinetic constant value. The data used to obtain the group contributions in the linear group contribution analysis was used for training the neural network.

In this study group contribution approach is selected because of its potential to predict rate constants for different classes of compounds. However, it should be emphasized that this approach requires large amount of data for any meaningful analysis. This study was started with 56 organic chemicals with this requirement in mind, but more than half of them did not degrade. The data obtained is not sufficient for calculating the contributions of the different groups and validation of the approach. The literature does not have the Monod constants for sufficient number of compounds, and hence literature data in conjunction with the data obtained in this study could not be used to calculate group contributions and then validate the approach. Hence, it was decided to use the first order biodegradation rate constants, which are available extensively in the literature to calculate group contributions and then validate the approach using the experimentally obtained first order rate constants. However, the group contributions were also calculated using the Monod constants obtained in this study and the study at the Clemson University (7,36).

Results and Discussion

First order rate constants. Urano and Kato (10,37) had obtained data [kinetic constant (k) values] for 74 compounds in their study. However, in our analysis using the linear group contribution method, it was necessary to ensure that each group, considered in the analysis, occurred in at least 5 compounds in the training set. This requirement prevented us from using the entire set of compounds studied by Urano and Kato (10,37) and only 18 compounds were used for calculating the group contribution values for 8 groups. The compounds used in our analysis (training set) were: ethyl alcohol, butyl alcohol, ethylene glycol, acetic acid, propionic acid, n-butyric acid, n-valeric acid, adipic acid, methyl ethyl ketone, hexamethylenediamine, n-hexylamine, mono ethanol amine, acetamide, benzene, benzyl alcohol, toluene, acetophenone, and aminophenol. The experimental values of the kinetic constants (-ln(k)) for these chemicals have been tabulated in Table II.

Table II.   Values of First Order Kinetic Constant for
            Training and Testing Set Compounds, used in
            the Linear Group Contribution Method

Compound	ln(k)		Avg. ln(k)
Training Set			
Ethyl alcohol	-2.9 ˜	-3.15	-3.02
Butyl alcohol	-3.08 ˜	-3.32	-3.19
Ethylene glycol	-3.35 ˜	-3.65	-3.49
Acetic acid	-2.60 ˜	-2.72	-2.66
Propionic acid	-2.67 ˜	-2.97	-2.81
n-Butyric acid	-2.70 ˜	-3.06	-2.87
n-Valeric acid	-2.63 ˜	-2.66	-2.65
Adipic acid	-2.81 ˜	-3.12	-2.96
Methyl ethyl ketone	-3.37 ˜	-3.69	-3.58
Hexamethylene-diamine	-4.34 ˜	-4.51	-4.43
n-Hexylamine	-2.86 ˜	-3.06	-2.96
Mono ethanol amine	-3.32 ˜	-3.38	-3.35
Acetamide	-3.00 ˜	-3.06	-3.03
Benzene	-2.86 ˜	-2.98	-2.92
Benzyl alcohol	-2.78 ˜	-3.17	-2.96
Toluene	-2.60 ˜	-2.86	-2.73
Acetophenone	-3.17 ˜	-3.54	-3.34
Aminophenol	-3.24 ˜	-3.30	-3.27
Training Set			
o-Cresol	-2.57 ˜	-2.81	-2.69
Phenol	-2.83 ˜	-3.17	-3.00
2,4-Dimethyl phenol	-2.63 ˜	-3.07	-2.85
Butylbenzene	-2.99 ˜	-3.27	-3.13
Acetone	-3.04 ˜	-3.20	-3.12
1-Phenylhexane	-3.30 ˜	-3.50	-3.40
Aniline	-3.00 ˜	-3.24	-3.12
Benzoic acid	-2.12 ˜	-2.20	-2.16

Using the training set values, the group contribution parameters for all the groups considered in the analysis are given in Table III (38). Note these contribution values are used for calculating the value of ln(k) rather than the kinetic constant (k) itself.

To validate the results, experiments were conducted by the authors using an electrolytic respirometer (Voith-Morden, Milwaukee, WI) for o-cresol, phenol, 2,4-dimethyl phenol, acetone, butyl benzene, 1-phenyl hexane, aniline and benzoic acid. These compounds, obtained from Aldrich chemical company, were of 99+% purity. Except for the source and nature of biomass, the experimental conditions were the same as used by and Kato (10,37).

The oxygen uptake data was analyzed using the kinetic model presented earlier, and the best fit kinetic constant (k) value was obtained for each compound in the testing set. The experimental k values obtained for the test compounds have been tabulated as -ln(k) in Table II.

Except for the source and the nature of biomass, the experimental conditions in this study are similar to that of and Kato (10,37). The kinetics for the above mentioned compounds are determined from oxygen consumption data in the similar way as and Kato (10) and are also predicted using the group contribution parameters given in Table III. The comparison of the experimental and predicted ln(k) values for training set are given in Table IV and for testing set are given in Table V. The percentage errors in Tables IV and V are calculated by

$$100*[\ln(k)_{pre} - \ln(k)_{exp}]/\ln(k)_{pre} \qquad (12)$$

For the training set compounds, the results of the linear group contribution analysis and the nonlinear group contribution or neural network approach have been presented in Table IV. It shows the goodness of fit [percentage error, defined as the absolute difference between the experimental and predicted values for ln(k) divided by the experimental value of ln(k)] between the experimental values and the computed values. It can be seen that the mean error in the linear group contribution method is generally larger than the neural network method. However, the fit for the linear and nonlinear methods are both good since the training set was used to obtain the group contribution values in the linear method and for converging on the neural network weights in the nonlinear method. The true test of these methods is to examine the percentage errors obtained for the testing set, consisting of compounds which were not in the training set.

Figure 2 provides a comparison between the experimental and predicted (calculated) biodegradation kinetic data by neural network and linear group contribution.

Table V lists the results for the testing set compounds. This shows the ability of the method to predict

**Table III. Groups and their Contribution Values for
First Order Rate Constants**

Group	Symbol	α_i
Methyl	CH_3	−1.37
Methylene	CH_2	−0.04
Hydroxy	OH	−1.71
Acid	COOH	−1.31
Ketone	CO	−0.51
Amine	NH_2	−1.46
Aromatic CH	ACH	−0.50
Aromatic carbon	AC	1.06

Table IV. Comparison of Experimental and Predicted ln(k) for Training Set using the linear and nonlinear (neural network) group contribution method

Compound	Experimental -ln(k)	Neural Network -ln(k)	Neural Network %Error	Linear Method -ln(k)	Linear Method %Error
Ethyl alcohol	3.02	3.01	0.33	2.97	1.43
Butyl alcohol	3.19	3.16	0.94	3.24	1.30
Ethylene glycol	3.49	3.45	1.15	3.39	2.87
Acetic acid	2.66	2.68	0.75	2.49	6.55
Propionic acid	2.81	2.81	0.06	2.65	5.84
n-Butyric acid	2.87	2.83	1.39	2.75	4.17
n-Valeric acid	2.65	2.70	1.89	2.88	8.86
Adipic acid	2.96	2.93	1.01	2.94	0.55
Methyl ethyl-ketone	3.58	3.63	1.40	3.31	11.90
Hexamethylene diamine	4.43	4.22	4.74	3.96	10.43
n-Hexylamine	2.96	2.97	0.33	3.52	19.11
Mono-ethanol amine	3.35	3.38	0.90	3.41	1.80
Acetamide	3.03	3.01	0.66	3.48	15.19
Benzene	2.92	2.94	0.68	2.87	1.62
Benzyl alcohol	2.96	2.94	0.68	3.12	5.57
Toluene	2.73	2.70	1.10	2.70	1.10
Acetophenone	3.34	3.31	0.90	3.33	0.38
Aminophenol	3.27	3.29	0.61	3.13	4.26

Table V. Comparison of Experimental and Predicted ln(k) for Testing Set using the linear and nonlinear (neural network) group contribution method

Compound	$-\ln(k)$ Experimental	$-\ln(k)$ Neural Network	%Error Neural Network	$-\ln(k)$ Linear Method	%Error Linear Method
o-Cresol	2.69	2.62	2.02	2.87	6.59
m-Cresol	2.37	2.62	10.74	2.87	20.92
p-Cresol	2.47	2.62	6.46	2.87	16.24
Phenol	3.00	2.91	3.17	2.99	0.29
2,4-Dimethyl phenol	2.85	2.58	9.29	2.74	3.82
Butylbenzene	3.13	3.18	1.53	3.10	5.84
Acetone	3.12	3.14	0.60	3.15	0.96
1-Phenylhexane	3.40	3.67	7.90	4.76	40.00
Aniline	3.12	3.00	3.85	4.03	29.00
Benzoic Acid	2.16	2.31	6.95	3.64	68.50

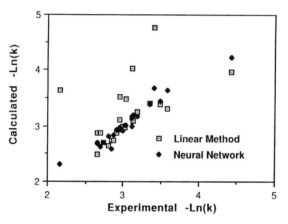

Figure 2. Comparison of experimental -ln(k) by neural
          network and linear group contribution method.

the kinetic constant value for compounds that were not in the training set.

The predicted values for the training set are within 10 % except for aliphatic amines. As can be seen from Table V the reported results for the testing set agree within 20 % with the predicted values except for o-cresol and benzoic acid. The model does not differentiate between ortho-, meta- and para-cresols. This is because the contribution of OH group is calculated assuming that all the three positions are equivalent. The contribution of COOH group, used to calculate ln(k) value for benzoic acid, is calculated using aliphatic acids. This can be the reason for large error in prediction of ln(k) value of benzoic acid. These problems could be alleviated if more data is available for each unique group.

The group contributions α_j can be used to estimate the rate constants for any organic chemical that consists of the groups listed in Table III. The values of α in Table III suggest that the larger aliphatic molecules will have smaller degradation rates and vice versa. This dependance on molecular size is not generally true since some groups have a positive contribution to the ln(k) values as evident from Table III.

Both the linear group contribution and the neural network methods are unable to distinguish between the ortho-, meta- and para-cresols since the position of the hydroxy group was not considered in the analysis. Hence only o-Cresol was used in the testing set. With availability of more biodegradation kinetics data for compounds with groups in ortho, meta and para positions, the effect of group position can be included in the analysis.

The absolute percentage error between the predicted and experimental values of -ln(k) are generally lower by the neural network method when compared with the linear group contribution method. The error by the linear group contribution method, which ignores the interaction between the groups, becomes significant for compounds such as Aniline (29%), 1-Phenylhexane (40%) and Benzoic acid (68.5%). However, the percentage error for the neural network predictions remains below 10%. This type of difference between the linear and nonlinear methods is expected for compounds, since the nonlinear method accounts for all the interactions between the groups. On examining the weights in the neural network we found that the weights interconnecting the aromatic, amine, alkane (greater than 4 carbons) and acid groups had large values in the converged network. Note that in the case of Butylbenzene, the error by the linear method is not large.

The Monod Constants. The compounds used to calculate the Monod parameters were: benzene, toluene, ethyl benzene, xylenes, cumene, butyl benzenes, 1-phenyl hexane, dimethyl phthalate, diethyl phthalate, dipropyl phthalate, dibutyl phthalate, phenol, cresols, 2,4-dimethyl phenol, catechol, resorcinol, acetone, 2-butanone, 4-methyl-2- pentanone and benzyl alcohol from this study and dichlorobenzenes, chlorophenol, 4-nitrophenol and 2,4-dinitrophenol from the study at Clemson University (7,36). These compounds form the training set. The group contribution parameters, $\alpha_{j,Monod}$, of the ten groups occurring in the above compounds are given in Table VI. The contribution for chloro and nitro groups should be used only when acclimated biomass is present.

The compounds from Urano and Kato's data set (10,37) which could be formed using the groups given in Table III, were selected to validate the group contribution approach for prediction of the Monod constants. Urano and Kato (10) have calculated the first order rate constant by fitting an empirical exponential form of equation to the increasing part of the oxygen uptake curve. The Monod relationship, used in this study, can be reduced to linear relationship under the assumption, $K_s \gg S$. Under this assumption equation (4) reduces to

$$dS/dt = -\mu_m\ S\ X/(Y\ K_s) = -(K\ X)\ S \qquad (13)$$

where $K=\mu_m/(YK_s)$ and S and X are expressed in BOD units. Also, if it is assumed that there is no appreciable change in biomass concentration then we get a pseudo first order constant 'KX'.

The Monod rate constants, um, Ks and Y were calculated for different compounds using group contribution parameters of Table VI. Comparisons of predicted Monod kinetic values for the training set are provided in Table VII and the testing set in Table VIII, using the linear group contribution approach. The percent error was calculated as the average error between the BOD values obtained from the experimental Monod parameters and predicted Monod parameters.

The experimental values agree within 25 % with the predicted values for most of the compounds. The error is high for most of the aliphatic compounds (alcohols and esters). This high errors may be because the contribution of all the groups is determined by using aromatic compounds except for ketones and assumptions made in reducing the Monod relationship to linear form may not be valid under the experimental conditions of Urano and Kato (10,37).

Most of the substrate degradation takes place in the exponential phase. The maximum specific rate constant μ_m, has the major impact on the exponential part of the oxygen uptake curve. Hence, the magnitude of this constant will indicate the extent of biodegradation that can be achieved for a particular compound. The Monod rate constant μ_m is calculated for different compounds using group contribution parameters of Table VI and are given in Table IX. These values are used to show that the different trends observed by researchers for different groups of compounds is also predicted by the group contribution approach.

Table VI. Groups and Their Contribution Values for the Monod Constants

Group	Symbol	$\alpha_{j, Monod}$		
		μ_m	K_s	Y
Aromatic carbon	AC	-0.033	0.19	-0.01
Aromatic CH	ACH	0.048	0.95	0.06
Methyl	CH_3	0.045	0.92	0.06
Methylene	CH_2	-0.028	0.51	0.01
Methelene	CH	-0.107	2.41	-0.03
Hydroxy	OH	0.173	0.80	0.07
Ester	COO	0.057	-0.11	0.00
Ketone	CO	0.182	2.87	0.33
Chlorine	Cl	-0.023	-0.29	0.09
Nitro	NO_2	-0.025	-0.13	0.09

Table VII. Comparison of Experimental and Predicted BOD Values For Training Set

Compound	Experimental			Predicted			Err %
	μ_m 1/hr	K_s mg/l	Y mg/mg	μ_m 1/hr	K_s mg/l	Y mg/mg	
Benzene	0.28	6.5	0.27	0.29	5.7	0.36	5.5
Toluene	0.52	7.7	0.36	0.25	5.9	0.36	12.3
Ethyl benzene	0.21	6.6	0.34	0.23	6.4	0.36	5.8
m-Xylene	0.12	4.7	0.26	0.22	6.0	0.35	21.1
p-Xylene	0.14	4.9	0.36	0.22	6.0	0.35	18.1
Cumene	0.19	10.0	0.23	0.19	9.2	0.39	3.9
Butyl benzene	0.11	8.8	0.23	0.17	7.4	0.38	15.0
sec-Butyl benzene	0.11	7.6	0.62	0.16	9.7	0.39	21.0
1-Phenyl hexane	0.14	7.5	0.53	0.12	8.4	0.39	11.9
Dimethyl phthalate	0.32	7.0	0.49	0.33	5.8	0.35	16.5
Diethyl phthalate	0.28	6.9	0.36	0.28	6.8	0.36	1.8
Dipropyl phthalate	0.22	6.5	0.23	0.22	7.9	0.37	10.4
Dibutyl phthalate	0.18	6.2	0.39	0.17	8.9	0.39	6.4
Phenol	0.28	3.0	0.30	0.38	5.7	0.36	12.0
o-Cresol	0.29	5.1	0.53	0.35	5.9	0.36	18.9
m-Cresol	0.54	6.5	0.44	0.35	5.9	0.36	14.7
p-Cresol	0.32	6.8	0.39	0.35	5.9	0.36	6.2
2,4-Dimethyl phenol	0.30	7.0	0.32	0.31	6.5	0.35	5.3
Catechol	0.49	5.4	0.28	0.47	5.8	0.36	9.6
Resorc-inol	0.39	5.8	0.28	0.47	5.8	0.36	15.2

Table VII. Comparison of Experimental and Predicted BOD Values
for Training Set (continued)

| Compound | Experimental | | | Predicted | | | Err |
	μ_m 1/hr	K_s mg/l	Y mg/mg	μ_m 1/hr	K_s mg/l	Y mg/mg	%
Acetone	0.22	1.0	0.55	0.27	4.7	0.46	7.5
2-Butanone	0.24	7.1	0.45	0.24	5.2	0.47	4.0
4-Methyl-2-pentanone	0.24	7.5	0.43	0.18	8.6	0.15	13.6
Benzyl alcohol	0.36	4.8	0.55	0.35	6.3	0.37	6.8
Chloro phenol	0.29	3.8	0.31	0.28	4.7	0.38	4.4
4-Nitro phenol	0.34	6.1	0.40	0.27	4.8	0.38	11.8
2,4-Dinitro phenol	0.17	3.8	0.41	0.17	4.0	0.40	3.4
Dichloro benzene	0.06	2.8	0.40	0.08	3.6	0.40	8.9

Table VIII. Comparison of Experimental and Predicted BOD
Values for Testing Set

Compound	Experimental rate const. k 1/hr	Predicted Monod constants			Error %
		μ_m 1/hr	K_s mg/l	Y mg/mg	
Ethyl alcohol	0.049	0.19	2.23	0.14	16.0
Butyl alcohol	0.041	0.13	3.25	0.15	17.2
iso-Butyl alcohol	0.017	0.13	5.57	0.16	27.6
Ethylene glycol	0.031	0.29	2.62	0.16	27.1
Acetone	0.018	0.27	4.70	0.46	15.1
Methyl ethyl ketone	0.028	0.24	5.20	0.47	14.2
Ethyl acetate	0.046	0.12	2.24	0.13	20.3
Butyl acetate	0.049	0.06	3.27	0.15	22.5
Benzene	0.054	0.29	5.70	0.36	6.8
Benzyl alcohol	0.052	0.35	6.30	0.37	5.1
Toluene	0.065	0.25	5.90	0.36	10.4
Phenol	0.026	0.38	5.70	0.36	13.5
Phenyl acetate	0.062	0.31	5.76	0.36	11.5
Acetophenone	0.035	0.44	8.74	0.69	8.4

**Table IX. The Monod Constants Predicted by Group
Contribution Method**

#	Compound	μ_m 1/hr	K_s mg/1	Y mg/mg
1	Benzene	0.29	5.7	0.36
2	Methyl benzene (Toluene)	0.25	5.9	0.35
3	Dimethyl benzene (Xylene)	0.22	6.0	0.34
4	Trimethyl benzene	0.18	6.2	0.33
5	Ethyl benzene	0.23	6.4	0.36
6	Propyl benzene	0.20	6.9	0.37
7	iso-Propyl benzene	0.16	9.7	0.39
8	(2-Ethyl propyl) benzene	0.14	10.2	0.40
9	Methanol	0.22	1.7	0.13
10	Ethanol	0.19	2.2	0.14
11	Propanol	0.16	2.7	0.15
12	iso-Propanol	0.15	5.0	0.16
13	Butanol	0.13	3.2	0.16
14	2-Butanol	0.12	5.6	0.17
15	Chlorobenzene	0.19	4.6	0.38
16	Dichlorobenzene	0.08	3.6	0.40
17	Phenol	0.38	5.7	0.36
18	Chlorophenol	0.28	4.7	0.38
19	Dichlorophenol	0.17	3.1	0.40
20	Trichlorophenol	0.07	2.6	0.42
21	Methyl phenol	0.35	5.9	0.35
22	Dimethyl phenol	0.31	6.1	0.34
23	Nitrophenol	0.27	4.8	0.38
24	2-Chloro-4-nitrophenol	0.16	3.8	0.40
25	Dimethyl phthalate	0.33	5.8	0.34
26	Diethyl phthalate	0.28	6.8	0.36
27	Dipropyl phthalate	0.22	7.8	0.38
28	Dibutyl phthalate	0.17	8.9	0.40
29	Dioctyl phthalate	-(0.06)		
30	Bis(2-ethyl hexyl) phthalate	-(0.07)		

Geating (15) has found that percentage degradation decreases with increase in molecular weight and branching and in presence of bulky side chains. The μ_m values in Table IX reflect similar trend with increase in molecular weight (compounds 1-4) and branching and size of side chain(s) (compounds 1,2,5-8). Babeu and Vaishnav (32) and Boethling (31) have reported that for alcohols percentage degradation decreases with increase in carbon number. The group contribution approach predicts similar trend for alcohols (compounds 9-14 in Table IX). Pitter (24) and Yonezawa and Urushigawa (39) reported that the degradation rate constant for normal alcohols is higher than alcohols with hydroxy group attached to any carbon atom other than the end carbon atoms. The values for compounds (11-14) in Table IX show that the Monod rate constant is higher for normal alcohols. Vaishnav et al. (25) has shown that addition of chlorine decreases the percentage degradation with respect to parent compound. The μ_m values of chlorinated compounds (15-16 and 17-20) in Table IX show that a similar trend is predicted by the group contribution approach.

Paris et al. (21,22) has studied the degradation of para-substituted phenols and has reported the second order rate constants for these compounds. The rate constant with respect to phenol decreases with substitution and magnitude of decrease depends on the type of substituent. The rate constants of substituted phenols are reported in the following decreasing order : phenol > p-methyl phenol > p-chlorophenol > p-nitrophenol. Pitter (24) has reported the following order : phenol > methyl phenols > dimethyl phenols > chlorophenols > nitrophenols > dichlorophenols > 2-chloro-4-nitrophenol > trichlorophenol. Pitter has reported the degradation as percentage of theoretical oxygen demand (ThOD) achieved. The values of the Monod rate constant for compounds (17-24) in Table IX agrees with this trend.

The predicted Monod rate constants decreases with increase in molecular weight of phthalates (compounds 25-30 in Table IX). Urushigawa and Yonezawa (39), Wolfe et al. (22) and Sugatt et al. (40) have reported the degradation rate constants for phthalates. Sugatt et al. (40) found that dibutyl phthalate was outlier and did not follow the trend but other two studies and experimental results of this study confirm the results predicted by the group contribution method. The molecular weight of both dioctyl phthalate and bis(2-ethyl hexyl) phthalate is 391 but bis(2-ethyl hexyl) phthalate is more branched than dioctyl phthalate. Hence, bis(2-ethyl hexyl) phthalate should be more refractory than dioctyl phthalate. The group contribution approach predicts negative Monod rate constant for both these phthalates, indicating that none of them would degrade. The predicted rate constant of bis(2-ethyl hexyl) phthalate is greater than dioctyl phthalate, which suggests that bis(2-ethyl hexyl) phthalate should be more difficult to degrade than dioctyl phthalate. The agrees with the findings of above three studies.

Conclusions

The results of this research study show that the group contribution approach can predict first order kinetics and Monod rate constants. The extent of degradation does follow the trends reported in literature for different groups of compounds. More data are required to predict the unique contribution due to positional effects of the groups and to include more functional groups in SAR.

The developed fate predictive group contribution model will closely predict the results on biodegradation kinetics found experimentally for many organic pollutants with varied molecular structure configurations and varied chemical groups composing the molecule.

It was demonstrated that both group contribution models can be used for estimating the biodegradation kinetic constants for chemically related compounds and that the nonlinear model gives better results than the linear approach.

It was shown that the non-linear model is able to provide a superior fit to the training set data (goodness of fit between the experimental and computed values) and that it produces lower per cent error between the predicted and experimental value than the linear group contribution approach.

Further research on biodegradation of organic compounds with different functional groups is needed to extend the applicability of the predictive biodegradation group contribution models.

Further work is needed to extend the nonlinear analysis to additional compounds. This would require electrolytic respirometric measurements for other compounds, since currently there is a lack of sufficient biodegradation kinetics data to extend our analysis to additional compounds. Eventually it would be possible to obtain estimates of biodegradation kinetic constants for a variety of compounds using the nonlinear method presented in this paper. Furthermore, this type of analysis would aid in the "designing" of molecules with favorable biodegradation kinetics.

Literature Cited.

1. Alexander, M. 1980. Biodegradation of chemicals of environmental concern. Science 211: 132-138.
2. Strier, M. P. 1980. Pollutant treatability : A molecular engineering approach. Environ. Sci. Tech. 14: 28-31.
3. Nirmalakhandan, N. and R.E. Speece. 1988. Structure-activity relationships. Environ Sci. Tech. 22: 606-615.

4. Free, S.M. and J.W. Wilson. 1964. A mathematical contribution to structure activity studies. J. Medical Chemistry. 7: 395-397.
5. Martin, Y.C. 1978. Quantitative Drug Design. Marcell Dekker, New York, New York.
6. Howard, P. H., S. Banerjee and A. Rosenburg. 1981. Review and evaluation of available techniques for determining persistence and rates of degradation of chemical substances in environment. U.S. Environmental Protection Agency, Report No. 560/5-81-011.
7. Grady, C.P.L.,Jr. 1985. Biodegradation : Its measurement and microbiological basis. Biotech. Bioengg. 27: 660-674.
8. 1983 OECD guidelines for testing of chemicals. EEC Directive 79/831, Annex V, part C: Methods for Determination of Ecotoxicity, 5.2 Degradation, Biotic Degradation, Manometric Respirometry, Method DGX1, Revision 5. pp. 1-22.
9. Tabak, H. H., R.F. Lewis and A. Oshima. 1984. Electrolytic respirometry biodegradation studies, OECD ring test of respiration method for determination of biodegradability. Draft manuscript, MERL, U S Environmental Protection Agency, Cincinnati, OH.
10. Urano, K. and Z. Kato. 1986. Evaluation of biodegradation ranks of priority organic compounds. J. Hazardous Mater., 13, 147-159.
11. Schroeder, E.D. 1977. Water and wastewater treatment. McGraw Hill Book Company, New York, NY.
12. Grady, C.P.L., Jr. and H.C. Lim. 1980. Biological Wastewater treatment, Theory and Application, Marcel Dekker, New York, NY.
13. Ludzack, F. J. and M.B. Ettinger. 1960. Chemical structures resistant to aerobic biochemical stabilization. J. Wat. Pollut. Control Fed. 32: 1173-1200.
14. Lyman, W. J., W.F. Reehl and D.H. Rosenblatt. 1982. Handbook of chemical property estimation methods : Environmental behavior of organic compounds, Mc-Graw Hill Book Company, New York, NY.
15. Geating, J. 1981. Literature study of the biodegradability of chemicals in water. U.S. Environmental Protection Agency, Report No. EPA-600/2-81-175, Cincinnati, OH.
16. Rothkopf, G. S. and R. Bartha. 1984. Structure-biodegradability correlation among xenobiotic industrial amines. Am. Oil Chemists Soc. J. 61: 977-980.
17. Yoshimura, K., S. Machida and F. Masuda. 1980. Biodegradation of long chain alkyl amines. Am. Oil Chemists Soc. J. 57: 238-241.
18. Paris, D. F., N.L. Wolfe and W.C. Steen. 1982. Structure-activity relationships in microbial transformation of phenols. Appl. Environ. Microbiol. 44: 153-158.
19. Paris, D. F., N.L. Wolfe, W.C. Steen and G.L. Baughman.

1983. Effect of phenol molecular structure on bacterial transformation rate constants in pond and river samples. Appl. Environ. Microbiol. 45: 1153-1155.

20. Paris, D. F., N.L. Wolfe and W.C. Steen. 1984. Microbial transformation of esters of chlorinated carboxylic acids. Appl. Environ. Microbiol. 47: 7-11.

21. Paris, D. F. and N.L. Wolfe. 1987. Relationship between properties of a series of anilines and their transformation by bacteria. Appl. Environ. Microbiol. **53**: 911-916.

22. Wolfe, N. L.,D.F. Paris, W.C. Steen and G.L. Baughman. 1980. Correlation of microbial degradation rates with chemical structure. Environ. Sci. Tech., 14, 1143-1144.

23. Reinke, W. and H.J. Knackmuss. 1978. Chemical structures and biodegradability of halogenated aromatic compounds : Substituent effects on 1,2-dioxygenation of benzoic acid. Biochimica et Biophysica Acta. 542: 412-423.

24. Pitter, P. 1984. Correlation between the structure of aromatic compounds and the rate of their biological degradation. Collection Czechoslovak Chem. Commum. 49: 2891-2896.

25. Vaishnav, D. D. R.S. Boethling and L. Babeu. 1987. Quantitative structure-biodegradability relationships for alcohols, ketones and alicyclic compounds. Chemosphere. 16: 695-703.

26. Banerjee, S., P.H. Howard, A.M. Rosenburg, A.E. Dombrowski, H. Sikka and D.L. Tullis. 1984. Development of a general kinetic model for biodegradation and its application to chlorophenols and related compounds. Environ. Sci. Tech. 18: 416-422.

27. Deardan, J. C. and R.M. Nicholson. 1986. The prediction of biodegradability by the use of quantitative structure-activity relationships : Correlation of biological oxygen demand with atomic charge difference. Pestici. Sci. 17: 305-310.

28. Randic, M. 1975. On characterization of molecular branching. J.Am. Chem. Soc.. 97: 6609-6615.

29. Kier, L. B. and L.H. Hall. 1976. Molecular connectivity in chemistry and drug research. Academic Press, New York, NY.

30. Govind, R. 1987. Treatability of toxics in wastewater systems. Hazardous Substances. 2: 16-24.

31. Boethling, R. S. 1986. Application of molecular topology to quantitative structure-biodegradability relationships. Environ. Toxicol. Chem. 5: 797-806.

32. Babeu, L. and D.D. Vaishnav. 1987. Prediction of biodegradability for selected organic chemicals. J. Industrial Microbiol. 2: 107-115.

33. Desai, S., R. Govind and H. H. Tabak. 1990. Development of quantitative structure-activity relationships for predicting biodegradation kinetics. Environ. Toxicol. Chem. 9: 473-477.

34. Rosenblatt, R. 1959. Principles of Neurodynamics, Spartan Books, New York, NY.

35. Rumelhart, D.E., G.E. Hinton and R.J. Williams. 1986. Learning Internal Representations by Error Propagation, in D.E. Rumelhart and J.J. McClelland (Eds.) Parallel Distributed Processing: Explorations in the Microstructure of Cognition. Vol 1: Foundations, MIT Press.
36. Grady, C.P.L., Jr., J.S. Dang, D.M. Harvy, A. Jobbagy and X.L. Wang. 1989. Determination of biodegradation kinetics through use of electrolytic respirometry. Wat. Sci. Tech. 21: 957-1003.
37. Urano, K. and Z. Kato. 1986. A Method to classify biodegradabilities of organic compounds. J. Hazardous Mater.,13, 135-146.
38. Tabak, H.H., S. Desai and R. Govind. 1989. The determination of biodegradability and biodegradation kinetics of organic pollutant compounds with the use of respirometry, in Proceedings of EPA 15th Annual Research Symposium: Remedial Action, Treatment and Disposal of Hazardous Waste.
39. Yonezawa, Y. and Y. Uruchigawa. 1979. Relation between biodegradation rate constants of aliphatic alcohols by activated sludge and their partition coefficients in a 1-octanol-water system. Chemosphere, 159-164.
40. Sugatt,R.H., D.P. O'Grady, S. Banerjee, S. Howard and W.E. Gledhill. 1984. Shaker flask biodegradation of 14 commercial phthalate esters. Applied Environ. Microbiol. 47: 601-609.

RECEIVED September 3, 1992

Chapter 10

Degradation of Cyanides by the White Rot Fungus *Phanerochaete chrysosporium*

Manish M. Shah and Steven D. Aust

Biotechnology Center, Utah State University, Logan, UT 84322–4705

Potassium cyanide and various other cyanide salts (Fe, Cu, Zn, Cd and Cr) were mineralized by the white rot fungus Phanerochaete chrysosporium. At 1.5 mM potassium cyanide, the rate of mineralization was about 0.17 ± 0.01 mmoles/lit/day. P. chrysosporium also mineralized [^{14}C]-cyanide contaminated soil (3000 ppm) using ground corn cobs as nutrient (10 ± 0.75 ppm/day). Cyanide was oxidized to the cyanyl radical by a lignin peroxidase from P. chrysosporium. We suggest that the ability of P. chrysosporium to mineralize cyanides may make it useful in the treatment of cyanide contaminated soils, sediments and aqueous wastes.

Lignin is a complex three dimensional biopolymer that provides structural support to plants. The structure of lignin is heterogeneous, non-specific and non-stereoselective. The complexity of its structure makes it resistant to most microbes. The white rot fungus Phanerochaete chrysosporium is known to degrade lignin to CO_2 (1). The lignin degradative process of P. chrysosporium is free radical based (2,3) which gives the advantages of being non-specific and non stereoselective. Some of the known important components of the lignin degradation system of P. chrysosporium include lignin peroxidases, H_2O_2 and vertryl alcohol (2,3). P. chrysosporium is also able to degrade variety of structurally diverse organopollutants (4,5). Results of various studies indicate that lignin degrading enzyme system is involved in the degradation of organopollutants by P. chrysosporium (4-6). Recent attention has focused on the possible usefulness of P. chrysosporium for biodegradation of hazardous and environmentally persistent organopollutants (3-8). Hydrogen cyanide is the form of cyanide used most by industry. It is used as a chemical intermediate in the production of methyl methacrylate, cyanuric chloride, sodium chloride, chelators and other chemicals (9). Potassium cyanide and sodium cyanide have

0097–6156/93/0518–0191$06.00/0

been designated as hazardous substances and priority pollutants by the EPA (9). Cyanide containing wastes are generated by various industries like metal plating, mining, pharmaceutical, refining, paint, electronic and others (9). The presence of cyanides in sewage plant upsets treatment processes and are toxic to fish at very low concentration (0.02 mg/l) (10). The available options for management of cyanide wastes include minimization, recycling and treatment by alkaline chlorination, ozonation, thermal destruction and biodegradation. Of all the treatment methods, alkaline chlorination is most widely used. Biological treatment of cyanide is an attractive process as the products are generally nontoxic and the process is inexpensive.

In this paper, we show that P. chrysosporium is able to mineralize potassium cyanide and salts of Fe, Cd, Cu, Cd, Cr(VI) in soil and liquid cultures. We also show that a pure lignin peroxidase from P. chrysosporium oxidized cyanide to the cyanyl radical.

Toxicity of Cyanide to P. chrysosporium

Cyanide is an inhibitor of heme proteins and metal containing oxidases which include cytochrome oxidases. Cyanide forms relatively stable complexes with many metals, reacts with keto groups to form cyanohydrin, and reduces thiol groups (10). Therefore, exposure of cells to cyanide results in inactivation of respiration. Figure 1a and 1b shows the effect of addition of cyanide on respiration (glucose metabolisms) of spores and six day old cultures of P. chrysosporium (11). Fifty percent inhibition of respiration for spores and six day old cultures occured at about 100 μM and 5 mM of cyanide, respectively.

Mineralization of [^{14}C]KCN by P. chryosporium

The ability of six day old cultures of P. chrysosporium to tolerate cyanide could be due in part to its ability detoxify cyanide. Figure 2 shows that P. chrysosporium mineralized cyanide to CO_2 and the rate of mineralization of cyanide followed first order kinetics (11). The inhibition of mineralization at 10 mM of cyanide could be partly due to its toxicity as this concentration inhibited fungal respiration by 65% (Figure 1b).

Degradation of Cyanide in the Presence of Metals

Table I shows that P. chrysosporium mineralized cyanide and its salts in the presence of Fe, Cu, Zn, Cd and Cr(VI) in liquid cultures. The rate of mineralization of cyanide was about 30% $\pm$ 10% in 30 days and was not effected significantly by various concentration (0-1 mM) of Cu, Cd, and Zn. In the case of iron, some (~10%) stimulation was observed at lower concentration (12 μM) while higher concentration of iron inhibited mineralization of cyanide with I_{50} occurring at about 200 μM (Figure 3). However, higher concentration of cyanide could reverse the inhibition of cyanide mineralization (Figure 4). It can be seen that the rate of mineralization increased with increasing in concentration of cyanide. In the case of Cr(VI), mineralization was also

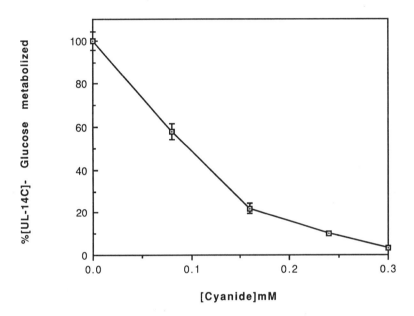

Figure 1a. Toxicity of cyanide to spores of P. chrysosporium. Sodium cyanide and [UL-^{14}C]-glucose were added to cultures of P. chrysosporium at day 0, and 3 days later evolved [^{14}C]-CO$_2$ was trapped and its radioactivity quantitated using liquid scintillation spectrometry. The controls were without cyanide. Glucose metabolism in the controls was considered as 100 percent. All incubations were carried out in quadruplicates in a closed system at room temperature and in stationary cultures. The data are average values with standard deviations, some of which are within the data points. Reproduced with permission from Reference 11. Copyright 1991 Academic.

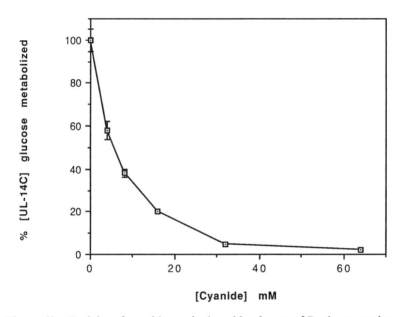

Figure 1b. Toxicity of cyanide to six day old cultures of <u>P. chrysosporium</u>. Conditions were the same as in Figure 1a except sodium cyanide and [UL-^{14}C]-glucose were added to six day old ligninolytic cultures of <u>P. chrysosporium</u>. Reproduced with permission from Reference 11. Copyright 1991 Academic.

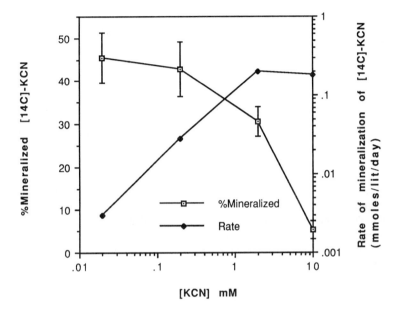

Figure 2.   Mineralization   of   cyanide   by   six   day   old   cultures   of   P. chrysosporium.   [^{14}C]-KCN   was   added   to   six   day   old   cultures   of   P. chrysosporium,   3 days later   evolved   [^{14}C]-CO$_2$ was   trapped   using   1M Ba(OH)$_2$.   [^{14}C]-BaCO$_3$ was separated   from   [^{14}C]-Ba(CN)$_2$ by centrifugation. Radioactivity   of   [^{14}C]-BaCO$_3$ was   quantitated   using   liquid   scintillation spectrometry.   All incubations   were carried out in quadruplicate   in a closed system at room temperature   and in stationary cultures. Open squares show the percent of the cyanide mineralized in three days and closed symbols show the rate of mineralization in mmoles per liter per day. The error bars (standard   deviations) are   within the   data   points   in   some cases. Reproduced   with   permission   from   Reference   11.   Copyright   1991 Academic.

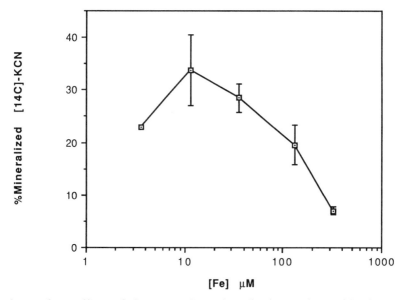

Figure 3. Effect of iron on the mineralization of cyanide by P. chrysosporium. Solutions of ferrocyanide [^{14}C]-K$_4$Fe(CN)$_6$ were added to 6 day old liquid cultures of P. chrysosporium and the amount of [^{14}C]-KCN converted to [^{14}C]-CO$_2$ determined as described in ref. (11). Data are the mean percent mineralization in 3 days ± standard deviations for triplicate incubations.

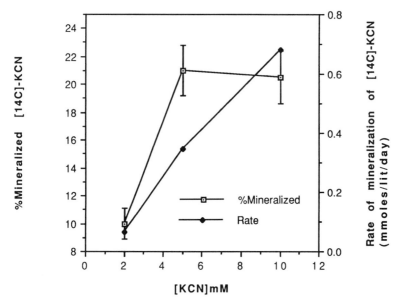

Figure 4. Effect of cyanide concentration on mineralization of cyanide at 200 μM Fe. Reaction conditions were similar to those described in Figure 3.

Table I. Mineralization of [^{14}C]-KCN in the Presence of Metals[a]

	% mineralization
KCN	25 ± 2
KCN + FeSO$_4$	33 ± 4
KCN + CuSO$_4$	30 ± 3
KCN + ZnSO$_4$	35 ± 2
KCN + CdSO$_4$	35 ± 3
KCN + CrO$_3$	40 ± 5

[a] Culture conditions were as described in ref. 11. Solution of cyanide complexes with metals were prepared by mixing various amounts of metals (0-1 mM) with 1.5 mM [^{14}C]-KCN and they were added to six day old cultures of P. chrysosporium. There was no significant change in mineralization of cyanide at different concentration of Cu, Zn and Cd. At 200 μM and 500 μM of Fe and Cr(VI), respectively, mineralization of cyanide was inhibited about 50%.

inhibited at higher concentration with I$_{50}$ occurring at about 500 μM. Cyanide might be partly oxidized by Cr(VI) as it is a good oxidizing agent. This might help in enhancing cyanide mineralization. Iron and Cr(VI) inhibited fungal respiration (glucose metabolism) at 1200 μM and 100 μM, respectively. The mechanism of inhibition of cyanide mineralization by iron needs more study due to the complexity of iron chemistry and reactivity of cyanide.

Mineralization of [^{14}C]-KCN from Contaminated Soil

Table II shows the properties of the soil used in studies of cyanide mineralization in soil. Mineralization of cyanide was observed when soil contaminated with cyanide at a final concentration of 3000 ppm was incubated with P. chrysosporium grown on corn cobs (1:1) as a nutrient (Figure 5). The rate of mineralization was 10 ppm $\pm$ 0.75 ppm/day. It can be seen that degradation of cyanide was continuous. Volatilization of cyanide was about 4% of the total cyanide mineralized over 30 days.

Oxidation of Cyanide by Lignin Peroxidases

Lignin peroxidases are the enzymes which are proposed to be important components of the lignin degrading system of P. chrysosporium (3). They are produced during idiophasic metabolism and the production of H$_2$O$_2$ also occurs at the same time (3). Lignin peroxidases are activated by H$_2$O$_2$ as are typical

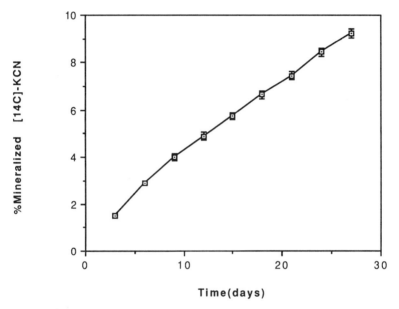

Figure 5.   Mineralization   of cyanide in soil by P. chrysosporium.   Five grams of corn cobs previously (15 days) inoculated with P. chrysosporium were incubated with 5 grams of an agricultural silt loam soil containing 3000 ppm [^{14}C]-KCN. The amount of [^{14}C]-KCN converted to [^{14}C]-CO$_2$ was determined every 3 days and plotted as a percent of the cyanide mineralized versus time. The data are means $\pm$ standard deviations for triplicate incubations. In most cases the standard deviation bars are within the data points.

Table II. Properties of the Agricultural Silt Loam Soil in Which the Metabolism of Cyanide by P. chrysosporium was Studied

pH	6.4
Nitrogen (ppm NO_3)	18.4
Sulfur (ppm SO_4)	6.7
Total nitrogen (%)	0.19
Organic matter (%)	4.0
Organic carbon (%)	2.3
Diethylenetriaminepentoacetic acid extractable metals (ppm)	
Zn	2.3
Fe	35.1
Cu	1.2
Mn	29.4
Pb	0.86
Ni	0.47
Cr	0.16
Water extractable ions (meq/100 g)	
Na	0.10
K	0.99
Ca	16.85
Mg	3.59

peroxidases (13) and the resultant activated enzyme intermediate is called as compound I. In the presence of suitable electron donors (ie. veratryl alcohol), compound I can undergo two single electron reductions to bring the enzyme to its resting state (3). On incubation of lignin peroxidase with cyanide in the presence of H_2O_2, cyanyl radicals were formed suggesting that lignin peroxidase is capable of oxidizing cyanide through a free radical based mechanism (11). Figure 6a shows the PBN (α-tert-butyl-N-tert-butyl nitrone) cyanyl radical adduct observed by ESR spin trapping upon incubation of lignin peroxidase isoenzyme H2 with H_2O_2 and cyanide. The formation of the cyanyl radical was confirmed by experiments with [^{13}C]-KCN (Figure 6b).

Advantages of P. chrysosporium for the Degradation of Cyanide Containing Wastes

Cyanide containing wastes usually contain various metals like Fe, Cd, Cu, Zn and Ni. The results of our present research suggest that P. chrysosporium has the ability to degrade cyanide in the presence of moderate concentrations of such metals. In addition to metals, the wastes sometimes also contain a variety of hazardous organopollutants and some of them are highly insoluble and recalcitrant (ie. polyaromatic hydrocarbons). The extracellular degradative

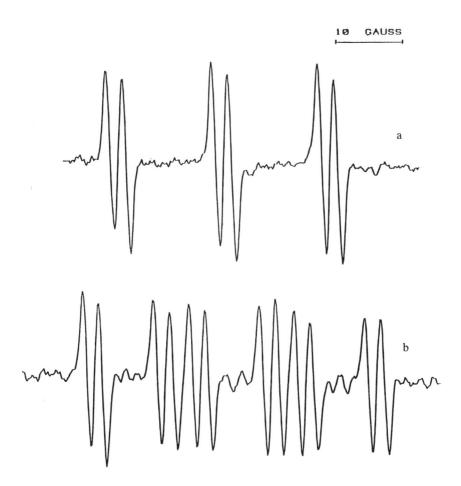

Figure 6. ESR Spectrum of PBN-cyanyl radical adduct formed by lignin peroxidase H2 and H_2O_2. Reaction mixtures contained 10 μM lignin peroxidase H2, 40 mM sodium cyanide, 500 μM H_2O_2, and 100 mM PBN in 100 mM sodium phosphate buffer, pH 6.0. (b) [^{13}C]-KCN was substituted for NaCN. The receiver gain was 4 x 10^4 in (a) and 2.5 x 10^4 in (b). Reproduced with permission from reference 11. Copyright 1991 Academic.

system of P. chrysosporium may make it suitable for the degradation of insoluble chemicals which might be present in cyanide containing wastes. Also the nonspecific nature of the degradation system may also be useful in the case of mixtures of organopollutants. Also, under nitrogen limiting condition, the organism may compete well with other microorganisms. And biodegradation is not dependent on prior exposure to the pollutants as the lignin degradation enzymes are expressed in response to nitrogen starvation, not prior exposure to the chemicals. Degradation of organochemicals can procede to essentially non-detectable levels as there is no true Km due to the free radical based nonspecific degradative process and degradation can result in mineralization. Furthermore, degradation can be supported by inexpensive lignocellulose based nutrients such as corn cobs.

Summary

The white rot fungus P. chrysosporium was shown to mineralize cyanide and cyanide complexes of Fe, Cu, Cd, Zn and Cr(VI). The fungus was also able to mineralize cyanide in contaminated soil continuously using ground corn cobs as nutrient. The results also suggest that lignin degrading enzymes might be important in degradation of cyanides by P. chrysosporium as cyanide was oxidized to cyanyl radical by a pure lignin peroxidase.

The degradation of pollutants by white rot fungi is a free radical process so the rate of reaction would be expected to be first order with respect to concentration of chemical. The rate of mineralization by the fungus was directly proportional to the concentration of cyanide unless it became toxic. Further, cyanide wastes generally contain metal cyanides, especially in the case of metal plating industries. In the present study, we showed that P. chrysosporium mineralized cyanide complexes of Fe, Cd, Cu, Zn and Cr(VI). It might be possible to use a white rot fungus based biological treatment system for detoxification of cyanide containing wastes under moderate concentration of cyanide and metals.

Acknowledgements

This work was supported by NIEHS grant no. ES04922. The authors would like to thank Terri Maughan for assistance in the preparation of this manuscript.

Literature Cited

1. Kirk, T.K.; Farrel, R.L., Ann. Rev. Microbiol. 1987, 41, 465-505.
2. Shoemaker, H.E., Recl. Trav. Chim. Pays - Bas 1990, 109, 255-272.
3. Tien, M., Crit. Rev. Microbiol. 1987, 15, 141-168.
4. Bumpus, J.A.; Tien, M.; Wright, D.S.; Aust, S.D., Science, 1985, 228, 1434-1436.

5. Fernando, T.; Aust, S.D., In <u>Biological Degradation and Bioremediation of Toxic Chemicals</u>; Chaudhry, G.R., Ed.; 1991, in press.
6. Bumpus, J.A.; Aust, S.D., <u>BioEssays</u> **1986**, <u>6</u>, 166-170.
7. Aust, S.D., <u>Microb. Ecol.</u> **1990**, <u>20</u>, 197-209.
8. Eaton, D.C., <u>Enzyme Microb. Techn.</u> **1985**, <u>7</u>, 194-196.
9. Palmer, S.A.K.; Breton, M.A.; Nunnon, T.J.; Sullivan, D.M.; Surprenant, N.F., <u>Pollution Technology Review No. 158</u> **1988**, 11-45, Noye Data Corp., Park, Ridge, NJ.
10. Knowles, C.J., <u>Bacteriol. Rev.</u> **1976**, <u>40</u>, 652-680.
11. Shah, M.M.; Grover, T.A.; Aust, S.D., <u>Arch. Biochem. Biophys.</u> **1991**, <u>290</u>, 173-178.

RECEIVED October 2, 1992

Chapter 11

Protocol for Determining the Rate of Biodegradation of Toxic Organic Chemicals in Anaerobic Processes

Nagappa Sathish[1], James C. Young[1], and Henry H. Tabak[2]

[1]Department of Civil Engineering, The Pennsylvania State University, University Park, PA 16802
[2]Risk Reduction Engineering Laboratory, U.S. Environmental Protection Agency, Cincinnati, OH 45268

A protocol is presented for determining the rate of biodegradation of toxic organic chemicals in anaerobic processes. The protocol involves use of ethanol-enriched cultures that have been acclimated to specific toxic chemicals in a controlled environment and operated at steady-state conditions. These cultures are transferred to serum bottles, dosed with toxicant, and monitored for toxicant degradation using batch techniques. The kinetic model uses the biomass specifically involved in the toxicant degradation reaction to estimate toxicant degradation kinetic parameters. These parameters are then used to estimate the fate of toxic organic chemicals in full-scale anaerobic reactors.

Anaerobic treatment processes frequently receive complex organic chemicals that are classified as being hazardous or toxic. One specific concern is that these chemicals will pass through wastewater treatment units and be deposited along with waste sludges where they will reach streams and aquifers and contaminate water supplies. In spite of recent advances in identifying the principal microorganisms involved in biodegradation reactions there is a lack of definition of the kinetics of toxicant degradation reactions, thereby making it difficult to predict the nature and extent of toxicant degradation in anaerobic processes.

Objective

Consequently, the authors conducted a study to develop a protocol for measuring the kinetic parameters needed to model the rate of biodegradation of toxic organic chemicals in anaerobic wastewater treatment processes (1). The primary objective of this study was to develop a toxicant degradation protocol as part of a broader multi-level protocol for assessing fate and effect of toxic organic chemicals in anaerobic processes (2).

Background

Toxic organic chemicals have been observed to degrade under methanogenic conditions as far back as the 1930s when Tarvin and Buswell (*3*) studied the biodegradation of common aromatic compounds. Among other organics subsequently found to degrade are benzoates (*4*), phenols and catechols (*5, 6*), chlorinated benzenes (*7, 8*), chlorinated phenols (*9, 10*), anilines and chlorinated anilines (*11, 12*), and polychlorinated biphenyls (*13*). The contribution of these researchers to our understanding of toxicant degradation is very important. Actual kinetic data, however, is scarce and often difficult to interpret because of the mixed culture composition and the complex metabolic interactions in anaerobic environments. And in most cases, rates of degradation have been expressed relative to a control, on a volumetric basis, or when using total biomass rather than the biomass specific to the degradation reaction under study.

Various biological and non-biological factors must be considered when determining kinetic parameters for toxicant degradation reactions. The utility of the kinetic parameters is greatest when the test results do not vary between analysts and between measurements conducted at different times and therefore are intrinsic, that is, they are not system dependent. Experimental methods for such tests must incorporate precise control over the factors that influence the test progress. These factors include, but are not necessarily limited to, the composition of the microbial culture, the ratio of active biomass to total biomass solids, environmental stresses, solid/liquid interactions such as adsorption and volatilization, presence of other toxic substances, culture history, nutrient availability, and the physical arrangement of the microbial solids matrix.

Another important factor that must be considered in estimating kinetic parameters from test data is the mathematical model to be used. Various models have been developed for describing toxicant degradation (*14-18*). The most popular of these is the Haldane model (*19*) and, in fact, many of the other models are essentially modifications of equations developed by Haldane. Recently, Han and Levenspiel (*20*) presented a generalized model for describing inhibition reactions in which:

$$\frac{dI}{dt} = -\frac{k_o\,(k^*)\,I\,M_I}{I + K_{so}(K_s^*)} \tag{1}$$

and

$$\frac{dM_I}{dt} = Y\left(\frac{-dI}{dt}\right) - K_d\,M_I \tag{2}$$

where

dI/dt	=	Rate of toxicant degradation, mg/L-hr,
dM_I/dt	=	Rate of growth of toxicant-degrading biomass, mg/L-hr,
I	=	Toxic organic chemical concentration, mg/L
M_I	=	Active biomass that degrades the toxicant, mg/L
Y	=	Biomass yield coefficient, mg VSS/mg toxicant converted
k_d	=	Biomass decay coefficient, hr^{-1}
k_o	=	Maximum toxicant conversion rate, mg toxicant/mg VSS-hr
K_{so}	=	Substrate concentration at which the reaction rate equals half the maximum rate, mg/L
k^*, K_s^*	=	Inhibition terms

Equation (1) encompasses the Haldane model when:

$$k^* = 1.0 \tag{3}$$

$$K_s^* = \left(1 + \frac{I^2}{K_{so} K_H}\right) \tag{4}$$

where K_H is the Haldane inhibition coefficient defined as the highest toxic substrate concentration at which the specific conversion is equal to half the maximum rate in the absence of inhibition.

Values of k_0, K_{so}, and K_H typically are found by fitting Equations 1 through 4 to toxicant conversion data measured in batch laboratory tests. Biomass yield coefficients for toxic organic chemicals usually are not known because of the lack of test data and lack of knowledge of the stoichiometry of the degradation reactions and must be estimated by means described below. The biomass decay rate, k_d, generally is considered to be constant; a value of 0.024/d was used in the authors' study (*1*).

Toxicant Degradation Protocol

The protocol is divided into three steps. The first step (STEP A) involves development of acclimated cultures in Master Culture Reactors (MCRs), or chemostats, that are operated under controlled conditions and receive a known amount of one or more toxicants in addition to a base organic co-substrate. These master culture reactors initially should be seeded from an active anaerobic digester and operated for at least three hydraulic retention times prior to using the cultures for test purposes. Ethanol generally is used as the base co-substrate but other organic chemicals may be used as required to meet test objectives. In STEP B, these acclimated cultures are transferred to test reactors that are dosed at various concentrations of one test toxicant and without adding a co-substrate. The residual toxicant concentration is tracked over time. In STEP C, the resulting toxicant degradation data serves as a basis for assessing the toxicant transformation kinetics.

A standardized Nutrient/Mineral/Buffer medium as listed in Table I is used in all cultures -- the MCRs as well as cultures transferred to serum bottles. This medium is patterned after work conducted by other investigators of anaerobic processes (*21-24*).

The steps of the protocol are described in greater detail below.

STEP A -- Development of Acclimated Master Culture Reactors (MCR). Acclimation is developed by adding one or more test toxicants to a steady-state ethanol-enriched MCR so that the concentration of toxicant builds up slowly. This procedure gives acclimation an opportunity to occur before toxic effects are expressed. The amount of toxicant added to the MCR will vary with the relative toxicity of the test chemical and the physical characteristics of the chemical -- solid vs. liquid, solubility, etc. A typical toxicant feed rate is 0.5 to 1.0 % of the ethanol feed rate or 5 to 10 mg toxicant/L-d, but lower concentrations may be required with highly toxic chemicals. The MCRs are operated at a 20-day hydraulic and solids retention time (SRT) and are fed one time per day using a draw-and-fill procedure. If acclimation does not occur, the toxicant concentration in the MCR will approach that in the feedstock, or 20 times the daily toxicant loading rate except for volatile chemicals which may be partially lost with the gas stream.

Table I.   Nutrient/Mineral/Buffer Medium Used in Batch Tests

Constituent	Concentration (mg/L)	Constituent	Concentration (mg/L)
Nutrients		Minerals	
KH_2PO_4	500	$CaCl_2 \cdot 2H_2O$	150
Na_2SO_4	150[a]	$MgCL_2 \cdot 6H_2O$	200
NH_4Cl	530	$FeCl_2 \cdot 4H_2O$	20
Cysteine	100[b]	$MnCl_2 \cdot 4H_2O$	0.5
		H_3BO_3	0.25
Buffer		$ZnCl_2$	0.25
$NaHCO_3$	6000	$CuCl_2$	0.15
		$NaMoO_4 \cdot 2H_2O$	0.05
		$CoCl_2 \cdot 6H_2O$	2.50
		$NiCl_2 \cdot 6H_2O$	0.25
		Na_2SeO_4	0.25

[a] 150 mg/L in NMB feedstock provides 5.0 mg $SO_4^=$/L-d to test cultures
[b] 100 mg/L in NMB feedstock provides 5.0 mg/L-d to test cultures

This situation is illustrated in Figure 1. In this case, nitrobenzene was fed to an ethanol-enriched test reactor. The nitrobenzene was reduced immediately to form aniline which began to accumulate as the feeding continued. Initially, the residual aniline concentration increased in direct response to the nitrobenzene input. About day 7, the measured rate of aniline accumulation began to fall below the rate of input. The rate of aniline degradation continued to increase until after day 36 when essentially no residual aniline remained. An additional 60 days of operation would be required for the reactor to reach steady state with respect to the toxicant degradation and biomass concentration.

Mixed liquor samples removed daily from acclimated Master Culture Reactors are analyzed periodically to determine the residual toxicant concentration and to verify that acclimation is occurring. If acclimation does not occur, toxicity to the acetogenic or methanogenic cultures may cause gas production rates to decrease. When inhibition does occur, toxicant feed should be stopped immediately and time given for the culture to recover while receiving ethanol as the sole substrate. Small amounts of seed from external sources -- leachates, municipal sludge digester, industrial digester, or contaminated soils -- should be added periodically to help ensure the presence of microorganisms that will acclimate to the test toxicant. With highly toxic chemicals, for example chloroform or pentachlorophenol, continuous feed using small constant-rate pumps may provide improved operation. However, continuous feed may not be possible if the toxicant is volatile or is not soluble in the feedstock solution.

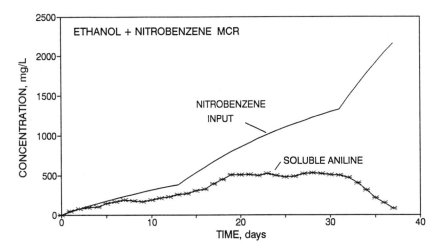

Figure 1. Toxicant input (calculated) and soluble residual (measured) versus time for a master culture reactor receiving nitrobenzene.

This method of developing acclimated cultures by adding small amounts of toxicant to ethanol-enriched cultures provides an approach that is expected to work with a large number of toxic organic chemicals. However, other means may be used to develop acclimated cultures. In some cases, use of co-substrates other than ethanol may be desired or necessary. Toxicants such as pentachlorophenol may require considerably longer times to develop acclimated cultures for complete degradation (25, 26). In some cases, mixtures of test toxicants may be added to the acclimated MCR, but this practice is risky. Mixtures can produce unexpected impacts and often are more toxic than either chemical alone (27).

The above acclimation procedure assumes that 1) a specific species or group of microorganisms is responsible for the toxicant degradation reaction for which measurements are made, 2) the toxicant-degrading reaction is not adversely affected by other anaerobic reactions, and 3) the rate-controlling step in the toxicant degradation pathway is known. Toxicants degraded by non-specific acetogenic microorganisms, those requiring co-metabolites, and those for which reaction products are toxic to the degradation of the parent compound may require a different testing and modeling approach than that described in this paper.

Step B -- Test Set Up. The protocol test program requires a minimum of six test reactors operated as low-rate batch units, that is, reactors in which the ratio of initial toxicant concentration to biomass concentration is greater than about 10:1. Test cultures are first transferred from the acclimated MCR under anoxic conditions to serum bottles that are then sealed and stabilized for 3 to 5 days by feeding ethanol-toxicant feedstock at the same rate as used in the MCR. One test reactor is used as a Control which is essentially a seed blank to provide a measure of background gas production. An Abiotic Control reactor receives 160 mg/L of mercuric chloride as a preservative and serves as a check against interference or false readings of toxicant residual. Toxicant is applied to test reactors at three to four concentrations, and one toxicant level is duplicated to provide a basis for quality control assessment. Toxicant test concentrations causing 10, 20, 40 and 60% reduction in gas production relative to that in a control sample, as determined by a separate screening protocol, generally give acceptable results (28). Higher concentrations may be used with some chemicals but can extend the tests beyond a reasonable time.

Changes in toxicant concentration are determined by measuring the total residual toxicant at daily or more frequent intervals by removing samples from the test reactors. Specific degradation products also are monitored to the extent possible within the limits of knowledge of the reaction stoichiometry and analytical procedures available. While a number of procedures can be used for measuring these residuals, no one method is applicable in all cases. The micro-extraction procedure used for analysis of volatile organic compounds has been found useful for PROTOCOL purposes because of the small sample volumes (29). Other methods may be used with demonstration of efficient compound recovery.

Methane production may be used as a surrogate measure of toxicant degradation in some cases but with some sacrifice in precision and accuracy. This approach requires careful monitoring of the headspace gas throughout the reaction and correction for gas produced by the seed culture and for changes in gas composition. Moreover, seed corrections are subject to considerable error if the toxicant affects the test cultures in a manner that is different from the effect in the seed blanks. Since methane production presents a measure of overall toxicant degradation, the rate of methane production may reflect the rate of conversion of an intermediate compound and not the toxicant added.

Step C -- Modeling Procedure. Modeling involves the use of a suitable mathematical relationship to describe the depletion of toxic organic chemical, formation of intermediates, and the generation of biomass that is responsible for the depletion of toxic organic. The Haldane model (*19*), as described by Equations 1, 3 and 4, was used in the authors' study. It should be noted that when the substrate concentration is low or the Haldane coefficient is high, the term $I^2/K_{so}K_H$ becomes negligible. Under these conditions, Equation 1 reduces to the classical Monod function (*30*). In batch cultures even for higher values of K_H, the term $I^2/K_{so}K_H$ may be significant because of the higher toxic substrate concentrations present during the early stages of growth.

Three approaches can be used for estimating the toxicant-degrading biomass to use for modeling purposes. The first is to calculate the total yield using bioenergetics methods followed by adjustment for the contribution of intermediates to the total biomass (*31*). While reasonably accurate, this method requires prior knowledge or a reasonable estimate of the stoichiometry of the reaction and knowledge of the free energies of the compounds involved in the reaction, and actual yields may differ considerably from theoretical yields. The second approach is to consider that the yield coefficient for the specific toxicant degradation reaction is the same as for the acetogenic conversion of compounds having similar structural composition, for example aromatics, alkanes, etc. A third approach is to consider that the toxicant degrading acetogenic biomass in the Master Culture Reactor is proportional to the contribution of the toxicant degradation reaction to the chemical oxygen demand (COD) of the ethanol-toxicant feedstock (*32*), that is, the yield coefficient for the critical conversion reaction is the same as for ethanol on a COD basis. This approach may be used when it is not possible to use either of the two methods cited above but is subject to considerable error because of differences in bioenergetics among toxicants. Each of the above approaches assumes that a specific culture is responsible for the initial toxicant conversion reaction.

Experimental Study

Toxicant degradation kinetics were analyzed for phenol using the above PROTOCOL. Phenol degrading culture was enriched in two MCRs. Reactor CP-1 was fed ethanol at a loading of 480 mg/L-d plus 5 mg/L-d each of phenol, 2-chlorophenol, 4-chlorophenol, 2,4-dichlorophenol and 2,4,6-trichlorophenol. Reactor P-1 was fed phenol as a sole substrate at a rate of 420 mg/L-d. Steady-state performance was confirmed by monitoring daily gas production, and weekly analysis of reactor contents for biomass, volatile fatty acids, and residual toxicant concentrations.

Low rate batch kinetic tests were conducted by transferring 300 or 400 mL of seed culture from the CP-1 MCR to each of a number of 500 mL serum bottles which previously had been flushed with a mixture of 70% methane and 30% carbon dioxide. Prior to the start of the test, the cultures were stabilized over a period of three to four days by daily removing 5% of the volume and adding an equal amount of stock feed solution. Total gas production from each serum bottle was monitored by means of an automated anaerobic digital respirometer (*33*) to verify the activity of the transferred cultures. Bottles not producing the expected amount of gas were discarded. The initial head space gas composition was determined for each reactor using gas chromatography. A one-time slug dose of toxicant was applied to the test units and test duplicates. The toxicant doses applied to the test units were selected from previous screening tests to represent low, intermediate, and moderately high levels of toxicity. A test duplicate was used to assess variability between replicates. Gas composition was monitored at frequent intervals (daily or hourly depending on the test progress) throughout the duration of the test.

The above procedure was repeated with frequent sampling of the aqueous content of the serum bottles to determine the conversion of the toxicant. Sampling was done as follows: five mL of the serum bottle contents were removed at frequent intervals using a syringe and transferred to a vial containing two drops of 16 g/L HgCl$_2$ solution and three drops of phosphoric acid to enhance extraction. The vial was covered with a Teflon^{TM} coated septum and capped tightly. One milliliter of methylene chloride that contained 300 mg/L of 2-nitrophenol as an internal standard was transferred through the septum which had an empty syringe (called a disposer syringe) inserted into it (34). The addition of methylene chloride displaced an equal volume (1 mL) of the contents of the vial into the disposer syringe. This volume was discarded. The contents in the vial were mixed on a rotary shaker for at least 15 minutes, after which the sample was ready for storage or analysis by gas chromatographic methods.

Similar tests and analyses were conducted using the P-1 culture to ascertain the degradation potential of phenol in high-rate batch tests, that is, tests in which the initial toxicant-to-biomass ratio is about 1:1. Phenol residue was measured throughout the test sequence. This test provided high-rate phenol conversion data for comparison to that obtained from low-rate tests performed using the CP-1 MCR. P-1 contained 271 mg/L of phenol-degrading biomass because of the higher phenol loading.

Results

Figure 2 shows the methane production and phenol disappearance profiles for one set of low-rate batch tests. Symbols indicate the measured parameter and lines indicate the model results. Modeling was accomplished using Equations 1 through 4 to produce a non-linear, best-fit of the data. The specific biomass yield was estimated for the acetogenic conversion reaction by using bioenergetics methods. In this case, the yield coefficient was 0.04 mg biomass/mg of phenol converted and the steady-state phenol degrading biomass concentration was 16 mg/L in CP-1 and 271 mg/L in P-1 (Table II).

Table II. Summary of Kinetic Parameters for Phenol Degradation

Parameter		Low-rate Batch Test (CP-1)	High-rate Batch Test (P-1)	Published Data:(Suidan, et al., 1988)
M_o,	mg/L	16	271	130
k_o,	hr^{-1}	0.17	0.17	0.10
K_{so},	mg/L	70	75	60
K_H,	mg/L	480	420	345
Y,	mg VSS/ mg phen.	0.04	0.04	0.04
k_d,	hr^{-1}	0.001	0.001	0.001

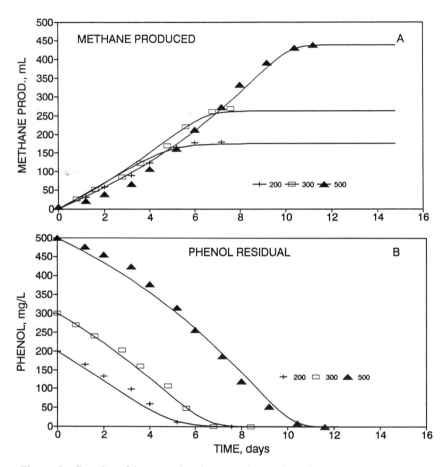

Figure 2. Results of low-rate batch tests when using phenol as a sole carbon source (MCR: CP-1). **A** shows measured (symbols) and modeled methane production (lines) while **B** shows measured (symbols) and modeled (lines) phenol residuals.

Results of high-rate batch tests using cultures transferred from the P-1 (phenol enriched) MCR are shown in Figure 3. The data were modeled using a technique called compositing. In compositing, data collected for individual test concentrations were projected onto the highest concentration data in that set by adding a time increment that was equal to the time difference between the first point in the data set to be projected and the corresponding concentration on the data set to be projected to. For example, in Figure 3A, T1 was added to the measurement time for the 500 mg/L test and T2 was added to the times for the 200 mg/L test to produce the combined data set shown in Figure 3B. This technique has the advantage of revealing outliers in a given test and provides an averaged, combined data set for modeling in the determination of kinetic parameters. Parameters evaluated from composited data sets are less biased by lack of data in sensitive areas of the model application than are parameters evaluated from individual data sets. For example, the data for the 500 mg/L test, shown in Figure 3A, would have been difficult to fit using Equations 1 through 4 because of the lack of measurements between 15 and 25 hours.

The above compositing procedure was applied to phenol conversion data reported by Suidan et al.(35) (Figure 4). In this case, the MCR was operated at a 30-day SRT at a phenol loading of 133.3 mg/L-d. The steady-state acetogenic biomass was estimated as 130 mg/L. The kinetic and inhibition coefficients by this analysis agree reasonably well with those obtained by the tests described above (Table II).

While effective for providing kinetic parameters for low and moderately toxic organics such as phenol, toxicant enriched cultures as represented by P-1 MCR cannot be developed and maintained easily for more severely toxic compounds. These compounds would require the use of low-rate tests using cultures transferred from toxicant-co-substrate enriched reactors as described above for the CP-1 MCR.

Discussion

The above observations indicate that the PROTOCOL methodology can be used for deriving kinetic and inhibition parameters for toxicant degradation reactions. A search of the literature showed that essentially no kinetic parameters are available for biodegradation of toxic organic chemicals under methanogenic conditions. This situation underscores the need for the evaluation of kinetic parameters for toxicant degradation reactions. As pointed out previously, the active biomass fraction selected for modeling toxicant degradation has a major bearing on the kinetic analysis. At present, there are no known microbiological techniques that can be used for directly measuring the biomass responsible for one particular reaction in a mixed culture environment. Many researchers have chosen to use total biomass in their analysis, thus including every bacterial group present in the mixed culture environment. This obviously leads to erroneous determinations of kinetic parameters for specific reactions. In estimating the toxicant degrading biomass for the analysis reported here, we have taken an indirect approach of separating the biomass by using reaction energetics (31,36). Major requirements for the use of this method are the knowledge of complete reaction stoichiometry and free energy for the starting, intermediate, and final products. This information is not available for a large number of toxic organic chemicals, However, a partial listing of yield coefficients for the priority pollutants has been compiled and presented elsewhere (1). While co-substrates other than ethanol can be used -- and may be required in some cases -- this is not expected to affect the test results since the kinetic analysis is based on specific toxicant-degrading biomass.

The parameters derived for describing phenol degradation kinetics can be used for analyzing the performance of full scale treatment processes. For example, the following equation:

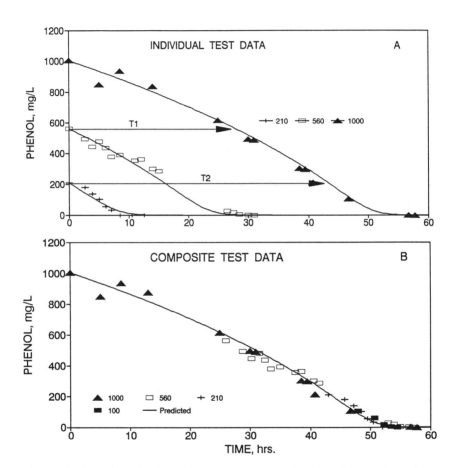

Figure 3. Results of high-rate batch tests when using phenol as a sole carbon source (MCR: P-1). **A** shows measured (symbols) and modeled (lines) phenol residuals for tests using three initial phenol concentrations. **B** shows a composite of the test data from **A**.

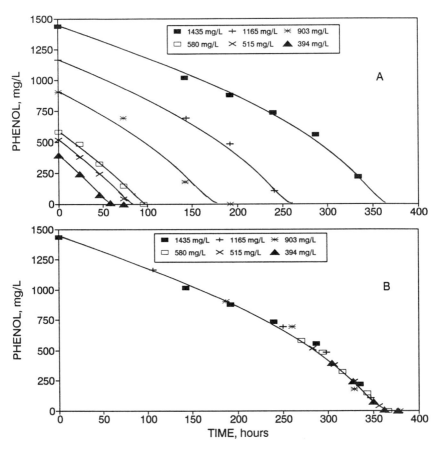

Figure 4. Kinetic analysis using data published by Suidan, et al. (35). **A** shows measured (symbols) and modeled (lines) phenol residuals for tests using three initial phenol concentrations. **B** shows a composite of the test data from **A**.

$$S_{et} = \frac{K_H}{2}\left[\frac{Yk_0SRT}{1+k_dSRT} - 1 - \left[\left(1 - \frac{Yk_0SRT}{1+k_dSRT}\right)^2 - 4\frac{K_{so}}{K_H}\right]^{0.5}\right] \tag{5}$$

relates the toxicant concentration, S_{et}, in contact with the biomass in a continuous flow, completely mixed reactor to the solids retention time and growth and toxicant degradation kinetic parameters (*16*).

In Figure 5, Equation 5 and the kinetic parameters for phenol are used to illustrate the anticipated ability of a digester to degrade toxic organic chemicals as a function of solids retention time and for various values of K_H. Predicted performance under varying influent toxicant concentrations is shown in Figure 6. If the kinetic parameters obtained from tests with laboratory cultures are truly intrinsic, such analyses should help reveal the cause of the differences between actual process performance and estimated performance. For example, adsorption can reduce the effective toxicant concentration, active biomass can be significantly less than indicated by the volatile solids measurement, diffusion limits can change the effective value of K_{so}, and interactions between toxicants and other co-substrates can change the effective value of the inhibition coefficient.

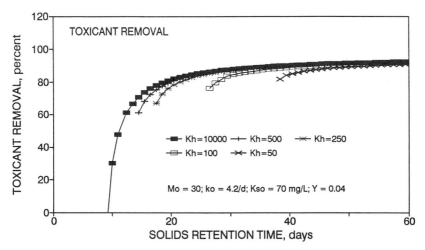

Figure 5. Predicted ability of continuous-flow anaerobic reactors to degrade toxic organic chemicals as a function of solids retention time and for different values of K_H.

Conclusions

The protocol presented above provides an organized and repeatable approach for analyzing the kinetics of biotransformation of toxic organic chemicals in anaerobic treatment processes. The protocol forms a part of a broader multi-level protocol for assessing the fate and effect of toxic organic chemicals in anaerobic processes (*28*). The toxicant degradation protocol presented here requires development of an acclimated culture and is designed to determine the kinetic and substrate inhibition parameters defining the biodegradation of the toxicant. It is based on prior work of a number of investigators of toxicant degradation. The novelty of the protocol is not in the biodegradation test methodology, but in the use of defined culture conditions and a specific reaction biomass in the mathematical modeling so that consistent and repeatable measurements can be made between test laboratories and between analysts.

The protocol needs to be verified for its ability to provide intrinsic measures of the parameters for describing the kinetics of biodegradation of a large number of toxicant organic chemicals. Additional research is especially needed for defining the biodegradation pathways and kinetics for compounds having more than one functional group (such as chloroanilines) and compounds that have two or more functional groups that are transformed sequentially (such as poly-chlorinated phenols). The methods given above are expected to be satisfactory for use with most toxic organic chemicals, but a number of other chemicals may require specialized techniques, specific co-substrates, or specialized culture development.

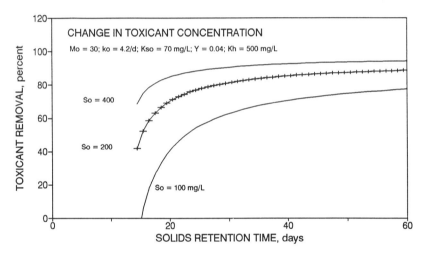

Figure 6. Anticipated behavior of anaerobic continuous-flow reactor under varying influent phenol conditions.

Acknowledgments

The research work described in this paper was supported in part by the U.S. Environmental Protection Agency through Cooperative Agreement R814488, in part by the University of Arkansas, Fayetteville, and in part by The Pennsylvania State University, University Park. This paper has not been subjected to the agency's peer and administrative review and therefore does not necessarily reflect the views of the Agency and no official endorsement should be inferred. The work was conducted while authors Sathish and Young were Graduate Research Assistant and Professor of Civil Engineering, respectively, at the University of Arkansas.

References

1. Sathish, N.; "Kinetics of Biodegradation of Toxic Organic Chemicals in Anaerobic Processes"; Ph.D. Dissertation, University of Arkansas, Fayetteville, AR, 1992.
2. Young, J. C.; Tabak, H. H. "Multi-Level Protocol for Assessing the Fate and Effect of Toxic Organic Chemicals in Anaerobic Reaction"; Final Report for Project R814488, U. S. Environmental Protection Agency, Cincinnati, OH, 1991.
3. Tarvin, D.; Buswell, A. M.; "The Methane Fermentation of Organic Acids and Carbohydrates"; *J. of American Chem. Soc.* **1934**, *56*, 1751-1755.
4. Clark, F. M.; Fina, L. B.; "The Anaerobic Decomposition of Benzoic Acid During Methane Fermentation"; *Arch. Biochem. Biophys.* **1952**, *36*, 26-32.
5. Healy, J. B.; Young, L. Y.; "Catechol and Phenol Degradation by a Methanogenic Population of Bacteria"; *Appl. Environ. Microbiol.* **1978**, *35*, 216-218.
6. Knoll, G.; Winter, J.; "Degradation of Phenol via Carboxylation to Benzoate by a Defined Obligate Syntrophic Consortium of Anaerobic Bacteria"; *Appl. Microb. Biotechnol.* **1989**, *30*, 318-324.
7. Suflita, J. M.; Horowitz, A.; Shelton, D. R.; Tiedje, J. M.; "Dehalogenation: A Novel Pathway for the Anaerobic Biodegradation of Haloaromatic Compounds"; *Science* **1982**, *218*, 1115-1117.
8. Horowitz, A.; Shelton, D. R.; Cornell, C. P.; Tiedje, J. M.; "Anaerobic Degradation of Aromatic Compounds in Sediments and Digested Sludge"; *Dev. Ind. Microbiol.*, **1982**, *23*, 435-444.
9. Boyd, S. A.; Shelton, D. R.; "Anaerobic Biodegradation of Chlorophenols in Fresh and Acclimated Sludge"; *Appl. Environ. Microbiol.* **1984**, *47*, 272-277.
10. Kohring, G-W. et al.; "Anaerobic Biodegradation of 2,4-Dichlorophenol in Fresh Water Lake Sediments at Different Temperatures"; *Appl. Env. Microb.* **1990**, *55*, 348-356.
11. Struijs, J. and Rogers, J. E.; "Reductive Dehalogenation of Dichloroanilines by Anaerobic Microorganisms in Fresh and Dichlorophenol-Acclimated Pond Sediment"; *Appl. Environ. Microbiol.* **1989**, *52*, 2527-2531.
12. Schnell, S., et al.; "Anaerobic Degradation of Aniline and Dihydroxybenzenes by Newly Isolated Sulfate Reducing Bacteria and Description of *Desulfobacterium Anilini*"; *Arch. Microbiol.* **1989**, *152*, 556-561.
13. Quinsen, J. F., III; Tiedje, J. M.; Boyd, S. A.; "Reductive Dechlorination of Polychlorinated Biphenyls by Anaerobic Microorganisms from Sediments"; *Science* **1988**, *242*, 752-754.
14. Andrews, J. F.; "Modeling of Biological Treatment Processes"; *Biotech. Bioeng.*, **1968**, *10*, 707-723.
15. Edwards, V. H.; "The Influence of High Substrate Concentrations on Microbial Kinetics"; *Biotech. Bioeng.*, **1970**, *12*, 679-712.
16. Gaudy, A. F., Jr.; Ekambaram, A.; Rozich, A. F.; "Practical Methodology for Predicting Critical Operating Range of Biological Systems Treating Inhibitory Substances"; *J. Water Poll. Control Fed.*, **1988**, *60*, 77-85.
17. Lewandowski, Z.; "Denitrification in Packed-Bed Reactors in the Presence of Chromium (VI)"; *Water Res.* **1985**, *21*, 147-153.
18. Brown, S. C.; Grady, C. P. L., Jr.; Tabak, H. H.; "Biodegradation & Kinetics of Substituted Phenols"; *Water Res.* **1990**, *24*, 853-861.
19. Haldane, J. B. S. *Enzymes*; MIT Press: Cambridge, MA, 1930.

20. Han, K.; Levenspiel, O.; "Extended Monod Kinetics or Substrate, Product, and Cell Inhibition"; *Biotech. Bioeng.* **1988**, *32*, 430-437.
21. Owen, W. F.; Stuckey, D. C.; Healy, J. B., Jr.; Young, L. Y.; McCarty, P. L.; "Bioassay for Monitoring Biochemical Methane Potential and Anaerobic Toxicity"; *Water Res.* **1979**, *13*, 485-492.
22. Parkin, G. F.; Speece, R. E.; "Modeling Toxicity in Methane Fermentation Systems"; *J. of Env. Eng.* **1982**, *108*, 515-531.
23. Shelton, D. R.; Tiedje, J. M.; "General Method for Determining Anaerobic Biodegradation Potential"; *Appl. and Env. Microbiol.* **1984**, *47*, 850-857.
24. American Society for Testing and Materials; "Standard Methods for Determining the Anaerobic Biodegradation Potential of Organic Chemicals"; *ASTM Method E 1196-7*; ASTM, Philadelphia, PA, 1987.
25. Kirsch, E. J.; Grady, C. P. L., Jr.; Wukash, R. F.; Tabak, H. H.; "Protocol Development for the Prediction of Fate of Priority Pollutants in Biological Wastewater Treatment Systems";   Report No. EPA-600/2-85/141, U.S. Environmental Protection Agency, Cincinnati, OH. **1986**.
26. Guthrie, M. A.; Kirsch, E. J.; Wukash, R. F.; Grady, C. P. L., Jr.; "Pentachlorophenol Biodegradation"; *Water Res.* **1984**, *18*, 451-461.
27. Nelson, M. A.; "Effect of Mixtures of Toxic Organic Chemicals on Anaerobic Treatment Processes"; Ph.D. Dissertation, University of Arkansas, Fayetteville, 1991.
28. Young, J. C. and Tabak, H.; "Multi-Level Protocol for Assessing the Fate and Effect of Toxic Organic Chemicals in Anaerobic Treatment Processes," In *Proc. Research Symp. 64th Annual Conference of the Water Pollution Control Federation*: Water Pollution Control Federation, Washington, D. C., 1991.
29. *Standard Methods for the Examination of Water and Wastewater*; 17th Ed.; American Public Health Assoc., Washington, D.C., 1979.
30. Monod, J.; "The Growth of Bacterial Cultures"; *Ann. Rev. Microbiol.* **1949**, *III*, 371-394.
31. McCarty, P. L.; "Bioenergetics and Bacterial Growth"; In *Anaerobic Biological Treatment Processes*, Gould, R. F., Ed., Advances in Chemistry Series 105, American Chem. Soc.: Washington, D.C. 1971.
32. Grady, C. P. L., Jr.; "Biodegradation of Toxic Organics: Status and Potential"; *J. of Environ. Eng.* **1990**, *116*, 805-828.
33. Young, J. C.; Kuss, M. L.; Nelson, M.; "Use of Anaerobic Respirometers for Measuring Gas Production in Toxicity and Treatability Tests"; In *Proc. 84th Annual Meeting of the Air and Waste Management Association*, Vancouver, B. C., Canada, 1991.
34. Henderson, J. E.; Peyton, G. R.; Glaze, W. H.; In *Identification and Analysis of Organic Pollutants in Water*, Keith, L. H., Ed.; Ann Arbor Science: Ann Arbor, MI, **1976**, pp. 105-111.
35. Suidan, M. T.; Najm, I. N.; Pfeffer, J. T.; Wang, Y. T.; "Anaerobic Biodegradation of Phenol; Inhibition Kinetics and System Stability"; *J. Env. Eng.* **1988**, *114*, 1359-1376.
36. Thauer, R. K.; Jungermann, K.; Decker, K.; "Energy Conservation in Chemotrophic Anaerobic Bacteria"; *Bacteriological Reviews*, **1977**, *41*, 100-180.

RECEIVED October 2, 1992

Chapter 12

Anaerobic Digestion of Industrial Activated Aerobic Sludge

J. C. Goodloe[1], H. M. Kitsos[2,4], A. J. Meyers, Jr.[3], J. S. Hubbard[1],
and R. S. Roberts[2]

[1]School of Biology and [2]School of Chemical Engineering, Georgia Institute
of Technology, Atlanta, GA 30332
[3]Tennessee Eastman Company, Kingsport, TN 37662

Two 750 milliliter (ml) anaerobic reactors (CSTR and percolating
bed) with a chemical industry wastewater sludge feed (at 7% total
solids) gave a volatile solids (VS) reduction of about 35% during
operation at a 6 day hydraulic retention time (HRT). The
biomass-retaining design of the percolating bed reactor
demonstrated no performance advantage over the CSTR and
accumulating solids in the percolating bed reactor eventually
plugged the bead bed. Poor mixing in both reactors (by single
flat-bladed paddles) caused a foam problem. Foam was not a
problem in a three liter conventional CSTR begun later (mixed by
two six-bladed turbines). VS reduction was also about 35% in the
three liter reactor (10 day HRT). However, an increase in free
ammonia concentrations seemed to be the cause of reduced biogas
production (ml biogas/g VS added).

Disposing of excess sludge from aerobic wastewater treatment facilities is a
growing problem in the chemical industry. These activated sludges are often
classified as hazardous wastes because the chemical wastewaters contain
residual amounts of hazardous chemicals, such as phenols. Sometimes sludges
contain high levels of heavy metals due to the sequestering and concentrating of
metals by many bacteria. These activated sludges are typically dewatered,
incinerated, and the resulting ash placed in hazardous waste landfills. Prior
treatment of the activated sludge by anaerobic digestion reduces the quantity of
sludge requiring final disposal. In addition, anaerobic digestion produces biogas
of high methane content which can be used to fuel conventional boilers.

[4]Current address: National Starch and Chemical Company, 1700 West Front Street,
Plainsfield, NJ 07063

0097–6156/93/0518–0219$06.00/0

Anaerobic digestion involves the complex interaction of several distinct bacterial groups acting through a series of metabolic steps (1-4):
1) hydrolysis of complex organic compounds such as carbohydrates, proteins and lipids to produce sugars, peptides, acetate, and H_2 and CO_2,
2) acidogenesis of these simple organic compounds to produce fatty acids such as propionate and butyrate,
3) acetogenesis to produce acetate, H_2, and CO_2 from fatty acids and also from H_2 and CO_2, and
4) methanogenesis to produce CH_4 from acetate and from H_2 and CO_2.

In the conventional anaerobic process these metabolic steps take place in a single reactor. Alternatively, Babbit and Baumann (5) suggested that separation of the hydrolytic and acidogenic steps from the acetogenic and methanogenic steps would produce more optimized reactor conditions for specific bacterial groups. Several researchers have used filter membranes to subdivide the anaerobic reactor (6-7). Pohland and Ghosh (8) suggested that increasing the dilution rate to washout the slower-growing methanogens from the acid-phase reactor would be an easier way to maintain metabolic separation. Such systems (termed "two-phase" for the separate acid phase and methanogenic phase reactors) have become popular at the laboratory and pilot-plant scale for the treatment of wastes with high solids content. These anaerobic reactor systems are reported to have increased process stability while operating at shorter hydraulic retention times (HRT's) than comparable single stage systems (9-11).

Anaerobic digestion of biological solids (such as cellulosic biomass or the bacterial cells in activated sludge) requires an initial solubilization step. This hydrolysis step is often the rate-limiting step in the production of methane from such materials (12-13). A study on the anaerobic digestion of activated sludge grown on synthetic feed revealed that the rate-limiting step was the hydrolysis of particulate protein rather than cell lysis of the activated sludge bacteria (14). However, for anaerobic digesters operating on high strength, soluble feed such as carbohydrates the rate-limiting step is the conversion of acetate to methane (15).

The specific aim of this research was to assess the effectiveness of anaerobic digestion in converting industrial wastewater sludge solids into biogas. Key parameters monitored (which affect the costs of such anaerobic treatment) included the volatile solids reduction achievable, the amount of biogas generated, the methane content of the biogas, and the HRT required. Experiments examined the effect of different pH levels and different initial solids levels on anaerobic digestion. In addition, the performance of a reactor design which retained biomass (percolating bed reactor) was compared to the performance of a continuously stirred tank reactor (CSTR).

Materials and Methods

Enrichment Cultures and Reactors.   Several different feeds were used for the experiments:
1) Settled activated sludge from a municipal wastewater facility, a new batch was taken at intervals of several weeks,
2) Tennessee Eastman activated sludge (TE sludge), batch 1 (45 gallons), and
3) TE sludge, batch 2 (20 gallons).

The settled municipal activated sludge was stored at 4°C and fed as a slurry. The TE sludges were divided into aliquots and frozen at -20°C. Aliquots of TE sludge batch 1 were fed both as a thawed slurry (to enrichment cultures and the three liter reactor) and also lyophilized and fed as reconstituted sludge. All aliquots of TE sludge batch 2 were lyophilized and fed as reconstituted sludge. After lyophilization several sludge aliquots were combined and then held at 4°C until use. This feed supply was replenished occasionally in order to minimize possible variation between feed aliquots. Lyophilized TE sludge was the feed for the 750 milliliter (ml) reactors, the three liter reactor and for the batch cultures. The use of lyophilized feed allowed accurate measurement of the feed solids (to 0.01 g) and enabled more precise solids monitoring.

The enrichment cultures, the 750 ml reactors, and the batch experiments were incubated in a 37°C warm room. The three liter reactor setup included an integral 37°C waterbath.

Enrichment Cultures at Different pH Levels. The pH 5.2, 6.3, and 7.1 enrichment cultures were two liter liquid volume in 2,800 ml Fernbach flasks. The medium (per 1,600 ml) contained 0.35 g $MgCl_2$, 0.43 g NaCl, 0.43 g NH_4Cl, 0.06 g $CaCl_2$, and 1.5 ml trace metals solution (*16*). Sodium phosphate (NaH_2PO_4-H_2O/Na_2HPO_4-$7H_2O$) was added in the following quantities for the pH 7.1, 6.3, and 5.2 cultures: 10.38 g/29.10 g, 46.56 g/12.76 g, and 46.56 g/6.92 g, respectively. These phosphate ratios gave the initial pH for the pH 6.3 and pH 7.1 cultures. Concentrated H_2SO4 was added to adjust the initial pH of the pH 5.2 culture.

The media was gas stripped with N_2 for approximately 2 hours prior to the addition of the sludge inoculum. The inoculum (400 ml per culture) was a 4:1 mixture of effluent from an anaerobic reactor operating on simulated alcohol stillage wastewater and anaerobic sludge from the South River Wastewater Treatment Facility, Atlanta, Georgia.

The enrichment cultures were initially fed settled activated sludge. TE sludge, batch 1, was fed after day 88. These cultures were fed once per day and operated for 155 days. Feed concentration was approximately 10% total solids. Concentrated sulfuric acid was added daily to maintain the initial pH in the pH 6.3 and pH 5.2 enrichment cultures (1.0 ml and 2.0 ml, respectively). The pH 7.1 enrichment culture required only infrequent acid addition.

These cultures were initially shaken at about 90 rpm (revolutions per minute) on a gyratory shaker. The shaking was discontinued on day 33 because foam problems resulted which hindered mixing.

The enrichment cultures were monitored daily for pH and the quantity of gas evolved. The volatile fatty acid (VFA) concentrations (acetate, propionate, isobutyrate, and butyrate) were also determined daily. Biogas composition and ammonia concentration were determined periodically.

Batch Cultures at Different Initial Solids Concentration. The batch culture experiments used 160 ml Wheaton serum bottles with 100 ml liquid volume. The bottles were sealed with butyl rubber stoppers secured with aluminum crimp-seal closures. The Tygon tubing gas exit lines were attached to 18 gauge needles which pierced the rubber stoppers.

The basal salts medium used was modified from McInerney et al (*17*). Composition (per liter) was 50 ml Pfennig's mineral solution, 1 ml Pfennig's metal solution, 10 ml trace vitamin solution (from Table 1, Standard Media, (*16*)), and 3.5 g $NaHCO_3$. Although anaerobic techniques were not used, the media was autoclaved to drive off oxygen before dispensing. Lyophilized TE sludge (batch 2) was added at levels of 3%, 7%, and 10% total solids (including inoculum solids).

The sludge feed and basal media (80 ml/bottle) were gas stripped sequentially with H_2, H_2/CO_2 (20%/80%), and N_2/CO_2 (20%/80%) prior to inoculum addition. The inoculum per bottle was 20 ml anaerobic sludge (from a reactor operating at a 10 day HRT on TE sludge feed). After inoculum addition the culture was again gas stripped with N_2/CO_2. Then 0.05 g cysteine-HCl (reducing agent) was added per bottle. The bottle headspace gas was then flushed with H_2, H_2/CO_2, and N_2/CO_2.

Triplicate cultures were used for each solids concentration. Gas production was monitored daily for 2 of these 3 cultures. Final gas composition was also determined. Control cultures of sludge inoculum without feed were run in duplicate. The experiment ran for 31 days, then all cultures were analyzed for total solids, volatile solids, and ammonia concentration.

750 ml Anaerobic Reactors. Each 750 ml anaerobic reactor was a glass cylinder, 40 cm x 10 cm diameter (dia), with a 10.5 cm conical base which contained 3 mm (dia) glass beads (Figure 1). Total volume of each reactor (without glass beads) was about 3.0 liters. Mixing in each reactor was by a single flat-bladed paddle, driven by a variable speed motor (0-300 rpm, typically operated at approximately 90 rpm). The reactors could be operated as either CSTR's (when aliquots were removed from the side port) or as percolating bed reactors (when aliquots were removed from the bottom port, below the bead bed). Two 750 ml anaerobic reactors (at pH 6.3 and pH 7.1) were established with aliquots of the corresponding enrichment cultures. The reactors were initially operated as two continuously stirred tank reactors (CSTR's) in series, with the pH 6.3 reactor as stage 1.

Feed for the 750 ml reactors was reconstituted TE sludge, batch 1. The reactors were typically fed twice a day. During operation as two CSTR's in series, the stage 1 feed concentration was always 7% total solids (loading varied only by volume fed per day). Stage 2 was fed only stage 1 effluent. Later, when each reactor was operated independently, the feed concentration remained at 7% total solids.

Solids concentrations in these reactors and gas production were monitored daily. VFA concentrations were determined several times a week. Biogas composition and ammonia concentration was monitored periodically.

Three Liter Anaerobic Reactor. A conventional CSTR (New Brunswick Scientific) was established in order to test a reactor design with better mixing than in the 750 ml reactors. The reactor dimensions were 46 cm x 15.2 cm (dia), with a total volume of 7.05 liters. Mixing was by two 6-flat bladed turbines which operated at about 300 rpm. The three liter liquid volume reactor was established using major portions of the two 750 ml reactors. To increase the liquid volume to three liters the reactor was fed for about two weeks with little or no sample removal. After this operation began at a 10 day HRT.

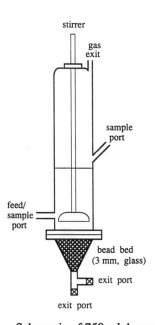

Figure 1.   Schematic of 750 ml Anaerobic Reactor.

Feed to the three liter reactor during the first several months of operation was a thawed slurry from TE sludge, batch 1 (typically at 8% to 10% total solids). Later the feed was changed to lyophilized TE sludge, batch 2, at 7% total solids.

Analytical Methods

Gas Chromatography. Volatile fatty acids (VFA's) were analyzed by gas chromatography (GC) using either a Poropak Q column (glass, 5' x 1/8" ID) or a Carbopack B-CA / 4% Carbowax 20M column (glass, 6' x 2 mm ID) and a flame ionization detector. Both columns used helium carrier gas (flow rate of 60 ml/min.) and were operated at 175°C. The injector and detector zones were both at 200°C for the Carbopack/Carbowax column and at 220°C and 230°C, respectively, for the Poropak Q column.

Reactor samples for GC analysis (1.5 ml) were centrifuged for 10 minutes at 13,000 rpm in a Eppendorf microcentrifuge. Before analysis on the Carbopack/Carbowax column the supernatants were adjusted to 30 mM oxalic acid with a 1 M stock solution. For the Poropak Q column, supernatants were acidified with 0.05 ml of concentrated H_2SO_4 and recentrifuged prior to injection. The Carbopack/Carbowax column resolved VFA's through valeric acid, with detection to the part per million level (reported to 0.01 mM). The Poropak Q column resolved VFA's through butyric acid, with detection to about the 0.5 mM level.

Methane and CO_2 concentrations in the biogas were determined by gas chromatography on a Poropak Q column (stainless steel, 5' x 1/8" ID) using a thermal conductivity detector and helium carrier gas. The column, injector, and detector were operated at room temperature.

Solids Determinations. Total solids (TS) and volatile solids (VS) were determined according to Standard Methods (18). Reactor samples (10 ml) were placed overnight at 105°C for total solids determination and then at 550°C for 30 minutes to determine inert solids. Volatile solids is the total solids sample weight minus the inert solids sample weight.

Determination of Ammonia Concentration. Ammonia concentrations were determined with an ion-specific probe (Corning). A standard curve ranging from 1 to 1,000 mg/l was run for each analysis. Reactor samples were diluted so that the concentration ranged from 400 to 600 mg/l.

Results and Discussion

Effect of pH on Anaerobic Digestion. To examine the effect of pH on anaerobic digestion, two liter enrichment cultures were begun and maintained at pH 5.2, 6.3, and 7.1. These cultures originally operated at a 20 day HRT and were reduced to a 10 day HRT on day 15. Feeding was stopped on day 24 due to rapidly increasing VFA levels. Feeding was resumed eight days later. A 15 day HRT was established on day 36 and used for the remainder of the 155 day experiment.

The composition of the biogas produced by the enrichment cultures varied with pH. Methane content was 68%, 61%, and 52% for the pH 7.1, pH 6.3 and pH 5.2 cultures, respectively. These differences were due to reduced methanogenic activity and lower CO_2 solubility at the more acidic pH levels.

The VFA levels for the pH 7.1 and pH 6.3 enrichment cultures are shown in Figures 2 and 3, respectively. The high propionate concentrations in the pH 6.3 culture while operating at the 10 day HRT indicated that the increased loading had a much greater effect on the pH 6.3 culture than on the pH 7.1 enrichment culture. Typical VFA levels in both the pH 6.3 and pH 7.1 enrichment cultures were equal to or less than 1.5 mM acetate and 0.5 mM propionate after the 15 day HRT was established. The low VFA levels indicated that operation at much shorter HRT's should be possible while maintaining similar VS reductions.

During the first two weeks of operation, biogas production from the pH 5.2 enrichment culture was only about 20 to 30 percent of that from the pH 6.3 and pH 7.1 cultures. Because of the reduced substrate conversion (as shown by the low gas production), the soluble chemical oxygen demand (COD) level in the pH 5.2 culture was five times the soluble COD level in the other cultures. This reactor showed very poor stability in comparison to the pH 6.3 and pH 7.1 cultures (stability was important as these cultures would be used later to start larger reactors). Operation of the pH 5.2 enrichment culture was discontinued on day 60 due to high VFA levels (13 mM propionate) and low gas production. Also, the cost for acid addition required by the pH 5.2 culture would have been prohibitive in any large scale system.

It has been reported that acid phase digesters should not operate at acid pH (*19-20*). Both these studies showed that hydrolysis was more effective at neutral pH levels (pH 7.0 and pH 6.8, respectively) and higher temperature (55°C and 50°C, respectively). Alternatively, an acid phase digester which operated at pH 5.6 to 5.9 but required no extra acid addition was recently reported (*21*). Perhaps a neutral pH, thermophilic acid phase digester or an acid phase digester requiring no acid addition for pH adjustment can be developed with the TE sludge.

The volatile solids (VS) reductions achieved in the pH 6.3 and pH 7.1 enrichment cultures are shown in Figure 4. The VS reduction was about 20% in both cultures after 30 days. By day 105 the VS reduction reached 35 % in both cultures and it remained at or exceeded this level for the rest of the experiment.

The equivalent ammonia concentrations for the pH 6.3 and pH 7.1 enrichment cultures are shown in Figure 5. From initial levels of about 400 mg/l, the ammonia concentrations in both cultures increased to about 1,400 mg/l by day 60. TE sludge feed was used after day 87. The equivalent ammonia levels in the pH 6.3 and pH 7.1 enrichment cultures ranged from 1,450 to 1,550 mg/l after about day 90. These cultures exhibited no apparent ammonia toxicity.

Free dissolved ammonia (NH_3) has been shown to be much more toxic to the microbial populations than the ammonium ion (NH_4^+)(*22*). At 35°C and pH 7.0 the free ammonia represents only about 1.1% of the total ammonia concentration. Typical reported values for the onset of ammonia inhibition include free ammonia concentrations of about 55 mg/l (*23*) and 80 to 100 mg/l (*24*). The equivalent ammonia concentrations in the pH 6.3 and pH 7.1 enrichment cultures corresponded to free ammonia concentrations of 25 mg/l or less.

Effect of Initial Solids Concentration on Anaerobic Digestion. The effect of initial solids levels on anaerobic digestion was studied with 100 ml batch cultures. Gas production was monitored daily, until gas production ceased. Total solids, volatile solids, and equivalent ammonia concentrations were determined after the experiment ended on day 31. Figure 6 shows the gas production from these "ultimate degradation" cultures. The total biogas produced averaged 227

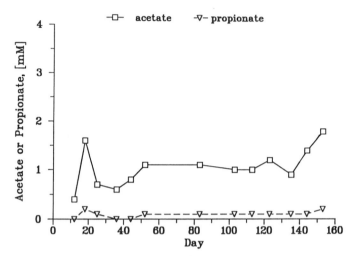

Figure 2 .   Acetate and Propionate Concentrations, pH 7.1 Enrichment
Culture.

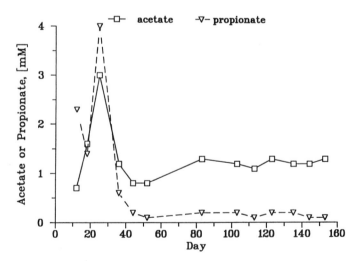

Figure 3.   Acetate and Propionate Concentrations, pH 6.3 Enrichment
Culture.

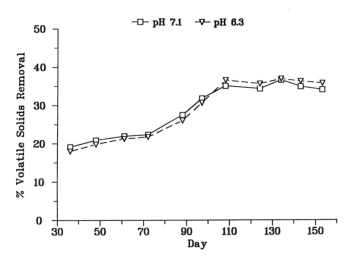

Figure 4. Volatile Solids Reduction, pH 7.1 and pH 6.3 Enrichment Cultures.

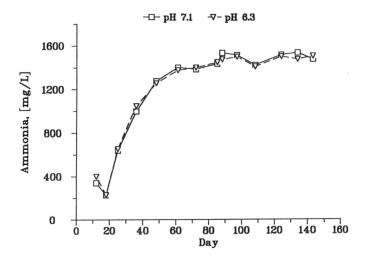

Figure 5. Equivalent Ammonia Concentrations, pH 7.1 and pH 6.3 Enrichment Cultures.

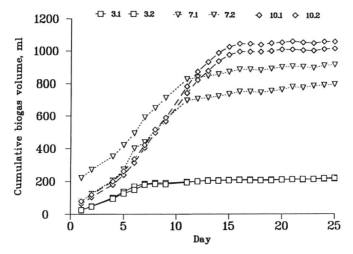

Figure 6.   Total Biogas Production, Batch Cultures at Different Initial
Solids Concentrations.

ml for the 3% solids cultures, 887 ml for the 7% solids cultures and 1,045 ml for the 10% initial solids cultures.   The volume of methane produced per gram of volatile solids added averaged 77 ml, 123 ml, and 94 ml for the 3%, 7%, and 10% initial solids cultures, respectively.   On the basis of volatile solids added, the 7% initial solids culture produced 60% more methane than the 3% solids culture and 31% more methane than the 10% solids culture.   Final equivalent ammonia concentrations averaged 1,200 mg/l for the 3% solids cultures, about 2,650 mg/l for the 7% solids cultures, and 3,500 mg/l for the 10% initial solids cultures.

The VS reductions averaged about 26.0%, 32.3%, and 33.9% for the 3%, 7%, and 10% initial solids cultures, respectively.   Although the VS reduction was slightly better in the 10% than in the 7% initial solids cultures, the total biogas production revealed the impact of the higher ammonia concentration in the 10% solids cultures: total biogas produced/ total VS was about 150 ml/g in the 7% cultures and only about 123 ml/g in the 10% initial solids cultures.   Additional experiments should be done using other intermediate initial solids levels, but it is clear that 3% initial solids is too dilute for maximum methane production.   The high equivalent ammonia levels in the 10% initial solids cultures showed that ammonia toxicity was the probable reason for reduced total biogas and methane production.

Operation of Two 750 ml CSTR's in Series.   The 750 ml reactors begun from the pH 6.3 and pH 7.1 enrichment cultures were operated as two CSTR's in series for 90 days.   The initial HRT was 10 days for the combined reactor system.   The first stage HRT, however, varied from a low of 4.4 days to a high of 15 days during series operation.   Although steady state conditions were not reached, stage 1 operation was stable at a 5.8 day HRT for 16 days before the loading was reduced due to an operator scheduling problem.   Later, the reduction to a 4.6 day HRT quickly resulted in high propionate levels.   Washout of either the methanogenic bacteria and/or the syntrophic propionate-degrading bacteria must have begun when operating at this low HRT.

Stage 2 was consistently underloaded throughout series operation.   This reactor showed essentially no propionate and significant acetate concentrations (5 to 15 mM) occurred only when reactor stage 1 was overloaded.   Stage 2 reactor required substantially increased loading.   If series operation were continued, the loading to stage 1 would have been increased from the 7% solids level in order to increase the VFA levels in the stage 1 effluent which stage 2 received.   Rather than continue operating two CSTR's in series, we decided to test performance of the CSTR reactor design versus the percolating bed reactor.

Two 750 ml Reactors: Independent Operation, Percolating Bed vs CSTR. The two 750 ml reactors were operated independently for more than 160 days after ceasing operation as two CSTR's in series.   Operation of the former stage 1 reactor was partially shifted to a percolating bed reactor mode on day 10 by taking two of three daily samples from the bottom rather than the side port.   The bottom port sampling was increased to 12 of 14 weekly aliquots on day 44 (thus the reactor functioned as 86% percolating bed and 14% CSTR).   The need to sample the reactor directly to monitor the solids concentration prevented operation as a 100% percolating bed reactor.   Reactor 1 (86% percolating bed) operated initially at a 7.3 day HRT, but foam problems and high propionate concentrations forced

HRT increases. The former stage 2 reactor continued to operate as a CSTR, but with gradually decreasing HRT.

Reactor 1 (86% percolating bed) began operation at a 6.1 HRT on day 108 and reactor 2 (CSTR) began operation at the same HRT on day 118. Both reactors remained at the 6.1 day HRT through day 148. Typical reactor 1 (86% percolating bed) VFA levels at this time were 3 to 5 mM acetate and 1 mM propionate. Reactor 2 (CSTR) VFA concentrations were substantially lower - only 0.5 to 0.7 mM acetate and 0.1 mM propionate.

Total solids concentration in reactor 1 (86% percolating bed) over the 160 day period are shown in Figure 7. Data from both side sampling (CSTR mode) and bottom sampling (percolating bed mode) are shown. There was a lot of variability in both sets of samples, but the solids levels in the bottom samples were usually 20 to 25% lower than side samples from about the same time (two samples taken at the same time required too large a volume for proper sampling and analysis). The difference in total solids between the side and bottom samples indicates that solids were retained by the bead bed - i.e., the reactor functioned as a percolating bed reactor.

Volatile solids (VS) reductions in reactor 1 (86% percolating bed) and reactor 2 (CSTR) are shown in Figures 8 and 9, respectively. During the 6.1 day HRT operation (day 108 to 148) of operation the reactor 1, VS reduction ranged from about 23 to 42%, with an average of about 35%. Reactor 2 (CSTR) VS reduction over the 118 to 148 day period at the 6.1 day HRT varied from 32 to 41% and averaged about 37%. The solids reduction values for reactor 1 (86% percolating bed) showed much greater fluctuation than the reactor 2 (CSTR) values. The variability in solids content of samples removed from the bottom of the percolating bed reactor resulted in variability in the solids which remained in the reactor.

Figure 10 shows volatile solids (VS) reduction versus the steady state hydraulic retention time (HRT) for both reactors. Because there was no steady state data for reactor 1 (86% percolating bed) at HRT's of more than 7.7 days, most of the data was from reactor 2 (CSTR) operation. The VS reduction at 6.1 days was very similar for both reactors, but there was much more variability with the percolating bed reactor than with the CSTR. Under our operating conditions there was no advantage to a percolating bed reactor design over a CSTR design. A more rigorous comparison could have been made if the anaerobic reactor cultures had been combined after series operation, redivided into new reactors and operated initially at a high HRT with gradual reductions only after steady state operation. Regardless of VS reduction, however, the percolating bed reactor had problems with solids buildup. These accumulating solids eventually plugged the bead bed and caused shutdown of the percolating bed reactor. Severe plugging problems due to the particulate nature of the feed would probably also have occurred with any type of packed bed reactor.

Three liter reactor. Foam problems occurred in the two liter enrichment cultures and with higher solids loading in both the 750 ml reactors (percolating bed and CSTR). The liquid level in the 750 ml reactors was typically about 11.5 cm high. When operating at an HRT of 6.1 days or less, the foam layer was 10 cm or more thick with both reactor types. Although the three liter reactor was operated at a longer HRT (10 days rather than 6 days), the feed concentration with the thawed slurry feed was often about 10% total solids. A large foam layer

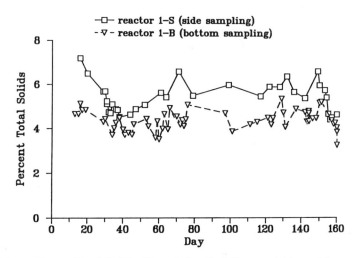

Figure 7. Total Solids, Percolating Bed Reactor (side and bottom sampling).

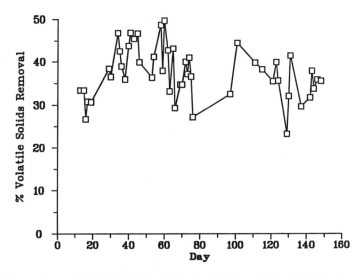

Figure 8. Volatile Solids Reduction in Percolating Bed Reactor.

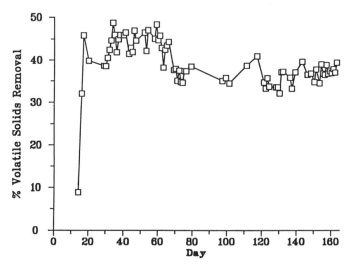

Figure 9.   Volatile Solids Reduction in CSTR.

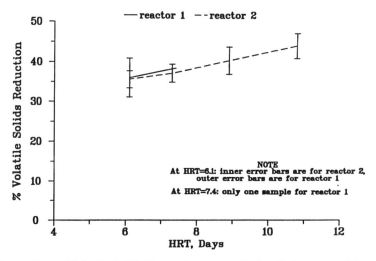

Figure 10.   Volatile Solids Reduction versus Hydraulic Retention Time, Percolating Bed Reactor and CSTR.

(15 cm thick) initially formed in the 3 liter reactor, but this layer decreased to about 2 cm after several weeks of operation. The foam problem thus seemed to have been a mixing problem rather than an inherent, unavoidable property of the TE sludge feed.

The equivalent ammonia concentrations in the three liter reactor ranged from about 2,000 mg/l to 2,150 mg/l. At a reactor pH of 7.45, the higher ammonia concentration corresponds to 75 mg/l of free ammonia. The three liter reactor showed VS reductions of about 35% while in stable operation at a 10 day HRT. The biogas produced had a methane content of 68% to 70%, as with the 750 ml reactors. However, on the basis of either total biogas volume or methane volume produced per gram VS added, the 3 liter reactor showed about a 20% decrease from the 750 ml reactor production levels. The 3 liter reactor produced about 175 ml biogas/g VS added compared to 215 ml biogas/g VS added for the 750 ml reactors. Other changes between the 750 ml reactors and the 3 liter reactor include different feed batches and the different prior histories for the bacterial populations. However, the increase in free ammonia levels with the three liter reactors seemed to be a more probable reason for the decrease in gas production. There have been several reports of reduced gas production from anaerobic digesters subjected to increased ammonia concentrations (25-27).

Conclusions

The increasing VFA concentrations (especially propionate) which occurred in the reactors as the HRT became too low indicates that polymer hydrolysis was not the rate-limiting step in the overall anaerobic conversion of the feed solids. Our practice of freezing TE sludge at -20°C may lyse microbial cells and thus constitute a pretreatment step. Longer hydrolysis times could be expected with non-frozen activated sludge feed.

The pH 5.2 enrichment culture showed that operation of a low pH hydrolytic reactor was impractical with the Tennessee Eastman activated sludge feedstock. The pH 6.3 and pH 7.1 enrichment cultures indicated that a VS reduction of about 35% was possible at an HRT of 15 days or less. The very low VFA levels in these cultures showed that stable operation at substantially greater loading levels was possible.

The VFA concentrations in stage 1 of the two 750 ml CSTR's operating in series indicated that the minimum HRT for stable operation was probably more than 4.6 days but less than 6.8 days. During independent operation both the 86% percolating bed reactor and the CSTR were capable of about 35% VS reduction while operating at about a 6 day HRT.

The three liter CSTR, with mixing greatly improved over the 750 ml reactors, demonstrated that the foam problem was due to inadequate mixing. Maximum equivalent ammonia concentrations increased to about 2,150 mg/l in this reactor. Along with the increase in pH to about pH 7.45, this corresponds to a free ammonia level of about 75 mg/l, a 3-fold increase over the free ammonia levels in the enrichment cultures. The three liter reactor also showed about a 20% decrease in the gas volume produced per gram VS added in comparison with the 750 ml reactor. The increased free ammonia levels probably caused the decreased gas production.

Acknowledgments

This research was funded by grants from the Department of the Interior, U.S. Geological Survey, through the Georgia Water Resources Research Institute (14-08-0001-G1556), the Tennessee Eastman Company (AGR DTD 890317) and the Georgia Tech Foundation.

Literature Cited

1. Novaes, R.F.V.; "Microbiology of Anaerobic Digestion"; *Water Sci. Tech.* **1986**, 18, pp. 1-14.
2. Mah, R.A.; Ward, D.M.; Baresi, L.; Glass, T.L.; "Biogenesis of Methane"; Starr, M.P.; Ingraham, J.L.; Balows, A., Eds.; Ann. Rev. Microbiol.; Annual Reviews: Palo Alto, CA, 1977, Vol. 31; pp. 309-341.
3. Wolfe, R.S.; Higgins, I.J.; "Microbial Biochemistry of Methane"; In *Microbial Biochemistry*; Quayle, J.R., Ed.; Vol. 21; University Park Press: Baltimore, MD, 1979, p. 270-300.
4. Wilkie, A.; Colleran, E.; "Microbiological Aspects of Anaerobic Digestion"; In Anaerobic Treatment of Industrial Wastewaters; Torpy, M.F., Ed.; Noyes Data Corporation: Park Ridge, NJ, 1988; pp. 35-48.
5. Babbit, H.E.; Baumann, E.R. Sewerage and Sewage Treatment, 8th ed.; John Wiley and Sons: New York, N.Y., 1958; p. 589.
6. Schaumburg, F.D.; Kirsch, E.J.; "Anaerobic Simulated Mixed Culture System"; *Appl. Microbiol.* **1966**, 14, pp. 761-766.
7. Hammer, M.S.; Borchardt, J.A.; "Dialysis Separation of Sewage Sludge Digestion"; *Proc. Am. Soc. Civil Eng.* 95 SA5 **1969**, p. 907.
8. Pohland, F.G.; Ghosh, S.; "Developments in Anaerobic Stabilization of Organic Wastes - The Two-Phase Concept"; *Environ. Lett.* **1971**, 1, pp. 255-266.
9. Ghosh, S.; Conrad, J.R.; Klass, D.L.; "Anaerobic Acidogenesis of Sewage Sludge"; *J. Water Pollution Control Fed.* **1975**, 47, pp. 30-45.
10. Ghosh, S.; "Comparative Studies of Temperature Effects on Single-Stage and Two-Phase Anaerobic Digestion"; *Biotechnology and Bioengineering Symposium;* Scott, C.D., Ed.; No. 17, (Proc. 8th Symp. Biotech. Fuels and Chemicals, May 13-16, 1986, Gatlinburg, TN); John Wiley and Sons: New York, NY, 1986, p. 365.
11. Ghosh, S.; "Improved Sludge Gasification by Two-Phase Anaerobic Digestion"; *J. Environ. Eng.* (Proc. 4th Int. Symp.on Anaerobic Digestion, November 11-15, 1985, Guangzhou, China) **1987**, 113, pp. 1265-1284.
12. Eastman, J.A.; Ferguson, J.F.; "Solubilization of Particulate Organic Carbon During the Acid Phase of Anaerobic Digestion"; *J. Water Pollution Control Fed.* **1981**, 53, pp. 352-366.
13. Boone, D.R.; "Terminal Reactions in the Anaerobic Digestion of Animal Waste"; *Appl. Environ. Microbiol.* **1982**, 43, pp. 57-64.
14. Pavlostathis, S.; Gossett, J.M.; "Preliminary Conversion Mechanisms in Anaerobic Digestion of Biological Sludges"; *J. Environ. Eng.* **1988**, 114, pp. 575-592.
15. Kaspar, H.F.; Wuhrmann, K.; "Kinetic Parameters and Relative Turnover of Some Important Catabolic Reactions in Digesting Sludge"; *Appl. Environ. Microbiol.* **1978**, 36, pp. 1-7.

16. Balch, W.E.; Fox, G.E.; Magrum, L.J.; Woese, C.R.; Wolfe, R.S.; "Methanogens: Reevaluation of a Unique Biological Group"; *Microbiol. Rev.* **1979**, 43, pp. 260-296.

17. McInerney, M.J.; Bryant, M.; Pfennig, N.; "Anaerobic Bacterium that Degrades Fatty Acids in Syntrophic Association with Methanogens"; *Arch. Microbiol.* **1979**, 122, pp. 129-135.

18. *Standard Methods for the Examination of Water and Wastewater*, 14th ed.; Greenberg, A.E.; Conners, J.J.; Jenkins, D., Eds.; American Public Health Association: Washington, D.C., 1980.

19. Henry, M.P.; Sajjad, A.; Ghosh, S.; "The Effects of Environmental Factors on Acid-Phase Digestion of Sewage Sludge"; *Proc. 42nd Purdue Univ. Ind. Waste Conf.* (May 12-14, 1987); Lewis Publishers: Chelsea, MI, 1988; pp. 727-737.

20. Perot, C.; Sargent, M.; Richard, P.; Phan Tu Luu, R.; Millot, N.; "The Effects of pH, Temperature, and Agitation Speed on Sludge Anaerobic Hydrolysis-Acidification"; *Environ. Tech. Lett.* **1988**, 9, pp. 741-752.

21. Ghosh, S.;"Pilot-Scale Demonstration of Two-Phase Anaerobic Digestion of Activated Sludge"; *Water Sci. Tech.* (Proc. 15th Biennial Conf. Int. Ass. Water Pollution Res. Control, July 29-August 3, 1990, Kyoto, Japan) **1991**, 23, pp. 1179-1188.

22. McCarty, P.L.; McKinney, R.E.; "Salt Toxicity in Anaerobic Digestion"; *J. Water Pollution Control Fed.* **1961**, 33, pp. 399-415.

23. Bhattacharya, S.K.; Parkin, G.F.; "The Effect of Ammonia on Methane Fermentation Processes"; *J. Water Pollution Control Fed.* **1989**, 61, pp. 55-59.

24. De Baere, L.A.; Devocht, M.; van Assche, P.; Verstraete, W.; "Influence of High NaCl and NH_4Cl Salt Levels on Methanogenic Associations"; *Water Res.* **1984**, 18, pp. 543-548.

25. van Velsen, A.F.M.; "Adaptation of Methanogenic Sludge to High Ammonia Nitrogen Concentrations"; *Water Res.* **1979**, 13, pp. 995-999.

26. Ripley, L.E.; Kmet, N.M.; Boyle, W.C.; Converse, J.C.; "The Effects of Ammonia Nitrogen on the Anaerobic Digestion of Poultry Manure"; *Proc. 39th Purdue Univ. Ind. Waste Conf.* (May 8-10, 1984); Butterworth Publishers: Stoneham, MA, 1985; pp. 73-80.

27. Shafai, S.; Oleszkiewicz, J.A.; Hooper, G.D.; "Anaerobic Treatment of High Nitrogen, High TDS Industrial Wastes"; *Proc. 41st Purdue Univ. Ind. Waste Conf.* (May 13-15, 1986); Lewis Publishers: Chelsea, MI, 1987; pp. 188-195.

RECEIVED September 3, 1992

SOIL REMEDIATION
AND TREATMENT

Chapter 13

Comparison of the Effectiveness of Emerging In Situ Technologies and Traditional Ex Situ Treatment of Solvent-Contaminated Soils

Sharon R. Just and Kenneth J. Stockwell

Engineering-Science, Inc., Suite 590, 57 Executive Park South, NE, Atlanta, GA 30329

This chapter examines the applicability of various treatment technologies to the remediation of contaminated soils in light of the Land Disposal Restrictions (LDRs) which were fully implemented for F001-F005 solvent contaminated wastes in November of 1990. Both traditional and emerging technologies are reviewed, including low temperature thermal treatment, radio frequency heating, steam stripping, vacuum extraction, aeration, in-situ bioremediation, and soil flushing/washing. In discussing the applicability of each technology, the feasibility, advantages, disadvantages, limitations, and performance of the treatment methods are reviewed. The treatment of soil contaminated with spent solvents such as trichloroethylene (TCE) is emphasized, and cleanup levels achieved during studies of different technologies are discussed. For some technologies, information concerning solvent removal was limited, and the treatment of other contaminants is summarized. The chapter concludes with a comparison of the removal efficiencies attained through emerging in-situ technologies and more traditional ex-situ treatment.

The ability of emerging in-situ treatment technologies to achieve satisfactory soil cleanup levels has increased in importance since November of 1990 when Land Disposal Restrictions (LDRs) were fully implemented for F001-F005 solvent contaminated wastes. Spent solvent contaminated wastes were prohibited from land disposal in November of 1986. However, exceptions were made to allow for continued disposal of certain Resource Conservation and Recovery Act (RCRA) corrective action and Comprehensive Environmental Response, Compensation, and Liability Act (CERCLA) response action wastes through November of 1990.

As outlined by EPA guidance documents, land disposal of restricted waste is allowed only if:
1) the wastes are treated to a level or by a method specified by EPA; or
2) it can be demonstrated that there will be no migration of hazardous constituents from the land disposal unit for as long as the wastes remain hazardous; or

0097–6156/93/0518–0238$11.00/0

3) the wastes are subject to an exemption or variance from meeting treatment standards *(1, pg. iv)*.

In general, once a spent solvent hazardous waste is excavated, it must meet one of the above requirements in order to be legally placed on the land.

Because of the long-term liability of future contaminant migration (Number 2) as well as the difficulties in obtaining exemptions or variances (Number 3), treatment (Number 1) is often the only viable or recommended solution for a hazardous waste once it is excavated. If a decision is made to excavate contaminated soil and LDRs apply, treatment standards are specified under 40 CFR 268 Subpart D. For most F001-F005 wastes, a technology-based standard is not stated. Instead, either total waste or waste extract concentrations are specified. For the TCE waste constituent, waste extract concentration limits are 0.062 mg/L for wastewaters and 0.091 mg/L for non-wastewaters (40 CFR 268.41).

Land disposal is defined in 40 CFR 268.2 as "placement in or on the land." The EPA has noted three general situations in which the land ban is not triggered since placement does not occur. These include:
1) when wastes are consolidated within a unit;
2) when wastes are treated in-situ, as long as the wastes are not picked up and treated in another unit; and
3) when wastes are capped in place *(1, pg. 5.2)*.

Of these three situations, only the in-situ treatment option may provide a permanent solution to the waste problem. The LDRs thus create an incentive for treating wastes in place, since excavation and placement restrictions are thereby avoided, while a permanent solution may be implemented.

Treatment Methods

Due to the impact of LDRs on site remediations, a knowledge of the relative effectiveness of in-situ and ex-situ treatment methods can be invaluable when selecting a soil remediation method. Two general approaches to treating contaminated soil exist. The first involves a phase transfer, where the contaminant is moved from one phase (either solid, liquid, or gas) into another, through a concentration gradient. The second involves destruction of the contaminants through a transformation process.

Transformation processes offer a significant advantage over phase transfer remediation alternatives, since soil contaminants are broken down instead of simply being moved from one phase to another. As a result, transformation is a more permanent solution to contamination than phase transfer remediation. Nonetheless, phase transfers can also reduce exposure risks, through altering the exposure path as well as decreasing contaminant concentrations through dispersal. Both transfer and transformation processes are reviewed in this paper, including low temperature thermal treatment, radio frequency heating, steam stripping, vacuum extraction, aeration, bioremediation, and soil flushing/washing.

Traditional methods of dealing with hazardous waste disposal in the past have included incineration and landfilling. With the LDRs, the landfilling of wastes has been curtailed. Although high temperature incineration has historically been proven to be the most reliable destruction method for the broadest range of wastes *(2, pg. 385)*, incineration is costly and can be difficult to permit *(3, pg. 1)*. On-site incineration can involve lengthy mobilization periods as the system is installed and trial burns are conducted. Off-site incineration increases costs and liability, due to waste transportation.

Recent developments have demonstrated the effectiveness of innovative technologies in the treatment of hazardous wastes. The following section discusses low temperature thermal stripping, to demonstrate the potential effectiveness of technologies which do not rely on high temperature removal processes. The remainder of this paper investigates the ability of several other in-situ and ex-situ technologies to treat contaminated soils, in light of the potential effectiveness of alternative treatment as well as the impact of LDRs.

Low Temperature Thermal Treatment

Both low and high temperature thermal processes are employed for soil remediation. In general, high temperature systems are viewed as those which operate at temperatures above 1,000°F (500-600°C), whereas low temperature systems operate below 1,000°F. High temperature processes include incinerators, electric pyrolysis, and in-situ vitrification. Low temperature thermal treatment systems are described as "soil roasting," "low temperature incineration," "low temperature thermal aeration," "infra-red furnace treatment," and "low temperature thermal stripping" processes.

Technology Description. Low temperature thermal treatment employs heat to increase the rate of contaminant volatilization and encourage the partitioning of the contaminant from soils. Thermal treatment systems vary by their soil heating methods, off-gas treatment, and contaminant removal abilities. During direct soil heating, the heating flames and/or gases directly contact the soil, thereby becoming contaminated. During indirect heating, the soil is warmed through contact with heated objects, such as an indirectly fired rotary kiln, oven, or hot oil pipes. The contaminated soil off-gas is thus segregated from the heat source gas during indirect heating, reducing the total volume of contaminated gas.

As a rule of thumb, volatile organic compounds (Henry's Law constant > 3.0×10^{-3} atm-m^3/mole) can be removed with these systems. Although some papers have stated that low temperature treatment is limited to volatile organic compounds (VOCs), significant removal efficiencies have been demonstrated with other compounds (4, pg. 78).

Several low temperature treatment systems exist which are capable of treating chlorinated solvents, such as the Desorption and Vaporization Extraction (DAVE) system, the Low Temperature Thermal Treatment (LT³) system, and the X*TRAX low temperature transportable treatment process. A brief description of each of these specific systems is included, in order to describe similarities and differences in treatment approach.

DAVE Description. The DAVE System claims to separate VOCs, semivolatile organic compounds, and volatile inorganic contaminants from soil, sludge, and sediments (5). Wastes are treated in a fluidized-bed reactor at a temperature of 160-170°C (320-340°F). A heated (1000-1500°F) gas containing 1-2% oxygen directly contacts and fluidizes the soil. Off-gases are treated with a cyclone separator and baghouse to remove particulates, and then proceed to a scrubber, counter-current washer, chill/reheat unit, and adsorption system. Condensed liquids are treated by a centrifuge, pressure filter, and carbon adsorption.

Large rocks must be crushed prior to processing (5, pg. 2), and heavy clays may require recycling through the system and/or mixture with other soils (5, pg. 6). A residence time of three minutes is provided for a feed rate of 10 tons per hour. The system is operated under a slight vacuum which aids in controlling emissions, and gases may be recycled within the system in the event of excessive contaminant levels or carbon bed breakthrough.

LT³ Description. The LT³ system uses hot oil-filled screw conveyors to indirectly heat the soil to approximately 204°C (400°F). Off-gases are passed through a fabric filter to remove dust, and then sent to a condenser. The liquid condensate is passed through an oil/water separator to skim off insoluble organics. The remaining water may then be treated through carbon adsorption to remove contaminants, and applied to treated soils for cooling and dust control purposes. Gases from the condenser may be sent to an afterburner or treated by carbon adsorption (Figure 1).

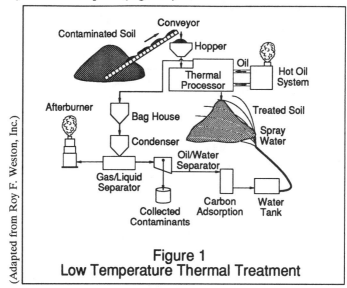

Figure 1
Low Temperature Thermal Treatment

(Adapted from Roy F. Weston, Inc.)

Three models of the LT³ system have been developed as follows:
- bench scale with 15 lbs/hr capacity
- pilot unit with 150 lbs/hr capacity
- full-scale with 15,000 lbs/hr capacity.

The full-scale unit is commercially available, and was designed to treat soil with a 20% moisture content and 1% VOC concentration *(6, pg. 139)*.

X*TRAX Description. In the X*TRAX process, soil is treated in an externally fired rotary kiln. Indirect heat is thus used, and the soil is mixed through the tumbling action of the rotating cylinder *(3, pg. 1)*. Soil temperatures are maintained at less than 900°F, usually in the 750-850°F range *(personal communication with Carl Swanstrom, X*TRAX)*. Exiting vapors pass through a liquid scrubber. Scrubber water is treated by a phase separator, and scrubbed gases are condensed in a heat exchanger. Approximately 90% of the carrier gas is reheated and recycled to the dryer, while the remainder is treated by carbon adsorption and released to the atmosphere.

As with the LT³ system, several scales of the X*TRAX have been tested, as follows:
- lab scale - 2-4 lb/hour capacity
- pilot scale - 5 ton/day capacity (30% moisture)
- full scale - 125 ton/day capacity (20% moisture).

The full scale system (Model 200) is available for commercial use.

Low Temperature Thermal Treatment Performance. Performance issues which have been associated with each system are summarized below.

DAVE Performance. Both waste characteristic and system operating parameters affect removal efficiencies. The vendor has stated that low porosity, non-adsorptive materials are the most easily processed. DAVE treatment categories are as follows:
- best - sand, sandy soils, non porous gravels
- acceptable - soils, clays, limestone, sandstone
- difficult - humus, organic decay products, wood, industrial adsorbents and catalysts *(5, pg. 6)*.

LT[3] Performance. A variety of soils have been tested with the LT[3] model, and the system has been successfully demonstrated on VOCs, semivolatile compounds (SVCs), and petroleum HC *(7, pg. 84)*. The full-scale system was first operated in Springfield, Illinois at a site contaminated with No. 2 fuel oil/gasoline. At a treatment temperature of 350°F and residence time of 70 minutes, over 99% of benzene, toluene, ethylbenzene, and xylene (BTEX) constituents were removed. Xylene removals exceeded 99.999%. Approximately 90 to 95% of carcinogenic PNAs were removed, while 63 to 95% of non-carcinogenic PNAs were removed *(6, pg. 141)*.

The LT[3] system was later demonstrated by the vendor at the Tinker Air Force Base in Oklahoma City, Oklahoma *(8)*. Clay soils at the site exhibited moisture contents of 13 to 23%, and were contaminated with a variety of volatile and semi-volatile compounds including chlorinated solvents. TCE removal efficiencies reported by the vendor to the U.S. Army Toxic and Hazardous Materials Agency are described in Table I *(8, pg. 9-1 to 9-6)*.

Table I. Low Temperature Thermal Treatment, LT[3] TCE Removal Efficiencies

Test	Soil Concentration Feed (mg/Kg)	Soil Concentration Processed (μg/L)[1]	Removal Efficiency
1	37.3	0.3	99.986
2	111.0	0.3	>99.995
4	10.6	1.2	99.78
8	8.5	2.3	99.46

SOURCE: Adapted from ref. 8.

[1]Processed Toxicity Characteristic Leaching Procedure (TCLP) extract concentrations were estimated based on dividing the processed soil concentrations by 20. The original processed soil values were not noted in the report summary.

The goal cleanup level was 70 ug/L, based on TCLP extract considerations. The above tests were conducted over a range of operating conditions such as oil temperature and soil residence time.

X*TRAX Performance. The pilot X*TRAX has been tested in several studies. Table II summarizes tests which were conducted on simulated waste streams, and Table III summarizes PCB removals which were achieved in various soils.

Table II. Low Temperature Thermal Treatment, Pilot X*TRAX Simulated Waste Stream

Compound	Feed Conc (ppb)	Product Conc (ppb)	Removal (%)
Methyl Ethyl Ketone (MEK)	100,900	<100.0	>99.90
Tetrachloroethylene (PCE)	91,000	15.6	99.98
Chlorobenzene (CB)	61,810	6.5	99.98
Xylene	56,365	2.8	99.99
1,4-Dichlorobenzene	78,400	1.4	99.99
1,2-Dichlorobenzene	537,000	74.1	99.99
Hexachlorobenzene	79,200	300.0	99.62

SOURCE: Adapted from ref. 3.

Although these tests failed to meet a treatment standard of 2 ppm for PCBs, they did comply with alternative treatment goals of 90 - 99.9% removal for soil concentrations initially greater than 100 ppm *(3, pg. 10)*.

Table III. Low Temperature Thermal Treatment, Pilot X*TRAX PCB Removals

Run #	Soil	Feed (ppm)	Product (ppm)	Removal (%)
0919	clay	5,000	24	99.5
1003	clay	1,600	4.8	99.7
0929	clay	630	17	97.3
0810	silty clay	2,800	19	99.3
0727	sandy	1,480	8.7	99.1

SOURCE: Adapted from ref. 3.

Low Temperature Thermal Treatment Advantages and Disadvantages. Advantages of low temperature thermal treatment systems include the following:
- can treat materials with moisture contents of 10 to 70% (vendor claim - DAVE System)
- mobile, on-site treatment
- screw conveyors mix and agitate soils, releasing volatiles from interstices (LT³)
- reduces heat requirements compared to incineration
- materials may be selectively recycled through the system for additional treatment
- treated soils can be used to backfill excavated areas
- avoids elevated temperatures which can cause thermal decomposition, for example, the formation of polychlorinated dibenzodioxins from PCBs (vendor claim - DAVE System)
- removed contaminants may be destroyed on-site through afterburner (LT³).

Low temperature disadvantages which have been noted include the following:

- fine solids are difficult to treat, due to preferential adsorption of contaminants and difficulties in removing fine particles from the gas stream
- limited to initial maximum total organic contaminant concentrations of 10% (DAVE System)
- pH extremes (pH <5, >11) have a corrosive effect on systems *(9, pg. 80)*
- lower treatment temperature may not volatilize certain compounds (e.g., metals, inorganics, less volatile organics)
- systems operate near the boiling point of mercury (356°C) *(9, pg. 80)*
- soil must be excavated, resulting in personnel exposure and LDR issues
- soil must be screened to remove materials over 2 inches in size (LT³)*(6, pg. 140)* or over 1.25 inches in size (X˙TRAX)*(9, pg. 78)*
- not economical for small volumes of soil, due to mobilization/demobilization costs (LT³)*(6, pg. 142)*
- not economical for high moisture contents, due to handling difficulties as well as increased energy requirements.

One major problem with ex-situ low temperature treatment systems are the volatilization losses which can occur during materials handling. During excavation, loading, and crushing operations, significant air emissions and personnel exposure can occur from VOC contaminated soils. These losses were tracked in a trial burn of clean sand contaminated with 3% oil and a 50/50 blend of xylene/toluene. Although a 99.25% removal efficiency was achieved, approximately 87% of the test hydrocarbons were lost in the loading process itself. Over 37% of the remaining HC were lost between the hopper and the conveyor *(Engineering-Science, August 1990)*. As indicated by such tests, the removal "effectiveness" of ex-situ processes must consider the losses attributed to volatilization during handling as well as treatment. When highly effective collection and off-gas treatment systems are designed, significant economical impacts to treatment costs can occur. When collection systems are insufficient, health and environmental concerns are raised due to the potential for simply transferring contaminants from the soil to the air. Due to such concerns, an in-situ technology may be preferred at sites which are contaminated with volatile organics.

Radio Frequency Heating

Technology Description. Radio Frequency Heating (RF Heating) is one method of achieving in-situ thermal decontamination of soil. This process was originally developed in the 1970s for use in recovering hydrocarbons from materials such as oil shales and tar sands, and was first tested on a pilot scale for the in-situ treatment of contaminated soils in 1987 *(13, pg. v)*.

RF heating uses electromagnetic energy applied in the radio frequency band to heat large volumes of soil, thereby encouraging the volatilization of contaminants. The process is similar to that of a microwave oven, except that the heated volumes are larger and the frequency of operation is different *(14)*. As a result, the soil is heated primarily by molecular agitation, rather than thermal conduction *(15, pg. 43)*.

The energy may be applied to the soil either through electrodes which are distributed horizontally over the contaminated area, or through a vertical electrode array which is inserted in the soil. If the contamination is shallow, the horizontal electrode array provides for contaminant removal while avoiding the exposure of drilling through contaminated materials *(16, pg. 58)*. However, the vertical electrode matrix is preferred for the treatment of deeper contamination. Table IV summarizes the applicability of each system *(17, pg. 337)*.

Table IV. Radio Frequency Heating, Electrode Matrix Systems

Electrode Array	Contaminated Soil Depth (feet)	Required Treatment Temperature (°C)
Horizontal	<3	<130
Vertical	>3	>130-150

SOURCE: Adapted from ref. 17.

Four subsystems are required for an in-situ RF heating system, as follows:
1) electrode array
2) RF power generation and control system
3) vapor containment
4) vapor treatment and handling *(18)*.

The electrode array is the most critical design parameter, since it affects the requirements of the remaining subsystems.

The Federal Communication Commission (FCC) has allocated several microwave frequencies for industrial, scientific, and medical (ISM) applications *(19, pg. 235)*. The frequencies used for radio frequency heating range from 6.78 MHz to 2.45 GHz *(14)*. A modified radio transmitter can serve as the power source for treatment of the contaminated soil *(20, pg. 83)*. The exact frequency is chosen based on the volume and dielectric properties of the soil.

Figure 2 describes the RF heating system which was used during a field study at the Volk Air National Guard Base (ANGB) in Wisconsin. The contaminated area was an abandoned fire training pit which contained waste oils, fuels, and solvents. The soil at the site consists of a medium grain sand, and the average total petroleum hydrocarbon concentration was 4000 mg/kg (ppm) to a depth of 12 feet *(13, pg. 26)*. A total of thirty-nine electrodes were placed in three parallel rows of drill holes, which simulated an enclosed, in-situ rectangular box. In this vertical array system, the outer rows form the outer walls of the box (ground plane), while the middle row represents the central conductor *(13, pg. 42)*.

Contaminated gases and vapors which are released during RF heating may be collected through a vapor barrier system, destroyed through incineration, or dispersed in the atmosphere. At the Volk ANGB site, vapors were collected through a vapor barrier system constructed of a silicon rubber sheet, cooled, and separated *(13, pg. 48)*. Uncondensed gases were then treated by a carbon bed.

RF Heating Performance.   RF heating decontaminates soils through mechanisms such as vaporization, steam distillation, and/or stripping. As the soil is heated, soil contaminants as well as moisture vaporize. Contaminants which are mixed with water exhibit lower boiling points than pure samples. For example, the boiling point of pure benzene is 80.1°C, while a benzene/steam mixture (0.092 lb steam/lb benzene) boiling temperature is 68.3°C *(16, pg. 59)*. Thus, components which exhibit vapor pressures close to that of water can be effectively removed through steam distillation.

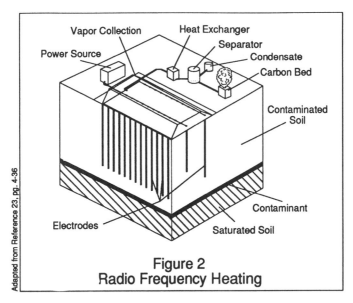

Figure 2
Radio Frequency Heating

For compounds with higher boiling points (two times greater than water), excessive amounts of saturated steam are required to achieve steam distillation. For these compounds, superheated steam can aid their removal by acting as a sweep gas. However, this requires increased temperatures, resulting in increased energy requirements and treatment costs. Finally, higher temperatures and increased residence times can result in thermal decomposition of contaminants, rather than simple vaporization *(16, pg. 58)*.

Bench, pilot, and field scale studies have been conducted by the Illinois Institute of Technology Research Institute (IITRI) to determine the effectiveness of RF heating. The bench scale studies involved heating small samples (75 grams, 2 to 3 inches deep) of sandy soil spiked with tetrachloroethylene (PCE) and chlorobenzene (CB) to confirm that thermal decontamination of such materials was possible *(13, pg. 6)*. The pilot scale studies were performed on six foot columns containing 3.5 to 4.0 kilograms of soil *(13, pg. 10)*. Both spiked soil samples and contaminated soil obtained from Volk ANGB were tested in the columns. Finally, the field scale studies were performed using the in-situ RF system described above.

In the pilot scale spiked soil experiments, difficulties were encountered due to volatilization losses of the spiked PCE and CB. Spiked soil concentrations ranged from 0 to 106.6% of the desired initial concentration *(13, pg. 7)*. Although high removal efficiencies were calculated in these tests (ranging from 98.4 to 100%), mass balance following treatment ranged from 66 to 240% (Table V).

Due to the spike losses, pilot scale studies were also conducted on soil contaminated with jet fuel which was obtained from the fire training pit at Volk. In one test the soil was heated to 157°C and maintained at that temperature for approximately 14 hours. In the second test, the soil was heated to similar temperatures (152-157°C), but maintained for a longer period of time (approximately 40 hours) and injected with heated water during the test to simulate an in-situ steam sweep. Overall removal efficiencies of 91% were achieved in the first test, while the steam sweep study resulted in a 99% removal efficiency. The steam sweep significantly increased the removal efficiency of higher boiling point compounds, such as pentadecane, which demonstrated a 75% removal without steam injection and a 94% removal with steam injection *(13, pg. 24)*.

Table V. Radio Frequency Heating, Pilot Scale Treatment of Spiked Soil

| Experiment | Initial Concentration[1] | | Removal from Soil | | Mass Balance | |
	PCE (ppm)	CB (ppm)	PCE (%)	CB (%)	PCE (%)	CB (%)
1	0.17	n.d.	-	-	-	-
2	0.58	n.s.	100	-	124	-
3	34.45	36.4	100	100	75.05	66.45
4	9.5	10.66	99.4	98.4	220.2	240.98

SOURCE: Adapted from ref. 13.
1. Experiments 1 & 2 were spiked to achieve 10 ppm initial concentrations, and Experiments 3 & 4 were spiked to achieve 50 ppm initial concentrations (n.d. = not detected; n.s. = not spiked)

In the field scale study, approximately 500 cubic feet of soil were heated over 12 days to final average temperatures of 150 to 160°C. Three different depths were studied. High removal efficiencies of contaminants were achieved, as described in Table VI.

Table VI. Radio Frequency Heating, Field Scale Removal Efficiencies

| Depth Interval (inches) | Moisture (%) | Volatiles | | Semivolatiles | | |
		Aliphatics (%)	Aromatics (%)	Aliphatics (%)	Aromatics (%)	Hexadecane (%)
6-72	97.2	99.3	99.6	94.3	99.1	82.9
6-12	96.1	98.2	99.2	88.1	98.1	67.5
30-42	96.1	99.7	99.6	97.6	99.6	91.4
60-72	98.9	99.8	99.9	98.5	99.9	93.6

SOURCE: Adapted from ref. 15.

It should be noted that identical sampling methods were not used to collect the initial and final soil samples in the field study. Specifically, initial samples were collected by split-core barrel sampling, while treated samples were collected by suctioning of a wet-dry shop-vac *(13, pg. 62)*. Although the revised sampling method was due to the dryness of the treated soil, which had caused it to become free-flowing, the vacuum method could cause additional volatilization resulting in further removal than had occurred in-situ.

The slightly lower removal efficiencies which occurred in the shallow soil layer were believed to have been caused by lower final temperatures near the surface and/or condensation of vapors in the surface zones after the gas collection fan system was turned off *(13, pg. 85)*. However, in general, uniform decontamination was achieved. The field study confirmed that high boiling point compounds such as hexadecane (b.p. 287.5°C) could be removed at lower treatment temperatures *(13, pg. 88)*.

RF Heating Advantages/Disadvantages. The advantages of RF heating include the following:

- reduces impact of LDRs since only 0.5% of the treated volume is excavated or drilled *(13, pg. 114)*
- removes both volatile and semivolatile organics
- reduces effluent gas production by avoiding the use of fuel combustion gases which mix with contaminated vapors
- achieves uniform heating which does not depend on conduction/convection
- minimizes exposure to personnel due to in-situ nature of treatment.

The disadvantages and limitations of RF heating include the following:

- does not treat nonvolatile organics, metals or inorganics
- costly when applied to very deep contamination *(20, pg. 85)*
- limited to soil types which allow for insertion of electrodes in drill holes and to soils which do not contain buried metal objects
- can destroy or inhibit microorganisms due to high temperatures
- not as cost effective in sands as other innovative technologies such as vacuum extraction *(personal communication with Doug Downey, ES)*.

Results of RF heating tests on clay soils have not yet been released. A second RF heating field test is currently planned to investigate the removal of chlorobenzene, trichloroethene, petroleum hydrocarbons, and RCRA hazardous wastes *(23, pg. 4-37)*. Further studies of RF heating are warranted, to determine the cost effectiveness of this technology under more adverse operating conditions.

Steam Stripping

Technology Description. Steam stripping is a variation of thermal stripping in which steam is used to contact the soil instead of hot air. As with RF heating, steam stripping can be applied as an in-situ process whereby air and steam are injected into the ground. The steam heats the ground, resulting in increased volatilization of the contaminants. A vacuum extraction process brings the air, steam, and contaminants to the surface, where they may be collected for treatment.

A variety of studies have reportedly used steam stripping to decontaminate soils and/or groundwater. These include research by Lord, et. al. and demonstrations by Novaterra's (formerly Toxic Treatments USA, Inc.) Detoxifier and AWD Technologies' integrated AquaDetox/soil vapor extraction technology *(20, 24-33)*.

The AWD system combines in-situ steam stripping for contaminated groundwater with vacuum extraction for contaminated soils *(24, 25)*. The systems are combined in a closed loop system. Specifically, each system's emissions are treated or re-injected within the closed loop, thereby greatly minimizing air emissions. The process was recently studied in a two-week long SITE demonstration at a location contaminated with VOCs such as TCE and PCE, and indicates promise as an effective combined soil and groundwater remediation system.

The Novaterra Detoxifier system is similar to the AWD system in that off-gases are treated and then reinjected. However, the Detoxifier is currently limited to treating more shallow depths, due to the use of an auger system for treatment. The Detoxifier system uses two hollow augers with 5 foot diameter blades to drill and mix the contaminated soil to depths of 30 feet *(20, pg. 82;* Figure 3). Treatment agents such as high pressure steam (200°C) and hot compressed air (135°C) are injected into the ground through blade-mounted jets *(30, 31).* The steam and hot air raise the soil temperature to approximately 80°C (170-180°F), increasing the vapor pressure of volatile contaminants *(26, pg. 31 and 36).* The steam, air, and vaporized contaminants are then recovered at the surface in a patented shroud and separated. Condensed organics are collected, while off-gases are treated by carbon adsorption and subsequently compressed and reinjected into the ground. Removal mechanisms appear to include volatilization, steam distillation, catalytic decomposition, hydrolysis, and binding to clayey soils *(26, pg. 37).*

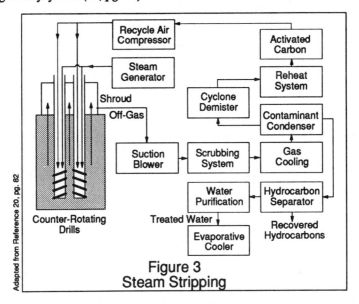

Figure 3
Steam Stripping

Steam Stripping Performance. Lord, et. al., conducted laboratory studies of vacuum-assisted steam stripping of organic contaminants from soil *(33).* The steam stripping of gasoline and kerosene from various mixtures of sand and silt were studied, as described in Table VII.

Table VII. Laboratory Scale Steam Stripping Approximate Removal Efficiencies[1]

Contaminant	Soil Mixture	Approximate Removal
Kerosene	50/50 sand/silt	50%
Kerosene	100 sand	80%
Gasoline	50/50 sand/silt	80%
Gasoline	100 sand	99%

SOURCE: Adapted from ref. 33.
1. After 150 minutes.

Removal efficiencies were measured based on volumetric separation of the collected organic and water fractions. Studies were also conducted over a range of kerosene saturation levels using a pilot scale, geosynthetic, vacuum cap to collect the steam and contaminants *(33)*.

One difficulty with these initial studies was the imprecise nature of volumetric calculations of removal efficiencies *(33, pg. 128)*. Subsequent studies were conducted using gas chromatographic (GC) analyses. Both alkanes (octane, decane, dodecane) and alcohols (butanol, octanol) were studied in various sand/silt mixtures. High initial concentrations (5-10%, e.g., >50,000 ppm) were reduced to the 1-1,000 ppm level, after five hours of steam stripping, resulting in the calculation of high (>99.8%) removal efficiencies. Removal efficiencies increased with increasing vapor pressure. Alcohols were removed more easily than alkanes, and removal did not appear to follow any trend as a function of silt content in this study *(33, pg. 128)*. However, the finer grain silt soils may have actually contained more contaminants after treatment, which were simply difficult to extract and therefore analyze by GC. The studies indicate that vapor pressure and polarity (alcohols are polar) appear to affect the ability to steam strip contaminants *(33, pg. 130)*. However, once again, the alcohols may have simply been more difficult to extract with the GC solvent *(33, pg. 130)*.

Steam stripping removal efficiencies have been reported in other tests, such as the SITE program demonstration of the Detoxifier. The Detoxifier was demonstrated at a former industrial chemical storage facility contaminated with forty-nine VOCs and SVCs. Heavy PCE and chlorobenzene (CB) contamination existed in portions of the site, in soils comprised of over 80% clays *(28, pg. 3)*. Other areas were composed of sandy/silty soil containing more heterogeneous contamination *(28, pg. 3)*.

Shallow groundwater existed at the site, and the Detoxifier was tested both above and below the water table. A 12-block area was tested to the water table (5 foot depth), and a 6-block area was tested to depths of 8-11 feet *(26, pg. 6)*. The 12-block area primarily contained CB, TCE, and PCE, while the 6-block area was contaminated with ketones *(26, pg. 28)*. An 85% average VOC removal was achieved in the vadose zone soils (Table VIII). The saturated zone removal efficiency could not be calculated, since sufficient pre-treatment data was unavailable. However, the average VOC concentration after treatment in the saturated zone (53 ppm) was less than that of the vadose zone (71 ppm) *(26, pg. 7)*. Thus, treatment in the saturated zone appears to achieve treatment times and removal efficiencies similar to treatment in the vadose zone.

It should be noted that high VOC removal efficiencies appear to be more easily attained with high initial concentrations of contaminants. For example, based on the above sampling results, an average 92% removal occurs in the sample blocks with the highest initial VOC concentrations (642-1133 ppm), while an average removal of only 68% occurs in the four blocks with the lowest initial concentrations (28-283 ppm).

Treatment time (and therefore costs) are affected by the type of contaminant, depth of contaminant, soil conditions, chemical reactions, soil adsorption, and cleanup requirements *(31)*. Clay soils require increased treatment times compared with sandy soils, and higher boiling point compounds also increase costs *(30)*. Although treatment time varies, a typical remediation rate of one minute per cubic yard has been reported *(29)*. However, SITE demonstration treatment times of twenty minutes per cubic yard were required, and the vendor has used treatment periods of three to twelve minutes in cost estimates *(26, pg. 27)*. The vendor states that soils with 70% clay require twice the treatment time as those with 30% clay *(26, pg. 27)*. Treatment rates in saturated sand and silt can be increased two to three times.

Table VIII. In-situ Steam Stripping, Detoxifier 12 Block (Unsaturated) Test Area

Block Number	Volatiles			Semivolatiles		
	Pre Treatment (μg/g)	Post Treatment (μg/g)	Percent Removal (%)	Pre Treatment (μg/g)	Post Treatment (μg/g)	Percent Removal (%)
A-33-e	1133	104	91	896	763	15
A-29-e	850	82	90	1310	726	45
A-31-e	788	61	92	781	610	22
A-27-e	642	29	96	1403	439	69
A-32-e	479	64	87	994	49	95
A-28-e	444	34	92	1040	576	45
A-34-e	431	196	54	698	163	77
A-30-e	421	145	65	1073	818	24
A-35-e	283	60	79	577	192	67
A-36-e	153	56	64	336	314	7
A-25-e	54	14	73	595	82	86
A-26-e	28	12	56	1117	172	85
Average	466	71	85	902	409	55
Std. Dev.	457	80	N/A	469	407	N/A

SOURCE: Adapted from ref. 26.

Compounds with boiling points (b.p.) below 80°C can often be removed to non-detect levels *(26, pg. 8)*. However, certain compounds may be difficult to remove due to their adsorption to soils. For example, cis 1,2-dichloroethene and PCE exhibited relatively high concentrations following treatment in recent tests, apparently due to adsorption, although the addition of auxiliary mixing blades improved their removal *(29)*. PCE can be expected to be difficult to remove due to its high stability and resistance to hydrolysis, as well as its elevated b.p. *(26, pg. 8)*.

In addition to the demonstration described above, several other Detoxifier studies have been conducted. One in-situ treatment of 10 blocks of soil achieved an 85% average reduction in SVCs and a 97% average reduction in VOCs. A VOC mass balance closure of 90% was achieved; however, SVCs demonstrated an extremely poor mass balance of 0.2%.

Table IX describes a bench-scale contained system test, whose purpose was to track final contaminant fate. VOC removal averaged 83%, but mass balance was poor. Higher mass balances were measured for SVCs, which achieved a 53% removal.

Table X summarizes mixing tests which were conducted using only air, and using air and steam. Consistent, high removal efficiencies were achieved only for VOCs treated with air and steam. Statistically, SVCs did not appear to be treated by either method *(26, pg. 35)*.

Table IX. Bench-Scale Steam Stripping, Detoxifier Contained System Tests

	VOCs		SVCs	
	Exp 3	Exp 4	Exp 3	Exp 4
Pre-treatment Soil (g)	9.376	9.376	30.987	30.987
Post-treatment Soil (g)	3.403	0	16.758	12.112
THC missing (g) [1]	5.436	7.470	1.655	11.612
Mass Balance (closure)	42.0%	20.3%	94.7%	62.5%
Percent Removal [2]	**63.7%**	**100%**	**45.9%**	**60.9%**

SOURCE: Adapted from ref. 26.
1. Note that VOC soil, water, and carbon bed concentrations were accounted for, but only soil and water SVC concentrations were measured.
2. VOC average removal = 83%; SVC average removal = 53%

Table X. In-situ Air and Steam Stripping Mixing Tests, Detoxifier Average Percent Removal

	VOCs			SVCs		
Soil Block	B48-g	B49-g	B50-g	B48-g	B49-g	B50-g
Air Removal[b]	16%	75%	79%	-122%[a]	93%	70%
Air and Steam Removal	89%	91%	100%	33%	-59%[a]	-9%[a]

SOURCE: Adapted from ref. 26.
a. Post-treatment concentration higher than pre-treatment concentration.
b. Note that B48-g was treated with air for 21 minutes, B49-g was treated for 11 minutes, and B50-g was treated for 100 minutes.

A project officer at EPA's Risk Reduction Engineering Laboratory has summarized that the Detoxifier's niche is the in-situ treatment of heavily contaminated, tight soil which cannot feasibly be excavated, since vacuum extraction is inadequate at such sites and thermal desorption is more timely and reliable when excavation is possible *(31)*.

Despite the mixing action of the augers, homogenous treatment does not occur and "hot spots" of contamination can remain after treatment. Mass balances could not be conducted in the Detoxifier demonstration tests, presumably due in part to the heterogeneous soil and wide standard deviations *(26, pg. 9)*. As a result, one in every six vadose zone cores and one in every nine saturated cores demonstrated post-treatment contamination levels above 100 ppm, in violation of operational objectives *(26, pg. 6)*. With increased treatment times, contaminant levels are reduced. Thus, the excesses appear to be due to premature termination of treatment. However, more lengthy treatment would have increased costs.

Steam Stripping Advantages/Disadvantages. Based on the demonstrated field scale application of the Detoxifier system in vadose zone soils, the advantages and disadvantages of this system are reviewed here. Several of the disadvantages are system specific (e.g. 30 foot depth limitation), and may not exist with alternative steam stripping techniques.

The advantages of steam stripping systems are as follows:
- minimizes exposure to personnel due to in-situ nature of treatment

- reduces impact of LDRs since augers drill through contaminated areas without requiring excavation
- removes volatile organics and reduces semivolatile organic concentrations, although semivolatile fate has not been determined *(26)*
- minimizes contaminated air emissions. Although recently treated soil blocks emit vapors, the placement of clean soil layers on newly treated blocks can reduce emissions. *(30)*
- achieves quick remediation when compared to other in-situ stripping technologies *(32, pg. 2)*
- potential to be used for injection of chemical/physical agents for solidification/stabilization; pH adjustment; oxidation/reduction contaminant destruction; and/or promotion of bioremediation (Detoxifier)*(32, pg. 3)*
- does not significantly alter soil structural properties, although slight increases in specific gravity (and decreases in porosity) have been noted (Detoxifier)*(26, pg. 11)*.

The disadvantages and limitations of steam stripping systems include the following:

- removal mechanisms and limitations are not fully understood *(20, pg. 83)*
- currently limited to 30 foot treatment depths, although vendor states system is being redesigned to achieve 100 foot depths (Detoxifier; personal communication with Robert A. Evangelists, Novaterra)
- not cost effective for sites with less than 2000 cubic yards of soil due to current system size (Detoxifier)*(31)*
- costly (>$300 per cubic yard at demonstrated treatment rate of 3 cubic yard/hour) compared with technologies such as vacuum extraction. However, lower costs could be expected in sandier soils with lower contamination levels. (Detoxifier)*(30)*
- not cost effective in sites with high clay contents, due to the clay's binding effect on compounds.
- treatment area must be greater than 0.5 acre in size, compacted to support the system's weight, free of hard pavement surfaces, and graded to a minimum slope of 1% (although vendor is designing new system for grades up to 5%) (Detoxifier)*(26, pg. 1)*
- underground objects larger than a one foot diameter must be removed, and an overhead clearance of 30 feet is required (Detoxifier)
- remaining "hot spots" of contamination can require further treatment resulting in increased costs to meet cleanup criteria
- elevated temperatures can adversely effect microbial populations *(26, pg. 37)*
- a variety of commercial steam stripping units for soil remediation are not yet available.

Steam stripping appears able to reliably achieve VOC (including TCE) removal efficiencies greater than 85%. Higher efficiencies are possible with greater treatment times; however, this results in increased costs and "hot spots" may nonetheless remain. SVC concentrations have also been reduced (by as much as 50-85%); however, mass balances for SVCs have generally not been demonstrated. In addition, some tests have demonstrated no statistical removal of SVCs, and the process should thus currently not be viewed as a SVC treatment technology.

Vacuum Extraction

Technology Description. Vacuum extraction (soil vapor extraction, SVE) is a modified aeration process, whereby partitioning is aided through application of a vacuum. Vacuum extraction is the most commonly used in-situ remedial technology *(20, pg. 78)*. Air is injected into contaminated soils, and then recovered through extraction wells. Vapor phase volatile organics in the air-filled pore spaces of the soil are removed. Through decreasing the air pore space concentration of contaminants, vacuum extraction encourages further partitioning of the volatile organics from the liquid/soil matrix into the air voids. Extracted contaminated air may then be dispersed, destroyed through incineration, or treated by an activated carbon system. Dispersion may reduce the risk to acceptable levels, since the health impacts of the resulting air concentrations are often much less severe than those posed by contaminated soils or groundwater *(34, pg. 7)*. In general, an activated carbon off-gas treatment system may be preferred *(35, pg. 151)*, due in part to concerns over the generation of chlorine gas and products of incomplete combustion during the direct incineration of chlorinated organics *(36, pg. 88)*.

SVE may be implemented with vertical wells, trenches (for shallow contamination), horizontal drilling (to reach under buildings), and above ground (under/in excavated soil piles). A typical vertical extraction well system is described in Figure 4. A PVC pipe is installed in a contaminated soil area. The screened (slotted) interval is located in contaminated soil layers, above or at the water table level. During the application of a vacuum, water table upwelling will occur, resulting in entrainment of water in the process stream. This effect can be minimized by pumping the water table down near the vapor extraction wells *(36, pg. 105)*.

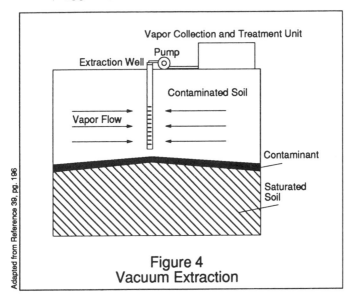

Adapted from Reference 39, pg. 196

Vapor Collection and Treatment Unit

Pump
Extraction Well

Contaminated Soil

Vapor Flow

Contaminant

Saturated Soil

Figure 4
Vacuum Extraction

Groundwater pump down can be especially important for immiscible contaminants with densities greater than water (i.e., halogenated compounds), compared with immiscible contaminants with densities less than one (i.e., petroleum hydrocarbons) *(37, pg. 289)*. The lighter compounds will tend to float on the water table, and therefore be easily exposed to vacuum extraction wells. If heavier compounds have sunk to the capillary and saturated zones, additional removal mechanisms will be required to achieve full site remediation (e.g., groundwater treatment). However, vacuum extraction can still aid in cleanup through removing residual contamination in the vadose zone which might otherwise serve as a continuing source of groundwater contamination. Some systems (e.g., AWD Technologies' integrated AquaDetox/SVE system), combine groundwater treatment and soil vapor extraction *(38)*.

The spacing and placement of the wells in a vacuum extraction system is site-specific, based on the radius of influence which the wells can achieve. The radius of influence has been defined as the distance to which an extraction well exerts a vacuum of 0.1 inch of water or more in a monitoring well *(36, Keech, 1989)*. Radii of influence increase with increasing suction head and increasing vadose zone depth *(36, pg. 58)*. An impermeable cap may be installed at the surface to prevent short-circuiting of air, thereby forcing a more horizontal flow path which intercepts greater volumes of contaminated soils. Surface seals also reduce fugitive emissions and prevent rainwater infiltration.

Passive air inlet and/or active air injection wells may also be used to control vapor flow, and reduce "dead spots" (Figure 5) *(36, pg. 59)*. As a rule of thumb, a spacing of wells twice the depth to which they are installed has been recommended *(36, pg. 58)*. Typical installation parameters are as follows:

- screened interval from 5-10 feet below land surface to the water table (with maximum screened interval of 20-30 feet) *(35, pg. 157)*
- flow rates of 10-1000 (typically 100) standard cubic feet per minute (scfm) *(39, pg. 26)*
- radius of influence of 30-100 feet *(39, pg. 205)*
- typical vacuum well pressures of 0.90-0.95 atm *(39, pg. 205)*.

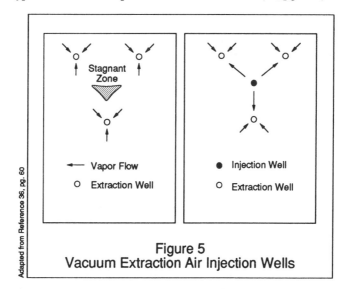

Figure 5
Vacuum Extraction Air Injection Wells

Vacuum Extraction Performance.   Site specific factors which affect the implementation of vacuum extraction include the following *(36, pg. 3)*:

- contaminant characteristics, including degree of weathering
- extent of contamination
- soil characteristics (moisture, air permeability)
- soil composition (organic carbon and clay content sorptive effects)
- groundwater depth
- emissions control requirements, and
- cleanup criteria.

Contaminants which have been released to the soil may exist as follows:
- vapor phase - affected by vapor pressure and Henry's law
- dissolved in pore/groundwater - affected by solubility
- sorbed to soil particles - affected by soil sorption coefficient (Kd); larger molecules have greater tendencies to sorb
- non-aqueous phase liquid (NAPL).

The volatility of a contaminant has been identified as the most important contaminant characteristic in determining the effectiveness of vacuum extraction (36, pg. 17-19). Vapor pressures increase with temperature increases, and higher vapor pressures reflect an increased tendency to volatilize. The vapor pressure represents the partioning between a pure substance and its vapor phase, while the Henry's law constant describes the volatilization of a contaminant dissolved in a solution (e.g., water). Vapor pressures greater than 0.5 mm Hg, and Henry's constants greater than 0.01 (dimensionless) represent significantly volatile compounds which should be amenable to vacuum extraction (36, pg. 17 & 21; 35, pg. 141). Terra VAC has reported that compounds with Henry's constants greater than 0.001 can be removed (34). Others have stated that vapor pressures of 1 mm Hg (20°C) are necessary (36, pg. 48; 41, pg. 165).

In addition to reducing vapor phase concentrations through removal, vacuum extraction causes a decrease in the pressure in soil pores. A compound's boiling point occurs when its vapor pressure equals the vapor pressure of the atmosphere. Decreases in atmospheric pressure (through increases in elevation or the application of a vacuum), reduce the boiling point of compounds and therefore encourage volatilization (36, pg. 21).

One major concern with vacuum extraction is its performance in various soil types. The U.S. EPA has noted that vacuum extraction is "typically limited to permeable unsaturated soils such as sands, gravels, and coarse silts" and that "clayey soils usually lack the conductivity necessary for effective vapor extraction, unless they are first fractured" (20, pg. 79). The report summarizes that soils must be unsaturated, permeable, and fairly homogeneous for the most effective vacuum extraction. However, vacuum extraction has been successfully demonstrated in other soil types.

Several reports have stated that the air-filled porosity is the most important factor in determining the effectiveness of vacuum extraction (36, pg. 27 and 34, pg. 1). The permeability to air flow represents the ability of vapors to flow through the soil, and is analogous to the permeability to water flow in saturated soils. Soil vapor migration occurs primarily through air-filled pore spaces, since movement through saturated pores relies on the slow process of diffusion (36, pg. 27). Air-filled porosity is affected by the moisture content and particle size distribution of the soil. The air-filled porosity increases with decreases in moisture content. However, contaminants tend to sorb more to drier soils (36, pg. 29). Thus, the optimal soil moisture content must be low enough to allow for air-filled pore spaces, but high enough to reduce electrostatic sorption forces (36, pg. 31). A 94-98.5% relative humidity in the soil gas has been reported as the critical moisture level for SVE (36, Davies, 1989), and an air-filled porosity of 40-50% has been suggested (36, Danko, 1989).

Numerous studies have demonstrated decreases in soil-gas concentrations over time (42, pg. 255; 39, pg. 236; 43, pg. 281). However, one difficulty raised in determining whether cleanup has occurred is the fact that soil-gas and soil concentrations do not necessarily correlate. For example, at one site soil-gas concentrations of 100 ng/cc were measured at locations with non-detectable soil concentrations, while gas levels of only 5 ng/cc were measured in soils contaminated at levels of 22 g/kg (44, pg. 183).

A pilot test conducted on a TCE spill *(45, pg. 30)* showed a combination air injection / air extraction system capable of removing TCE to a 20 feet depth. A removal efficiency of 75% was calculated based on a mass balance of estimated initial volumes and effluent air concentrations. Soil testing showed up to 99% reductions of TCE from initial concentrations around 2 ppm. The total test time was about 10 weeks. A second test of extraction on highly contaminated soils (up to 5,000 ppm TCE) resulted in calculations of 25% removal in 14 weeks, a rate of about 20 pounds/day. Final clean-up levels were not assessed. The highest cost was associated with cleanup of the effluent air stream.

At a site containing silty/clayey gravels and gravelly sand, a vacuum extraction system was operated for 440 days and removed approximately 30 kg of TCE *(46, pg. 794-795)*. Soil-gas concentrations exhibited a steady drop from approximately 640 to 1.5 ppm-volume. Average soil concentrations (0-27.5 m depth) dropped from approximately 0.11 mg/kg to 0.034 mg/kg.

Terra VAC's In-situ Vacuum Extraction System was tested through the U.S. EPA SITE program on TCE contaminated soils in Groveland, Massachusetts. Contaminant removal was measured through flow rates and vapor stream concentrations which indicated that 1297 lbs of VOCs were removed. The amount of VOCs adsorbed in the activated carbon was used to verify this analysis, and indicated a removal of 1353 lbs of VOCs *(47, pg. 3-4)*. However, as with other studies, VOC concentrations in the soil did not correlate well with VOC well head gas concentrations *(47, pg. 72)*.

The system was operated for one week, shut down due to excessive water extraction, and resumed one month later for a 56-day demonstration *(47, pg. 16)*. The average reduction of TCE over the 24 foot depth of treated soil is presented in Table XI *(47, pg. 68)*.

Table XI. Terra VAC In-situ Vacuum Extraction, Weighted Average Soil TCE Concentrations/Removals

	Pretreatment (mg/kg)	Post treatment (mg/kg)	Removal (%)
Extraction Well (EW)			
4	96.10	4.19	96
1	33.98	29.31	14
3	6.89	6.30	9
2	3.38	2.36	30
Monitoring Well (MW)			
3	227.31	84.50	63
2	14.75	8.98	39
1	1.10	0.34	69
4	0.87	1.05	negative

SOURCE: Adapted from ref. 47.

Each of the wells (MW and EW) was installed in two sections (one deep-D, one shallow-S), above and below an impermeable, horizontal clay lens *(47, pg. 43)*. One difficulty which was encountered during the Groveland *Terra VAC* tests was siltation blockage in one well. Attempts to clear the blockage by pulse-pumping water down the well were unsuccessful. Should such a blockage occur during a full-scale cleanup, it may be necessary to drill a new well adjacent to the blocked one *(34, pg. 7)*.

It should be noted that the highest removal efficiency (average 95.64%) was calculated in EW4. However, the deep portion of EW4 (EW4D) was the well which had experienced a siltation blockage and subsequent zero flow. It is thus unclear why a well which experienced zero flow (and subsequent zero extraction) for over half of the study would have achieved the highest removal efficiency. Further investigation demonstrates that removal efficiencies within individual wells were also variable, as demonstrated in Table XII.

Table XII. Terra VAC In-situ Vacuum Extraction, MW3 TCE Removal Efficiencies

Depth (ft)	Description of strata	Permeability (cm/s)	TCE conc. ppm		% removal
			pre	post	
0-2	M. stiff brn. fine sand	10^{-5}	10.30	ND	100% (ND)
2-4	M. stiff grey fine sand	10^{-5}	8.33	800.0	negative
4-6	Soft lt. brn. fine sand	10^{-4}	80.0	84.0	negative
6-8	Lt. brn. fine sand	10^{-4}	160.0	ND	100% (ND)
8-10	Stiff V. fine brn. silty sand	10^{-4}	ND	63.0	negative
10-12	Silty sand	10^{-4}	NR	2.3	negative
12-14	Soft brown silt	10^{-4}	316.0	ND	100% (ND)
14-16	Wet green-brown silty clay	10^{-8}	195.0	ND	100% (ND)
16-18	Wet green-brown silty clay	10^{-8}	218.0	62.0	71.6%
18-20	Wet green-brown silty clay	10^{-8}	1570.0	2.4	99.8%
20-22	Silt, gravel, and rock frag.	10^{-4}	106.0	ND	100% (ND)
22-24	M. stiff lt. brn. med. sand	10^{-4}	64.1	ND	100% (ND)
0-24	Average	-	227.3	84.5	62.8%

SOURCE: Adapted from ref. 47.

As described in the table, removal efficiencies of 71.6 to 99.8% were achieved in a relatively impermeable, wet clay at depths of 14 to 20 feet. However, a broad range of treatment results occurred in the other soil strata.

The variable removal efficiencies can necessitate lengthy treatment times to achieve cleanup criteria. A pilot *Terra VAC* vacuum extraction system which was installed at the Upjohn facility in Puerto Rico to remove carbon tetrachloride (CCl_4) was operated for over five years before closure criterion were met. In this instance, the regulatory agency had required non-detect levels of CCl_4 in all the exhaust systems for three consecutive months before considering treatment to be complete. With less stringent closure requirements, treatment operations could have been suspended at an earlier date.

A pulsed venting scheme has been suggested as the most effective removal mechanism in terms of energy expended. Pulsed venting allows the soil/soil-gas system to requilibrate. Without pulsing, removal rates can significantly decrease once relatively quick advection has removed initial vapor contamination, and further removal has become dependent on the relatively slow process of diffusion *(42, pg. 252)*. Over time, removal rates will slow as the remaining contaminant mixtures are composed of less volatile compounds *(48, pg. 269)* and the contamination level decreases *(39, pg. 211)*.

Vacuum Extraction Advantages/Disadvantages.   The advantages of vacuum extraction include the following:
- reduces impact of LDRs due to in-situ nature
- can be implemented under buildings and in other inaccessible contaminated areas
- air pathways can be controlled through placement of inlet wells *(36, pg. 59)*
- minimizes exposure to personnel, especially when air controls are installed
- treated air may be reinjected in air injection wells, further reducing emissions and acting to increase contaminant removal (of soil, if injected in vadose zone, and groundwater, if injected beneath water table)
- can treat soils to depths exceeding 75 feet *(34, pg. 35)*
- can treat large volumes of soil at reasonable costs *(36, pg. 2; 49, pg. 240)*
- a potential for product recovery exists *(35, pg. 136)*
- uses standard, available equipment, which allows for rapid mobilization and installation *(50, pg. 1)*
- demonstrated effectiveness in removing VOCs from unsaturated soils *(36, pg. 3)*
- enhances bioremediation *(48, pg. 269)*
- relatively inexpensive treatment method (approximately $50/ton for 10,000 cubic yards of TCE contaminated clay soils), although dewatering and off-gas treatment requirements result in higher costs *(34, pg. 7)*.

The disadvantages and limitations of vacuum extraction include the following:
- siltation effects can cause blockage of wells, potentially necessitating the drilling of new wells
- lengthy treatment times
- limited to unsaturated soils *(51)*
- highly soluble organics are more mobile and difficult to remove in dissolved phase *(9, pg. 84)*
- potential requirement for soil dewatering
- not applicable for less volatile organics, metals, inorganics, and cyanides *(9, pg. 84)*
- variable removal efficiencies.

If removal requirements are stated as low, compound-specific soil concentration levels and/or are required to be achieved in short periods of time, vacuum extraction should not be considered a reliable treatment. However, if the required removal is based on a total limit (e.g. total VOC concentration) and/or is to be verified based on gas stream concentrations, vacuum extraction can achieve significant removal efficiencies at relatively low costs *(41, pg. 168)*. An additional important issue with vacuum extraction is the amount of in-situ contaminant biodegradation which can be encouraged through aeration *(52)*. The effect of aeration on biodegradation is explored in the following sections.

Aeration

Technology Description.   Aeration encompasses a number of processes by which soil contaminants are removed from or transformed in the soil matrix by the presence of air (oxygen) or air flow (evaporation). Soil vapor extraction is one aeration mechanism. Other methods include air injection and "land farming" approaches. Land farming is used to remediate shallow soils and is accomplished by tilling the soil to expose surfaces to the air to enhance volatilization or oxygen availability. This technique is a low cost approach since the removal is accomplished primarily by natural processes (controlled by wind speed, temperature and contaminant vapor pressure). The oxygen supply provided by aeration mechanisms can encourage bioremediation, and removal efficiencies are thus achieved by a combination of processes.

Ex-situ aeration approaches may also be implemented. In such an application, aeration is accomplished mechanically by mixing/shredding similar to the shallow in-situ process, but exposure of contaminant surfaces to the air stream can more easily be affected. Shredding systems break the soil into smaller particles, thereby exposing more surface area. Waste piles may be aerated either through blowing air into piles (injection) or drawing air through piles (extraction). Extraction provides a more effective method for controlling emissions of the contaminated air stream. Passive evaporation can be accomplished through the spreading and mechanical mixing of contaminated soil on a prepared surface.

Requirements for air emission controls will be decided on a site specific basis and will be a function of site location, contaminant types, and vapor concentrations. Air stream controls that may be implemented include carbon adsorption and thermal treatment.

Aeration Performance. Aeration processes are most effective on the more volatile, less soluble contaminants such as halogenated solvents. However, success has also been reported on general category parameters, such as total petroleum hydrocarbons, aromatic volatile compounds, and other solvents such as acetone and 2-butanone.

The shredding and aeration of halogenated volatile organic contaminated soils at a spill site have been successfully demonstrated *(53, pg. 10)*. Approximately 1,300 cubic yards of soils were excavated and stockpiled for treatment. Concentrations of total volatile organics ranging up to 3,640 ppb were reduced below 2.5 ppb after up to seven passes through the shredder.

A passive evaporation bench-scale test was conducted on soils contaminated with spilled liquid refinery wastes to evaluate the effectiveness of this process on the treatment of phenol and cresoles *(54)*. A simulation was run for up to 21 days. Reduction of phenol ranged between 58 and 66% for initial concentrations in the 1,000 ppm range. Reduction of cresoles showed a wider range of reduction from 36 to 80%, with initial concentrations in the 100 ppm range *(54, pg. 346)*.

A pilot study conducted on soils contaminated with acetone, 2-butanone and other volatile organic compounds achieved significant removal with mechanical mixing and aeration *(55)*. Mechanical mixing was effected by a rototiller. Acetone concentrations were reduced 72-95% (range of 84-220 ppb reduced to 3-61 ppb) and 2-butanone concentrations were reduced by 70% (81 ppb to 24 ppb).

In another study, the effect of aeration on biodegradation was studied. Fuel contaminated soils were vented, and a mass balance for the fuel was calculated based on oxygen and carbon dioxide levels in the vented off-gas *(56, pg. A-8)*. Based on the extracted gas, approximately 62% of the fuel was removed through volatilization and 28% was biodegraded, for an overall 90% removal efficiency.

Aeration Advantages/Disadvantages. The advantages of aeration technologies are as follows:
- can be combined with bioremediation to enhance overall treatment effectiveness
- can be conducted in-situ in some applications, thereby avoiding LDRs
- relatively low cost when distinctly applicable (i.e. solvent contamination)

- assists in lowering cost, by increasing efficiency, when augmenting other technologies (i.e. biodegradation).

The disadvantages and limitations of aeration include the following:

- major effectiveness is limited to volatile compounds
- in-situ, and to some extent ex-situ, application is limited by soil types and soil homogeneity. Loose, sandy soils are more amenable than tight clays or organic soils. Lack of homogeneity may cause incomplete treatment due to short circuiting of the air flow.
- control of air emissions may be difficult in some applications.

Aeration can be applied in a number of applications but typically augments other processes. The majority of in-situ aeration applications are currently tied to soil vapor extraction systems (as opposed to land farming), due to the greater treatment depths and better emissions controls which can be effected through SVE. Aeration continues to be a chosen remedy for hazardous waste site clean-ups, typically for less contaminated soil and/or compound-specific treatment (i.e. VOC or halogenated solvent removal). Future application of aeration will be in its continued use for ex-situ bioremediation and greater future use to augment in-situ bioremediation.

Bioremediation

Technology Description. Bioremediation involves enhancing the natural biological transformation of chemical contaminants into less toxic and/or less mobile forms. The technologies are gaining increased attention due to their cost and effectiveness. As a transformation mechanism, bioremediation achieves permanent treatment of contamination. These two criteria (permanence and treatment) are important for remedy selection under the National Contingency Plan.

Some biodegradation occurs naturally at most hazardous waste sites. Bioremediation focuses on enhancing natural transformation by augmenting or optimizing conditions for its occurrence. In order for bioremediation to occur, organisms capable of degrading the compounds must be present, enzymes required for the process must be able to be synthesized, and environmental conditions must be favorable for the reaction to occur (75, pg. 661). Bioremediation technologies are currently being applied both in-situ and ex-situ.

Both anaerobic and aerobic approaches to bioremediation have been considered. Anaerobic degradation may only be effective as a partial approach, since complete degradation is not typically effected. However, anaerobic reduction reactions can degrade some compounds to others that are more amenable to further aerobic degradation (20, pg. 43). Dehalogenation of some compounds is also noted to occur anaerobically. However, the implementation and control of in-situ anaerobic conditions may be difficult (57, pg 306). Therefore, the majority of the bioremediation activities currently being conducted use aerobic processes.

There are four possible approaches for bioremediation of hazardous contaminants (20, pg. 38). These include the following:

- enhancement of naturally occurring biochemical mechanisms
- augmentation of natural systems with exogenous acclimated or specialized microorganisms
- application of cell free enzymes
- application of vegetation.

For deeper soils, in-situ enhancement of nutrients or microorganisms can be accomplished through well/nozzle injection, infiltration trenches, pits, or other irrigation approaches. For shallow soils, enhancement may be implemented via spray irrigation/infiltration, plowing, or other mechanical mixing.

The availability of the contaminants to the microbes is a significant factor in the potential effectiveness of bioremediation. For example, the volatility, sorptability, and solubility of a contaminant influence its bioavailability. Oxygen is often a limiting factor in bioremediation, and a number of different approaches have been considered to deliver oxygen to the soil (Figure 6). One approach is to inject air directly into the vadose zone while encouraging circulation through air extraction wells. This process has been described as bioventing, and is a variant of aeration and vacuum extraction. Another approach is to inject oxygen rich liquids such as hydrogen peroxide, ozone, or colloidal gas aphrons (CGAs). Unfortunately, hydrogen peroxide and ozone can be toxic to microorganisms, clog soil pores through bubble formation, and deplete rapidly. In addition, hydrogen peroxide may mobilize some metals such as lead *(20, pg. 43)*. As a result, the usefulness of these approaches may be limited.

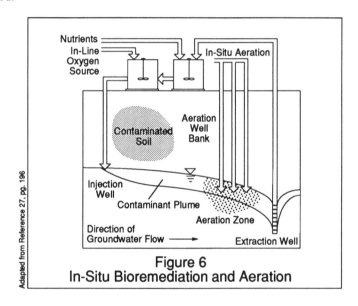

Figure 6
In-Situ Bioremediation and Aeration

Colloidal gas aphrons are microscopic gas bubbles formed with a surfactant. These bubbles can be formed with air, oxygen or nitrogen and do not tend to coalesce. Therefore, they can flow more readily through the soil matrix. Bench and pilot scale studies have shown a greater oxygen retention in the soil matrix using CGAs *(20, pg. 39)*.

For ex-situ bioremediation, a more controlled environment may be developed. In addition to optimizing oxygen, nutrient and microbial populations, ex-situ processes can better control pH, moisture, and temperature which significantly impact the viability of the remediation. Optimal conditions vary for different microorganisms, and a variety of optimal conditions have been reported *(58, pg. 728)*.

Biodegradation of some compounds can be enhanced by adding additional electron receptors, such as nitrates, but this does not appear to help degrade chlorinated ethanes and ethanes. These compounds may be degraded by cometabolism, a condition in which one compound is degraded beneficially while the microorganism is metabolizing another. There are indications that trichloroethylene and 1,1,1-trichloroethane are cometabolated with methane by methanotrophs *(20, pg. 51)*.

At some hazardous waste sites, the indigenous microorganisms may not be capable of bioremediating the compounds, or at least not in a reasonable length of time. In these instances, the natural system may be augmented with exogenous acclimated microorganisms and/or genetically engineered organisms. Such organisms are available commercially and have shown success in some laboratory trials. Pentachlorophenol (PCP) was rapidly degraded in the laboratory using Arthrobacter incubated at 30°C, and 2,4,5-trichlorophenoxy-acetic acid (2,4,5-T) was essentially removed from soil using genetically engineered Psendomonas cepacia *(20, pg. 52)*. However, in one reported testing, native microbial cultures did better in removing chlorinated hydrocarbons than did acclimated commercially available cultures.

Although some research on white rot fungus has demonstrated the ability to degrade some halogenated organic compounds, particularly pentachlorophenol, halogenated solvents are not typically addressed by commercially available microbial augmentation products. Products which are available are often limited in their application to natural soil environments, due to viability issues *(20, pg. 54)*.

The application of cell-free enzymes is another potential treatment method for hazardous compounds. Enzymes produced by microorganisms may be extracted and applied directly to the hazardous constituent. In theory, the enzymes might be effective in environmental extremes in which the microorganism may not thrive *(20, pg. 56)*. The development of engineered enzymes or extraction of adequate quantities of enzymes has apparently not yet become commercially viable. This technology is therefore not commonly used for bioremediation.

Vegetative uptake is another biological process with potential applicability to the remediation of hazardous contaminants. Some plants are capable of bioaccumulating or transforming contaminants either through the root system or through vegetative uptake of vapors. Several studies have shown that plant roots can adsorb lipophilic contaminants from the soil. Specifically, 2,3,7,8-TCDD has been known to be absorbed *(20, pg. 58)*.

Ex-situ processes include "prepared bed" processes such as windrow/composting, land farming, and bioreactor/bioslurry mechanisms. Composting processes typically are conducted in long piles (windrows) or in reactors. The biodegradable waste is mixed with a highly biodegradable bulking material, such as wood chips, at a ratio on the order of 9 to 1 bulking to waste material. Aeration is accomplished by turning the windrows/piles and/or by forced aeration of static piles via a grid of perforated pipes. Moisture, nutrients, and microbial populations are enhanced and temperature is controlled as necessary, based on pilot studies. A similar process is conducted in a composting reactor, except that mixing is provided by mechanical means such as tumbling *(9, pg. 99)*.

Bioreactor (bioslurry) treatment processes are conducted with the waste materials in an aqueous slurry, which typically contains approximately 50% solids. The optimal percent solids is determined based on the types and concentrations of contaminants. The slurry is enhanced with oxygen and nutrients, seeded with microbial cultures as necessary *(9, pg. 99)*, and treated typically through a batch operation *(60, pg. 3)*.

Wastes treated by bioreactors include PCBs, PNAs, pesticides, and wood treatment compounds such as cresol and pentachlorophenol. This method is geared toward the treatment of high concentrations (up to 250,000 mg/kg) of soluble organic contaminants in soils and sludges *(60, pg. 1)*. Bioslurry treatment processes produce sludge/soil residues that must be dewatered, potential air emissions, and wastewaters. The solids may still require additional treatment for residual organic or metal contaminants prior to disposal. The process water is typically treated and discharged or recycled. Air emissions, if any, can be controlled by vapor phase carbon treatment *(60, pg. 3)*. A typical bioslurry process is depicted in Figure 7 *(60)*.

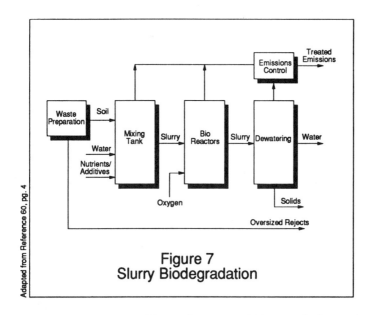

Figure 7
Slurry Biodegradation

Bioremediation Performance. The performance of bioremediation technologies vary considerably with their application and the type of target contaminants. In-situ bioremediation applications are typically combinations of groundwater and soil treatment, and performance data specific to soil cleanup is thus limited. The application of bioremediation to the treatment of halogenated volatiles is considered to be potentially effective *(9, pg. 13)*, but limited experience is available on in-situ soil treatment.

Microbial degradation of TCE and other related solvents has been studied in both aerobic and anaerobic laboratory experiments. Microbial degradation of 1,1,1-trichloroethane under anaerobic conditions was demonstrated using ethanol as a substrate *(61)*. Anaerobic degradation of TCE by dechlorination to 1,2 DCE in soil has also been demonstrated *(62)*. Degradation of TCE by cometabolism is evident by studies of TCE degradation with aromatic compounds (toluene) and phenols *(63-64)*. Recent work has also demonstrated the effectiveness of aerobic methanotrophic organisms to biodegrade TCE to DCE, vinyl chloride and CO_2 *(65-67)*. Stimulation of methanotrophs in an aquifer by injecting methane and oxygen resulted in degradation of TCE by 20%, cis-DCE by 40%, trans-DCE by 85%, and vinyl chloride (VC) by 95% within a one day detention time *(67, Roberts et al, 1989)*. Studies at Stanford University demonstrated a consistent 75-80% vinyl chloride (VC) removal in methane fed columns with influents ranging from 175-775 µg/L. Little or no removal occurred in a control column which received dissolved oxygen and VC but no methane *(67, pg. 9)*. VC removals of 95% were achieved later in the study. Columns which were fed TCE, trans-DCE, and VC initially failed to demonstrate TCE removal compared to the control column. However, after the columns were fed and left for an additional three weeks, 30% TCE, 80% trans-DCE, and 95% VC removal efficiencies were achieved *(67, pg. 10)*. Some data indicate an upper limit for microbial viability at 300 mg/kg TCE with optimum temperatures of 22-37°C and a pH of 7 to 8.1 *(65, pg. 1709, 1713)*. In groundwater studies of methanotrophic degradations, half lives for TCE were assessed in the range of hours to days *(67, pg. 3)*. These studies indicate that in-situ anaerobic and aerobic degradation of TCE and other halogenated solvents is possible, but further study for full scale application is necessary.

Field application data for the in-situ treatment of other compounds provides some insight on how such processes may succeed on chlorinated solvent contaminated sites. Using a soil bioventing approach on a pilot scale, treatment of fuel contaminated soil (JP4, diesel fuel and gasoline) after seven months showed biodegradation removal of about 30% of the total hydrocarbons and an additional 20% removal due to soil venting *(56, pg. A-9)*. Biodegradation was dominant over vapor extraction when the air flow rate was reduced to about 1/4 pore volume exchanges/day. Soil temperature was noted as a dominant treatment factor, while nutrient variations showed little impact. Since fuel related compounds biodegrade more rapidly than TCE, vapor extraction may be expected to dominate as a removal mechanism in TCE bioventing.

In a more conventional in-situ bioremediation approach, a gasoline contaminated soil site was biodegraded from 245 ppm volatile fuel hydrocarbons to 0.8 ppm after 200 days *(69, pg. 143)*. Injection and recovery wells were used to cycle nutrients and hydrogen peroxide through the soils.

The performance of ex-situ composting of solvent wastes has been demonstrated by laboratory studies *(70)*. Full scale operation of a windrow type operation was simulated for soils contaminated with up to 6,000 ppm of total volatile organic compounds (primarily F003 and F005 wastes). A 94% reduction was achieved over a 56 day period. Based on available data; however, it is not certain how much was due to biological activity and how much to aeration.

In a similar composting approach which used rototilling to provide aeration/mixing, a full scale biotreatment of PAHs achieved a 95% reduction of 2- and 3-ring PAHs and 70% reduction of 4- and 5-ring PAHs during the first year of operation *(69, pg. 142)*. A similar pilot scale test under controlled "green house" type conditions demonstrated success on oil field soils contaminated with styrene, still-bottom tars, and chlorinated hydrocarbon solvents *(69, pg. 142)*. Volatile compounds were reduced by more than 99%, probably by aeration, and semivolatiles by 89%, most likely by biodegradation.

The success of ex-situ slurry biotreatment is based primarily on vendor claims *(60, pg. 3 and 5)*. Full scale slurry biodegradation systems claim 99% removal efficiency for PCPs and PAHs in wood processing sludges. In an oil refinery sludge which was treated with a residence time of 60 days, 30 to 80% of the carcinogenic (5 and 6 ring PAHs) and virtually 100% of the non-carcinogenic PAHs were claimed to have been removed *(60, pg. 5)*.

A mobile slurry biotreatment system was used to treat approximately 1,700 cubic yards of contaminated silty-sand soils in California. The vendor claimed reduction of TCE from a maximum of 180 ppm to 25 ppm in four days. Retreatment of the soils was required to reach non-detectable levels *(71)*.

Bioremediation Advantages/Disadvantages.   The advantage of various bioremediation systems to the treatment of halogenated solvent contaminated soils are as follows:

In-situ
- avoids impacts of LDRs for mixtures of F001 - F005 listed waste solvents and soils
- treatment system is typically designed to treat contaminated groundwater as well as soils
- increases overall removal efficiency of soil vapor extraction system

Ex-situ
- allows for control of biodegradation parameters to enhance effectiveness of biodegradation
- provides a higher degree of treatment than in-situ bioremediation.

The disadvantages and limitations of bioremediation of halogenated solvent contaminated soils are as follows:

In-situ
- detailed information is required on hydrogeologic and soil conditions
- technical difficulties in providing in-situ nutrients and oxygen have not been resolved, thereby creating conditions of nonhomogeneous treatment
- effectiveness of bioremediation of TCE and related compounds has not been fully tested and cleanup requirements may not be reached

Ex-situ
- higher cost for treatment of soils and residuals
- potential difficulties in achieving required treatment levels
- LDRs may not be avoided without a treatability variance
- control of emissions during soil removal and handling.

For both in-situ and ex-situ processes, biodegradation presents a disadvantage due to its potential for generating unwanted end-products. For example, the biodegradation products may be more toxic, more persistent, and/or more susceptible to bioaccumulation than the original contaminants (*75, pg. 662*).

Soil Flushing/Washing

Technology Description. Soil flushing and soil washing are processes whereby solutions are used to remove organic or inorganic contaminants from soils. Soil flushing/washing differ in solids handling as well as solution application techniques.

In-situ application of this process is usually called soil flushing, but terms such as soil leaching may also be applied. In this application the fluid is used to wash contaminants from the saturated zone by injection and recovery systems. Removal from the vadose zone can be effected by spray irrigation, leaching and recovery systems (Figure 8). In either case a solvent, surfactant solution, or water with or without additives is applied to the soils to enhance contaminant release and mobility, resulting in increased recovery and decreased soil contaminant levels (*9, pg. 72*). This technology can be applied to volatile organics, such as halogenated solvents. However, it has been more often applied to heavier organic compounds such as oils or wood treating compounds.

The soil flushing treatment process is ideally a closed-loop system. The solution is applied by injection wells, sprinklers, or ponds and the soils are flooded (*20, pg. 7*). The specific solution used, whether a chemical solvent or water with various additives, is chosen based on the compounds of concern. The solution and released contaminants are recovered downgradient by an extraction system, typically recovery wells. The solution is then treated by standard wastewater treatment methods selected to address the particular contaminants and required level of treatment. The treated wastewater is enhanced with additional additives, if necessary, and then reinjected into the system. Depending upon the physical site setting, in-situ barriers such as slurry walls may be needed to control losses of solvents and contaminants from the system to surrounding soils. This type of soil flushing process may also be combined with enhanced bioremediation, as discussed previously, by augmenting the flushing solution with nutrients to encourage the biodegradation of organics remaining in the soil.

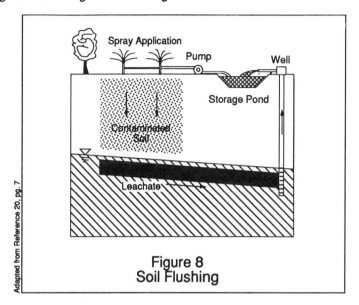

Figure 8
Soil Flushing

Adapted from Reference 20, pg. 7

Several unique variants to this basic process also exist which are considered to be soil flushing technologies. One method developed in Germany uses high pressure water injection to wash soils *(72)*. The high pressure water is applied by a rotating lance vertically down a boring that is jacketed by a casing. The soil/water mixture flowing out the top is collected, dewatered and treated. Subsidence in the borehole is compensated with clean sand. Boreholes are overlapped to ensure complete treatment of the site.

A second unique process evolved from the oil industry and uses steam and hot water to "sweep" oily waste to the point of extraction. This process, the Contained Recovery of Oily Wastes (CROW) process *(9, pg. 148)*, may be applicable to wood treating sites and sites with soils containing dense organic liquids such as coal tars, pentachlorophenol, or creosote. Further in-situ soil treatment such as additional flushing or bioremediation will be required following oil removal.

Ex-situ processes are typically termed soil washing. However, one approach that has been employed was actually a flushing/leaching procedure applied ex-situ to soils in a prepared bed. The bed was prepared with a double lining, as well as leachate collection and spray irrigation systems, and the soil was treated in batches.

Soil washing systems mechanically scrub contaminated soils in a solution to remove contaminants (Figure 9). The contaminants are removed from the bulk of the soils by either solubilizing or suspending them in a solution which is removed for separate wastewater treatment, or by separating the fractions of the soil matrix by particle size and concentrating those particles upon which the contaminants are adsorbed *(73, pg. 1)*. This latter approach is based on the concept that most organic contaminants are associated with silt and clay sized particles which adhere to larger soil particles. Separating the soil particles allows the larger fractions to be removed, dried, and returned to the source.

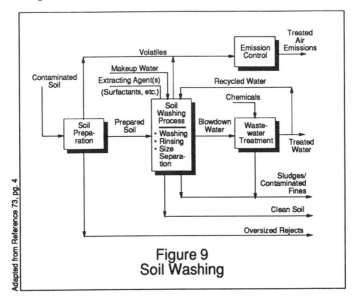

Figure 9
Soil Washing

Adapted from Reference 73, pg. 4

Soil washing may be applied as a stand-alone process or in conjunction with other treatment processes. As a stand-alone process, washing may not always be successful in attaining clean-up criteria. It may, however, provide a useful pre-processing step to reduce the volume for other treatment, or homogenize the soil for further treatment by processes such as bioremediation. Excessive amounts of silt and clay represent problems to the use of soil washing because of the difficulty of removing them from solution.

The soil washing system is comprised of an initial materials handling step, the soil washing process, emissions controls, wastewater treatment, and disposal of cleaned soils, residual sludge and treated water. The initial materials handling step uses various physical screening processes to remove large objects and break up the soils to enhance wetting. The soil washing process typically includes steps for washing, rinsing and particle size segregation. Any extraction agents to be used in the process are applied at this step. After washing, the clean soil is rinsed with water and removed. Suspended solids are separated from the wastewaters, which are then treated by conventional means and recycled. Residual sludges in which the majority of the original contamination will usually reside are removed and treated or disposed.

Soil Flushing/Washing Performance. The following summarizes several soil flushing and soil washing tests.

Soil Flushing. A test was conducted using surface soil flushing on sandy soils of an Air Force fire training pit contaminated with chlorinated hydrocarbons and other hydrocarbons (53, pg. 13). Initial concentrations of chlorinated hydrocarbons were up to 3.5 ppm and other hydrocarbons totaled hundreds of ppm. After about 14 pore volume washings, no measurable decrease in contamination was found, possibly due to high soil sorption values. Similar poor results were seen from in-situ flushing of fire training area soils using surfactants *(74)*. Although 75-94% removal of hydrocarbons was seen in laboratory column tests, in-situ application was not effective. Reduction of soil permeability due to clogging of soil pores was seen after repeated application of surfactants.

Application of high pressure in-situ soil washing on soils contaminated with PAHs showed relatively good removal. Pilot tests using the vertical cased borehole approach showed reductions of PAHs from 20,000 ppm to 10 ppm *(72)*.

Ex-situ soil leaching in prepared beds was effective in reducing phenol concentrations by 99.9% (from 980 ppm) and cresol concentrations 99.7% *(54, pg. 354)*.

Soil Washing. Soil washing performance data shows consistent success and removal of contaminants from soils. Table XIII summarizes soil washing performance by various vendor procedures on a variety of selected soil types and contaminants *(73, pg. 8,9)*.

Table XIII. Soil Washing Removal Efficiencies

Contaminant	Removal (%)	Residual Concentration (ppm)
Oil and Grease	50-83	250-600
	99	<5
	90	2,400 (small soil particle size)
Phenol	80-90	1-96
	<99.8	<0.01
Chlorinated hydrocarbons	98	<1
	>99	0.5
	>75	<0.01
Volatile organics	98-99	<50
Aromatics	>81	>45
	99.8	<0.02
Semivolatiles	98-99	<250
PAH/PNA	95	15
	86-90	91.4-97.5
	98	15-20
	95.4	15-16

SOURCE: Adapted from ref. 73.

Soil Flushing/Washing Advantages/Disadvantages. The advantages and disadvantages of soil flushing/washing technologies are as follows.

Soil Flushing. Soil flushing advantages include:
- avoids LDRs
- lower treatment costs unless surfactants/special additives are needed
- can be combined with enhanced biodegradation.

Soil Washing. Soil washing advantages include:
- proven effective on most contaminants
- lower treatment cost than thermal approaches
- can reduce volume for follow-on treatment.

Soil Flushing. Soil flushing disadvantages include:
- limited by soil characteristics. May have slow rates or incomplete coverage due to permeability and homogeneity issues.
- need to control application area to capture solvent/contaminants
- may introduce new contaminants either through transformation products or due to the solvents used
- cannot address complex mixtures of contaminants
- may not achieve clean-up objective.

Soil Washing. Soil washing disadvantages include:
- develops additional waste streams for treatment
- not highly effective on soils with high silt or clay content
- use of surfactants or other additives for washing may inhibit subsequent wastewater treatment.

Conclusions

This paper has presented summaries of selected technologies and their applications and limitations towards treating soils contaminated with halogenated solvents. Both in-situ and ex-situ technologies were addressed, including low temperature thermal treatment, radio frequency heating, steam stripping, vacuum extraction, aeration, bioremediation, and soil flushing/washing. Advantages and disadvantages of the technologies as well as specific applications were summarized.

As demonstrated by the technology reviews, a clear choice of the most appropriate technology to use for treating soils contaminated with F001 to F005 solvent listed wastes does not exist. Both soil and contaminant characteristics affect achievable removal efficiencies, while LDRs influence treatment requirements. Table XIV summarizes comparative criteria which can be used in a relative assessment of these technologies. Representative removal efficiencies which have been demonstrated in bench, pilot, and/or field scale studies are included. However, these results are based on specific applications, and other situations or treatment durations could result in higher or lower removal efficiencies.

As discussed in this paper, site-specific issues impact treatment effectiveness. For example, a process may be reliable in contaminated sands, and ineffective in a clay. Some of these effects can be compensated for by extending treatment duration; however, this increases costs. A detailed comparison of treatment costs was not made in this paper, since quoted costs vary widely even among vendors of similar technologies and site-specific issues (such as off-gas treatment trains) contribute significantly to any per unit cost. Some representative issues which should be considered in obtaining comparable costs for different technologies include the following: site preparation (excavation of soil, grading requirements), mobilization/demobilization (availability of power/water, set-up time, trial tests), treatment train (intended use of system, particle size limitations, operating temperature, through-put rate, cost of low vs. high volume, waste streams generated, moisture content effects, permits, off-gas treatment, analyses of contaminated and treated soil), disposal issues (soil, air, wastewater), man power requirements (number of operators, hours per day of operation), length of treatment (effect on operating costs,

Table XIV.
Comparison of Technology Performance on Contaminated Soils

Compound	Representative Range of Removal (%)	Comments
Low Temperature Thermal (ex-situ)		
TCE, MEK, PCE	>99	Not economical for small
non-carcinogenic PNAs	63-95	volumes. Lower efficiencies
carcinogenic PNAs	90-95	on fine soils. Good
BTEX	>99	reliability, proven. Higher cost. Volatilization losses in handling.
Radio Frequency Heating (in-situ)		
PCE	>99	Not economical for deep
semivolatile aliphatics	88-98	contamination. Not fully
semivolatile aromatics	98-99.9	tested.
Steam Stripping (in-situ)		
TCE, PCE, CB	85 (avg)	More costly for clay soils.
semivolatiles	55 (avg)	Increased speed of cleanup. Not economical for small volumes. High removals (>99%) possible.
Vacuum Extraction (in-situ)		
TCE	8-100	Can treat to greater depths (>75 feet). Proven technology. More cost effective for sandy soils. Applicable for large volumes. Relatively inexpensive. Variable removal efficiencies. Limited to unsaturated soils.
Aeration (ex-situ)		
volatile organics	99.9	Limited by soil types and to
phenol	58-66	volatile compounds. Proven
cresol	36-80	in limited application. In-
acetone	72-95	situ application typically tied to vacuum extraction.
Bioremediation (in-situ)		
TCE	20-30	Limited by soil permeability.
Vinyl chloride	95	Effectiveness on TCE not
trans-DCE	80-85	fully tested. Cleanup
cis-DCE	40	requirements may not be
total hydrocarbons	30-97	reached.

Continued on next page

Table XIV. - Continued
Comparison of Technology Performance on Contaminated Soils

Compound	Representative Range of Removal (%)	Comments
Bioremediation/Composting (ex-situ)		
F003/F005 solvents	94	Proven technology.
volatile compounds	99	Applicable to a wide
PAHs	70-95	variety of organics.
semivolatile compounds	89	
Bioremediation\Bioslurry (ex-situ)		
TCE	86-100	Higher cost. Treatment of
PCPs	99	residuals may be required.
carcinogenic PAHs	30-80	
non-carcinogenic PAHs	100	
Soil Flushing (in-situ)		
PAHs	99.9	Limited data on performance on TCE. Generally lower cost. Limited by soil conditions. May release contaminants. Lab scale 75 - 94% HC removal not reproducible in field study.
Soil Washing (ex-situ)		
chlorinated hydrocarbons	75-98	Proven effective. Higher
volatile organics	98-99	cost than in-situ, lower than
aromatics	81-99.8	thermal. Achieves volume
phenol	80-99	reduction. Not as effective
semivolatiles	98-99	on high silt or clay soils.
PAH/PNA	86-98	
Soil Leaching (ex-situ)		
phenol	99.9	Issues similar to ex-situ soil
cresol	99.7	washing.

community perceptions), site characteristics (water table level, extent/location of plume, climate), soil characteristics, and waste characteristics.

The reliability of any technology is also a consideration in treatment selection. Bench and pilot scale tests can be used as an indicator of a treatment technology's effectiveness for a specific site, although care must be taken to ensure testing of a representative portion of the site. Full scale tests from similar sites (if available) can aid in estimating treatment time and costs. As mentioned, vacuum extraction is currently the most widely used innovative technology. However, other technologies may be preferable for specific sites. In the final analysis, each potential application of the technologies considered here must be evaluated for its effectiveness, reliability, and relative cost based on soil characteristics, contaminants, bench scale tests, and engineering judgment.

Literature Cited

1. USEPA, Implementing The Land Disposal Restrictions, Solid Waste and Emergency Response (05-343), October, 1989.
2. Lee, C.C., Huffman, George L., and Sasseville, Sonya M., "Incinerability Ranking Systems for RCRA Hazardous Constituents." Hazardous Waste & Hazardous Materials, Vol. 7, No. 4, 1990, pp. 385-394.
3. Swanstrom, Carl, and Palmer, Carl, "X*TRAX Transportable Thermal Separator for Organic-Contaminated Soil." Presented at the Second Forum on Innovative Hazardous Waste Treatment Technologies: Domestic and International, May 1990.
4. Cudahy, James J., and Eicher, Anthony R., "Thermal Remediation Industry: Markets, Technologies, and Companies." Pollution Engineering, November 1989, pp. 76-80.
5. Recycling Sciences International, Inc., Desorption and Vaporization Extraction System (DAVES), Chicago, Illinois, 1990.
6. Nielson, Roger K., and Cosmos, Michael G., "Low Temperature Thermal Treatment (LT3) of Volatile Organic Compounds from Soil: A Technology Demonstrated." Environmental Progress (Volume 8, No. 2), May, 1989, pp. 139-142.
7. Johnson, Nancy P., and Cosmos, Michael G., "Thermal Treatment Technologies for Hazwaste Remediation." Pollution Engineering, October 1989, pp. 66-85.
8. Roy F. Weston, Inc., "Demonstration of Thermal Stripping of JP-4 and Other VOCs from Soils at Tinker Air Force Base." Prepared for the Army Toxic and Hazardous Materials Agency, Report Number CETHA-TE-CR-90026, March 1990.
9. Holden, Tim, et. al., "How to Select Hazardous Waste Treatment Technologies for Soils and Sludges: Alternative, Innovative, and Emerging Technologies." Noyes Data Corporation, 1989, pp. 1-149.
10. U.S. EPA, Innovative Treatment Technologies: Semi-Annual Status Report. EPA/540/2-91/001, January 1991.
11. U.S. EPA, The Superfund Innovative Technology Evaluation Program: Technology Profiles. EPA/540/5-90/006, November 1990.
12. Grosshandler, William L., "Research Needs for the Thermal Treatment of Hazardous Waste." Hazardous Waste & Hazardous Materials, Vol. 7, No. 1, 1990, pp. 1-6.
13. Dev, H., Enk, J., Stresty, G., Bridges, J., and Downey, D., "In Situ Soil Decontamination by Radio-Frequency Heating Field Test." Illinois Institute of Technology Research Institute, September 1989.

14. Dev, Harsh, and Downey, Douglas C., "Radio Frequency Thermal Soil Decontamination Pilot Test." Case Study for the NATO/CCMS Pilot Study on Remedial Action Technologies for Contaminated Land and Groundwater, November 1988.

15. Dev, Harsh, and Downey, Douglas, "Zapping Hazwastes." Civil Engineering, August 1988, pp. 43-45.

16. Dev, Harsh, Bridges, Jack E., and Sresty, Guggiliam C., "Decontamination of Hazardous Waste Substances From Spills and Uncontrolled Waste Sites by Radio Frequency In Situ Heating." Proceedings of the Hazardous Materials Spill Conference, April 1984.

17. Dev, H., Condorelli, P., Bridges, J., Rogers, C., and Downey,D., "In Situ Radio Frequency Heating Process for Decontamination of Soil." Solving Hazardous Waste Problems: Learning from Dioxins, American Chemical Society Symposium Series 338, April 1986.

18. Dev, Harsh, and Downey, Douglas, "Field Test of the Radio Frequency In Situ Soil Decontamination Process." Superfund '88: HMCRI 9th National Conference and Exhibition, November 1988.

19. Fang, C.S., Lai, Peter M.C., and Chang, Bruce K.L., "Oil Recovery and Waste Reduction by Microwave Radiation." Environmental Progress. Vol. 8, No. 4, November 1989, pp. 235-238.

20. U.S. EPA, Handbook on In Situ Treatment of Hazardous Waste-Contaminated Soils. EPA/540/2-90/002, January 1990.

21. Dev, H., Bridges, J., Sretsty, G., Enk, J., Mshaiel, N., and Love, M., "Radio Frequency Enhanced Decontamination of Soils Contaminated with Halogenated Hydrocarbons." EPA Project Summary, EPA/600/S2-89/008, October 1989.

22. Dev, Harsh, "Radio Frequency Enhanced In Situ Decontamination of Soils Contaminated With Halogenated Hydrocarbons." Proceedings of the 12th Annual Research Symposium, U.S. EPA, EPA/600/9-86/022, April 1986.

23. "A Guide to Innovative Thermal Hazardous Waste Treatment Processes." The Hazardous Waste Consultant, Vol. 8, Issue 6, November/December 1990.

24. AWD Technologies Integrated AquaDetox/SVE Technology. Draft Applications Analysis Report, U.S. EPA, April 1991.

25. AWD Technologies AquaDetox/SVE Technology. Draft Technology Evaluation Report, U.S. EPA, July 1991.

26. U.S. EPA, "Toxic Treatments, In Situ Steam/Hot Air Stripping Technology." Applications Analysis Report, EPA/540/A5-90/008, March 1991.

27. Ghassemi, Masood, "Innovative In Situ Treatment Technologies for Cleanup of Contaminated Sites." Journal of Hazardous Materials, 17, 1988, pp. 189-206.

28. LaMori, Phil, and LaMori, Jon, "Development of Post-Treatment Cleanup Criteria for In Situ Soils Remediation." Presented at the Second Annual West Coast Conference on Hydrocarbon Contaminated Soils and Groundwater, March 1991 (Reprint).

29. Diaz, Rudy, and Guenther, Jeff, "Detoxifier for Contaminated Soils." Pollution Engineering, January 1991 (Reprint).

30. James, Stephen, "In Situ Steam/Air Stripping." Haznews, April 1991, p. 17.

31. Roy, Kimberly A., "In Situ Detoxifier." Hazmat World, April 1990 (Reprint).

32. NovaTerra Corporate Summary, Torrance, California.

33. Lord, Arthur E., Koerner, Robert M., Hullings, Donald E., and Brugger, John E., "Laboratory Studies of Vacuum-Assisted Steam Stripping of Organic Contaminants from Soil." Presented at the 15th Annual Research Symposium for Land Disposal, Remedial Action, Incineration, and

Treatment of Hazardous Waste, U.S. EPA Risk Reduction Engineering Laboratory, Cincinatti, OH, 1989, pp. 124-136.

34. U.S. EPA, "Terra Vac In Situ Vacuum Extraction System." Applications Analysis Report, EPA/540/A5-89/003, July 1989.

35. Hutzler, N.J., Murphy, B.E., and Gierke, J.S., "Review of Soil Vapor Extraction System Technology." Soil Vapor Extraction Technology: Reference Handbook, Camp Dresser & McKee, February 1991, pp. 136-162.

36. Pedersen, Tom A., and Curtis, James T., "Soil Vapor Extraction Technology: Reference Handbook." Camp Dresser & McKee, February 1991, pp. 1-135.

37. Hoag, G.E., "Soil Vapor Extraction Research Developments." Soil Vapor Extraction Technology: Reference Handbook, Camp Dresser & McKee, February 1991, pp. 286-299.

38. Derammelere, Ron, and Hlegerson, Ron, "Integrated Zero-Emission Groundwater and Soil Remediation Facility at Lockheed, Burbank." AWD Technologies.

39. Kemblowski, M.W., Colthart, J.D., Byers, D.L., and Stanley, C.C, "A Practical Approach to the Design, Operation, and Monitoring of In Situ Soil Venting Systems" Soil Vapor Extraction Technology: Reference Handbook, Camp Dresser & McKee, February 1991, pp.195-239.

40. Sims, Ronald C., "Soil Remediation Techniques at Uncontrolled Hazardous Waste Sites: A Critical Review." Journal of the Air and Waste Managment Association, Vol. 40, No. 5, May 1990, pp. 704-732.

41. Danko, J., "Applicability and Limits of Soil Vapor Extraction for Sites Contaminated with Volatile Organic Compounds." Soil Vapor Extraction Technology: Reference Handbook, Camp Dresser & McKee, February 1991, pp. 163-170.

42. Silka, L.R., Cirpili, H., and Jordan, D.L., "Modeling Applications to Vapor Extraction Systems." Soil Vapor Extraction Technology: Reference Handbook, Camp Dresser & McKee, February 1991, pp. 252-259.

43. Mutch, R.D., Jr., Vlarke, A.N., Wilson, D.J., and Mutch, P.D., "In Situ Vapor Stripping Research Project: A Progress Report." Soil Vapor Extraction Technology: Reference Handbook, Camp Dresser & McKee, February 1991, pp. 273-285.

44. Kerfoot, H.B., "Soil Gas Surveys in Support of Design Vapor Extraction Systems." Soil Vapor Extraction Technology: Reference Handbook,   a mp Dresser & McKee, February 1991, pp. 171-185.

45. Koltuniak, Donna L., "In Situ Air Stripping Cleans Contaminated Soil." Chemical Engineering, August 18, 1986, pp. 30-31.

46. Ellgas, Robert A., and Marachi, N. Dean, "Vacuum Extraction of Trichloroethylene and Fate Assessment in Soils and Groundwater: Case Study in California." pp. 794-801.

47. U.S. EPA, "SITE Program Demonstration Test, Terra Vac In Situ Vacuum Extraction System, Groveland, Massachusetts." Technology Evaluation Report, EPA/540/5-89/003a, April 1989.

48. DePaoli, D.W., Herbes, S.E., and Elliot, M.G., "Performance of In Situ Soil Venting System at Jet Fuel Spill Site." Soil Vapor Extraction Technology: Reference Handbook, Camp Dresser & McKee, February 1991, pp. 260-272.

49. Marley, M.C., Richter, S.D., Cliff, B.L., and Nangeroni, P.E., "Design of Soil Vapor Extraction Systems--A Scientific Approach." Soil Vapor Extraction Technology: Reference Handbook, Camp Dresser & McKee, February 1991, pp. 240-251.

50. U.S. EPA, "Soil Vapor Extraction Technology Reference Handbook." Project Summary, EPA/540/S2-91/001, June 1991.

51. Bennedsen, Magnus B., Scott, Joseph P., and Hartley, James D., "Use of Vapor Extraction Systems for In Situ Removal of Volatile Organic Compounds from Soil." AWD Technologies.
52. Hinchee, R.E., Downey, D.C., and Miller, R.N., "In Situ Biodegradation of Petroleum Distillates in the Vadose Zone." Soil Vapor Extraction Technology: Reference Handbook, Camp Dresser & McKee, February 1991, pp. 186-194.
53. University of Wisconsin - Madison, The College of Engineers, "Underground Tank Technology Update," Volume 3, No. 4, August 1989.
54. Evangelista, Robert A., et. al., "Treatment of Phenol and Cresol Contaminated Soil," presented at Symposium on Characterization and Cleanup of Chemical Waste Sites, American Chemical Society 197th National Meeting, Dallas, TX, April 10, 1989.
55. Pilot Study Report, Triangle Chemical Company. USEPA Region VI (ATTIC data base).
56. Downey, Douglas C., Michael G Elliott, and Catherine M. Vogel, "Performance of Selected In-Situ Soil Decontamination Technologies." Environmental Progress, August 1990.
57. Lepore, J. V., et. al., "Process Development for In-situ Remediation of a Highly Contaminated Groundwater Plume," Journal of Hazardous Materials, 25 (1990).
58. Sims, Ronald C., "Soil Remediation Techniques at Uncontrolled Hazardous Waste Sites, A Critical Review," Air and Waste Management Association, Volume 40, No. 5, May 1990.
59. Koltuniak, Donna L., "In-situ Air Stripping Cleans Contaminated Soil," Process Technology, Chemical Engineering, August 1986.
60. EPA/540/2-90/016, Engineering Bulletin- Slurry Biodegradation; September 1990.
61. Boyer, Jean D., Robert C. Ahlert and David S. Kosson, "Degradation of 1,1,1-Trichloroethane in Bench Scale Bioreactors," Hazardous Waste & Hazardous Materials, Volume 4, No. 3, 1987, p. 241.
62. Kleopfer, Robert D., Diane Easley, Bernard Haas, Jr., and Trudy Delhi, "Anaerobic Degradation of Trichloroethylene in Soil," American Chemical Society, Environmental Science Technology, Volume 19, No. 3, 1985, p. 277.
63. Nelson, Michael J. K., Stacy O. Montgomery, and P. H. Pritchard, "Trichloroethylene Metabolism by Microorganisms That Degrade Aromatic Compounds," Applied and Environmental Microbiology, Feb 1988, p. 604-606.
64. Folsom, B. R., P. J. Chapman, and P. H. Pritchard, "Phenol and Trichloroethylene Degradation by Pseudomonas capacia, G4: Kinetics and Interactions between Substrates," Applied and Environmental Microbiology, May 1990, p. 1279-1285.
65. Fliermans, C. B.,T. J. Phelps, D. Ringelberg, A. T. Mikell, and D. C. White, Mineralization of Trichloroethylene by Heterotrophic Enrichment Cultures," Applied and Environmental Microbiology, July 1988, p. 1709-1714.
66. Vogel, Timothy M. and Perry L. McCarty, "Biotransformation of Trichloroethylene to Trichloroethylene, Dichloroethylene, Vinyl Chloride, and Carbon Dioxide under Methanogenic Conditions," Applied and Environmental Microbiology, May 1985, p. 1080-1083.
67. McCarty, Perry L., Lewis Semprini, Mark E. Dolan, Thomas C. Harmon, Sharon Just, Claire Tiedeman, Steven M. Gorelick, and Paul V. Roberts, "Evaluation of In-situ Methanotrophic Bioremediation for Contaminated Groundwater, St. Joseph, Michigan," Technical Report No. WR-1, Western Region Hazardous Substance Research Center, September, 1990.

68. Ostendorf, David W., and Don H. Kampbell, "Bioremediated Soil Venting of Light Hydrocarbons," Hazardous Waste & Hazardous Materials, Volume 7, No. 4, 1990.
69. Sims, Judith L., Ronald C. Sims, and John E. Matthews, "Approach to Bioremediation of Contaminated Soil," Hazardous Waste & Hazardous Materials, Volume 7, Number 2, 1990.
70. Kabrick, Randolph M., and June R. Coover, "Composting as BDAT for Solvent (F Wastes) Contaminated Soils," Environmental Engineering: Proceedings of the 1989 Specialty Conference, ASCE.
71. ENSITE, Case History on Channel Gateway Development Site, Report of Pilot Testing and Full Scale Remediation.
72. Balthaus, I.H., "High Pressure In-Situ Soil Washing," Phillip Holzmann Ag, West Germany (Attic Data Base).
73. EPA/540/2-90/017, Engineering Bulletin-Soil Washing Treatment, September 1990.
74. Nash, J., R.P. Traver and D.C. Downey, "Surfactant-Enhanced In-Situ Soils Washing," Final Report, Engineering and Services Laboratory, Air Force Engineering and Services Center, Tyndall AFB, FL, 1987.
75. Grady, C.P. Leslie, Jr., "Biodegradation: It's Measurement and Microbiological Basis," Biotechnology and Bioengineering, Vol. XXVII, 1985, pp. 660-674.

RECEIVED October 1, 1992

Chapter 14

Constant Capacitance Model

Chemical Surface Complexation Model for Describing Adsorption of Toxic Trace Elements on Soil Minerals

Sabine Goldberg

U.S. Salinity Laboratory, Agricultural Research Service, U.S. Department of Agriculture, Riverside, CA 92501

The constant capacitance model provides a molecular description of adsorption phenomena using an equilibrium approach. Unlike empirical models, the constant capacitance model defines surface species, chemical reactions, equilibrium constant expressions, surface activity coefficients and mass and charge balance. The model was able to describe cadmium, lead, copper, aluminum, selenite, arsenate, and boron adsorption on oxide minerals, cadmium, lead, copper, selenite, arsenate, and boron adsorption on clay minerals, and selenite, arsenate, and boron adsorption on soils. Additional research into model application to heterogeneous natural systems such as clay minerals and soils is needed. The constant capacitance model has been incorporated into transport models. Validation of these models is necessary.

Specific ion adsorption produces surface complexes which contain no water between the surface functional group and the adsorbing ion. Such surface complexes are quite stable and are referred to as inner-sphere (1). Aluminum and iron oxide surfaces and clay mineral edges often constitute the primary sinks for retention of specifically adsorbed ions in soil systems. The present study will describe the behavior of the following specifically adsorbing and toxic trace elements: cadmium, lead, copper, aluminum, selenium, arsenic, and boron.

Trace element adsorption reactions on soil minerals have historically been described using the Freundlich and Langmuir adsorption isotherm equations. Although these equations are often very good at describing adsorption, they are strictly, empirical numerical relationships (2). Chemical meaning cannot be assigned to Langmuir and Freundlich isotherm parameters without independent experimental evidence for adsorption (1). Since the use of the Langmuir and Freundlich adsorption isotherm equations constitutes essentially a curve-fitting procedure, the equation parameters are only valid for the conditions under which the experiment was carried out.

Chemical modeling of ion adsorption at the solid-solution interface has been successful using surface complexation models. These models use an equilibrium approach with mass-action and mass-balance equations to describe the formation of surface complexes. Unlike the empirical Langmuir and Freundlich equations, these models attempt to provide a general molecular description of adsorption reactions. Because surface complexation models take into account the charge on both the adsorbate ion and the adsorbent surface, they can have wide applicability and predictive power over changing conditions of solution pH, ionic strength, and ion concentration.

The Constant Capacitance Model

The constant capacitance model is a chemical surface complexation model developed by the research groups of Stumm and Schindler (*3-5*). The constant capacitance model explicitly defines surface species, chemical reactions, equilibrium constant expressions, and surface activity coefficients. The surface functional group is defined as SOH, an average reactive surface hydroxyl ion bound to a metal ion, S (Al or Fe), in the oxide mineral or an aluminol group on the clay mineral edge. The adsorbed ions are considered to form inner-sphere surface complexes, located in a surface plane with the protons and hydroxyl ions. A diagram of the surface-solution interface in the constant capacitance model is provided in Figure 1.

Assumptions. The constant capacitance model is characterized by four essential assumptions: (1) all surface complexes are inner-sphere complexes, (2) anion adsorption occurs via a ligand exchange mechanism, (3) the Constant Ionic Medium Reference State determines the aqueous species activity coefficients in the conditional equilibrium constants and therefore no surface complexes are formed with ions in the background electrolyte, (4) the relationship between surface charge, σ (mol_c L^{-1}), and surface potential, ψ (V), is linear and given by:

$$\sigma = \frac{C\,Sa}{F}\,\psi \qquad (1)$$

where C (F m^{-2}) is the capacitance density, S (m^2 g^{-1}) is the specific surface area, a (g L^{-1}) is the suspension density of the solid, and F (C mol_c^{-1}) is the Faraday constant.

Reactions. In the constant capacitance model the protonation and dissociation reactions of the surface functional group are:

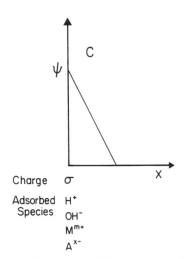

Figure 1. Placement of ions, potential, charge, and capacitance density for the constant capacitance model. (Adapted from ref. 6.).

$$SOH + H^+ \rightleftharpoons SOH_2^+ \tag{2}$$

$$SOH \rightleftharpoons SO^- + H^+ \tag{3}$$

The surface complexation reactions for specific ion adsorption are defined as:

$$SOH + M^{m+} \rightleftharpoons SOM^{(m-1)} + H^+ \tag{4}$$

$$2SOH + M^{m+} \rightleftharpoons (SO)_2M^{(m-2)} + 2H^+ \tag{5}$$

$$SOH + H_xA = SH_{(x-i)}A^{(1-i)} + H_2O + (i-1)H^+ \tag{6}$$

$$2SOH + H_xA = S_2H_{(x-j)}A^{(2-j)} + 2H_2O + (j-2)H^+ \tag{7}$$

where M is a metal ion, m+ is the charge on the metal ion, A is an anion, x is the number of protons in the undissociated form of the acid, $1 \leq i \leq n$, and $2 \leq j \leq n$ where n is the number of anion surface complexes and is equal to the number of dissociations undergone by the acid.

Equilibrium constants. The intrinsic conditional equilibrium constant expressions describing these reactions are:

$$K_+(int) = \frac{[SOH_2^+]}{[SOH][H^+]} \exp[F\psi/RT] \tag{8}$$

$$K_-(int) = \frac{[SO^-][H^+]}{[SOH]} \exp[-F\psi/RT] \tag{9}$$

$$K_M^1(int) = \frac{[SOM^{(m-1)}][H^+]}{[SOH][M^{m+}]} \exp[(m-1)F\psi/RT] \tag{10}$$

$$K_M^2(int) = \frac{[(SO)_2M^{(m-2)}][H^+]^2}{[SOH]^2[M^{m+}]} \exp[(m-2)F\psi/RT] \tag{11}$$

$$K_A^i(int) = \frac{[SH_{(x-i)}A^{(1-i)}][H^+]^{(1-i)}}{[SOH][H_xA]} \exp[(1-i)F\psi/RT] \qquad (12)$$

$$K_A^j(int) = \frac{[S_2H_{(x-j)}A^{(2-j)}][H^+]^{(j-2)}}{[SOH]^2[H_xA]} \exp[(2-j)F\psi/RT] \qquad (13)$$

where R is the molar gas constant (J mol^{-1} K^{-1}), T is the absolute temperature (K), and square brackets represent concentrations (mol L^{-1}).

Activity coefficients. The conventional thermodynamic Standard State for solid phases is pure solid. In the constant capacitance model the solid phase Standard State refers to the solid phase in a chargeless environment (7). The intrinsic conditional equilibrium constants are directly proportional to the thermodynamic equilibrium constants (7). The molecular basis of the intrinsic conditional equilibrium constant expressions derives from the concept that charged surface complexes create an average electric potential field at the surface. These coulombic forces provide the dominant contribution to the solid-phase activity coefficients; the contribution from other forces is assumed equal for all surface complexes (7). Thus the exponential term can be considered as a solid-phase activity coefficient.

Mass balance. The mass balance for the surface functional group is:

$$[SOH]_T = [SOH] + [SOH_2^+] + [SO^-] + [SOM^{(m-1)}] + [(SO)_2M^{(m-2)}]$$

$$+ \sum_{i=1}^{n} [SH_{(x-i)}A^{(1-i)}] + \sum_{j=2}^{n} [S_2H_{(x-j)}A^{(2-j)}] \qquad (14)$$

Charge balance. The charge balance equation is:

$$\sigma = [SOH_2^+] - [SO^-] + (m-1)[SOH^{(m-1)}] + (m-2)[(SO)_2M^{(m-2)}]$$

$$+ \sum_{i=1}^{n} (1-i)[SH_{(x-i)}A^{(1-i)}] + \sum_{j=2}^{n} (2-j)[S_2H_{(x-j)}A^{(2-j)}] \qquad (15)$$

This set of equations can be solved using the mathematical approach outlined by Westall (8).

Parameter Determination

Values of the intrinsic protonation and dissociation constants can be obtained experimentally from alkimetric or acidimetric titration curves carried out in the absence of specifically adsorbing metals or anions (4). The assumption is made that σ, the net surface charge, is equal to $[SOH_2^+]$ below the zero point of charge, ZPC, and is equal to $-[SO^-]$ above the ZPC. The ZPC is defined as the pH value at which σ = zero. In the constant capacitance model the ZPC is also defined by the equation:

$$ZPC = 1/2[\log K_+(int) - \log K_-(int)] \qquad (16)$$

A plot of the logarithm of the conditional equilibrium constant, $^cK_+$, versus surface charge, σ, will yield the logarithm of the intrinsic conditional equilibrium constant, $K_+(int)$, upon linear extrapolation to $\sigma = 0$. The conditional equilibrium constants for protonation and dissociation of the surface functional group are:

$$.^cK_+ = \frac{[SOH_2^+]}{[SOH][H^+]} \qquad (17)$$

$$.^cK_- = \frac{[SO^-][H^+]}{[SOH]} \qquad (18)$$

By combining equation (8) with equation (17) and equation (9) with equation (18), the intrinsic conditional protonation and dissociation constants and the conditional protonation and dissociation constants can be related:

$$K_\pm(int) = {}^cK_\pm \exp(\pm F\psi/RT) \qquad (19)$$

where + represents the protonation constant and - represents the dissociation constant. By substituting for surface potential, ψ, from equation (1), taking the logarithms of both sides, and solving for $\log^cK_\pm$, one obtains:

$$\log^cK_\pm = \log K_\pm(int) \pm \frac{\sigma F^2}{CSaRT(\ln 10)} \qquad (20)$$

Upon plotting the titration data as $\log^cK_\pm$ versus σ, an estimate of $\log K_\pm(int)$ can be obtained from the y-intercept, where $\sigma = 0$. The slope of such a plot provides an estimate of the capacitance density.

Protonation-Dissociation Constants. Values of $\log K_\pm(int)$ have been obtained experimentally for titanium oxide (3), aluminum oxide (9-12), iron oxide (13-

19), and silicon oxide (20, 21). Figure 2 provides an example of the linear extrapolation technique for the iron oxide, goethite (4). Table I presents values of $logK_{\pm}(int)$ obtained by various investigators using the linear extrapolation technique. Large differences between the value of $logK_{+}(int)$ and the absolute value of $logK_{-})int)$ result from the assumptions in the linear extrapolation technique.

Values of $logK_{\pm}(int)$ can also be obtained by optimization of alkimetric and acidimetric titration data using a computer program. The computer program FITEQL (23) uses a nonlinear, least squares optimization technique to fit equilibrium constants to experimental data and contains the constant capacitance model to describe surface complexation of ions.

Average values of $logK_{\pm}(int)$ can be obtained from a literature compilation of experimental $logK_{\pm}(int)$ values for aluminum and iron oxide minerals (24). Values of $logK_{\pm}(int)$ obtained in this fashion were not statistically significantly different for aluminum and iron oxide minerals.

Capacitance. A weakness of the constant capacitance model is that the value of the capacitance, C_{+}, obtained from linear extrapolations below the ZPC is usually not equal to the value, C_{-}, obtained from linear extrapolations above the ZPC. Values of $C_{\pm}$ obtained from linear extrapolations of titration data for the iron oxide, goethite and the aluminum oxide, boehmite are provided in Table II. The capacitance density exhibits great variability for two batches of goethite and even for three experiments on the same batch of boehmite. Because of this variability, applications of the constant capacitance model have often used single values of capacitance density. Values considered optimum are $C = 1.8$ F m^{-2} for goethite (14) and $C = 1.06$ F m^{-2} for aluminum oxide (25). The constant capacitance model was found to be insensitive to changes in capacitance density from 1.06 to 1.8 F m^{-2} for amorphous iron hydroxide (24). Thus a capacitance density of 1.06 F m^{-2} has been used for aluminum and iron oxides, clay minerals, and soils (26).

Surface Site Density. The total number of reactive surface hydroxyl groups, $[SOH]_T$, is an important parameter in the constant capacitance model. The value of $[SOH]_T$ has been determined experimentally by potentiometric titration (9, 10), fluoride adsorption (14), maximum adsorption (24, 26-28), or calculated from crystal dimensions (29). Initial sensitivity analyses showed that the constant capacitance model was relatively insensitive to changes in $[SOH]_T$ value for phosphate adsorption on a soil in the range of 1.25 to 2.5 sites per nm^2 (30). More detailed investigation of the constant capacitance model has indicated that for adsorption of phosphate, arsenate, selenite, silicate, and borate the ability to describe adsorption was sensitively dependent on the $[SOH]_T$ value (29).

The constant capacitance model contains the assumption that ion adsorption occurs at only one or two types of surface sites. Clearly this assumption is a simplification for soils which are complex, multisite mixtures.

Table I. Values of intrinsic protonation and dissociation constants obtained using the linear extrapolation technique for oxide minerals

Solid	Ionic Medium	log K_+(int)	log K_-(int)	Reference
Aluminum Oxides				
γ-Al_2O_3	0.1 M $NaClO_4$	7.2	-9.5	(9)
γ-Al_2O_3	0.1 M $NaClO_4$	7.4	-10.0	(10)
γ-$Al(OH)_3$	1 M KNO_3	5.24	-8.08	(11)
γ-AlOOH	0.001 M KNO_3	5.6	-8.6	(12)
Iron Oxides				
α-FeOOH	0.1 M $NaClO_4$	6.4	-9.25	(14)
α-FeOOH	0.1 M $NaClO_4$	5.9	-8.65	(14)
α-FeOOH	0.1 M $NaNO_3$	7.47	-9.51	(19)
Fe_3O_4	0.1 M KNO_3	5.19	-8.44	(17)
Fe_3O_4	0.01 M KNO_3	4.66	-8.81	(17)
Fe_3O_4	0.001 M KNO_3	4.40	-8.97	(17)
$Fe(OH)_3$(am)	0.1 M $NaNO_3$	6.6	-9.1	(18)
Silicon Oxides				
SiO_2(am)	0.1 M $NaClO_4$		-6.8	(20)
SiO_2(am)	1.0 M LiCl		-6.57	(13)
SiO_2(am)	1.0 M $NaClO_4$		-6.71	(13)
SiO_2(am)	1.0 M CsCl		-5.71	(13)
SiO_2(am)	1.0 M $NaClO_4$		-5.7	(15)
SiO_2(am)	1.0 M $NaClO_4$		-6.56	(16)
Quartz	0.1 M $NaClO_4$		-8.4	(21)
Titanium Oxides				
Anatase	3.0 M $NaClO_4$	4.98	-7.80	(3)
Rutile	1.0 M $NaClO_4$	4.46	-7.75	(15)
Rutile	1.0 M $NaClO_4$	4.13	-7.39	(16)

SOURCE: Adapted from ref. 22 and expanded.

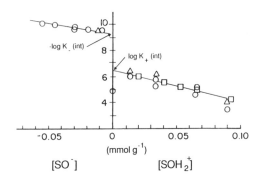

Figure 2. The logarithms of the conditional protonation and dissociation constants as a function of surface charge for goethite, α-FeOOH, LogK$_+$(int) = 6.4, logK$_-$(int) = -9.25, C$_+$ = 2.7 F m^{-2}, C$_-$ = 4.4 F m^{-2}. (Reproduced with permission from ref. 4. Copyright 1980.)

Table II. Values of capacitance obtained with the constant capacitance model using the linear extrapolation technique

Solid	C$_+$ (F m^{-2})[a]	C$_-$ (F m^{-2})[b]
goethite[c]		
α-FeOOH I	1.5	3.5
α-FeOOH II	2.7	4.4
boehmite γ-AlOOH[d]		
Experiment 1	1.3	0.7
Experiment 2	2.8	0.9
Experiment 3	2.9	2.0

[a] C$_+$ is obtained from the slope of extrapolation for log K$_+$(int).
[b] C$_-$ is obtained from the slope of extrapolation for log K$_-$(int).
[c] Calculated from ref. (14) data for two batches of goethite.
[d] Calculated from ref. (12) data for three experiments of one batch of boehmite.

However, experimental evidence indicates that even oxide mineral surfaces contain several sets of trace element reactive sites (*31, 32*). Thus surface complexation constants determined for soils and even for pure mineral systems likely represent average composite values for all sets of adsorbing sites.

Modeling of Trace Metal Adsorption Reactions on Oxides and Clay Minerals

Adsorption edges are curves that characterize metal ion adsorption behavior as a function of solution pH. The constant capacitance model has been used to describe trace metal adsorption edges on silicon oxide (*33*), titanium oxide (*15*), aluminum oxide (*9*), iron oxide (*19*), and the clay mineral, kaolinite (*34*). The conditional equilibrium constants for metal adsorption are:

$$.^{c}K_{M}^{1} = \frac{[SOM^{(m-1)}][H^{+}]}{[SOH][M^{m+}]} \tag{21}$$

$$.^{c}K_{M}^{2} = \frac{[(SO)_{2}M^{(m-2)}][H^{+}]^{2}}{[SOH]^{2}[M^{m+}]} \tag{22}$$

To allow graphical evaluation of the trace metal surface complexation constants, the simplifying assumption is made that ψ = zero (*33*). With this assumption, the intrinsic conditional equilibrium constants become equal to the conditional equilibrium constants:

$$K_{M}^{i}(int) = .^{c}K_{M}^{i} \tag{23}$$

For the special case of divalent metal ions, M^{2+}, equation (23) holds true without any simplifying assumption, when neutral bidentate surface complexes are formed. Excellent fits to trace metal adsorption data have been obtained despite the simplification (*9, 33*); however, the lack of dependence of the conditional equilibrium constants on surface charge indicates a self-consistency problem in these applications of the constant capacitance model (*1*). Figure 3 presents the ability of the constant capacitance model to describe trace metal adsorption on silicon oxide. The model describes the adsorption data very well.

Computer Optimization.   The above limitation can be overcome by optimizing the intrinsic conditional equilibrium constants using a computer program such as FITEQL (*23*). The computer approach negates the requirement for simplifying assumptions and has been used by Lövgren et al. (*19*) to describe aluminum adsorption on the iron oxide, goethite. These authors postulated the surface complexes $SOAlOH^{+}$ and $SOAl(OH)_{2}$ with equilibrium constants $\log K_{AlOH^+}(\text{int}) = -1.49$ and $\log K_{Al(OH)2}(\text{int}) = -9.10$. Table III provides values for trace metal surface complexation constants obtained with the constant capacitance model for various oxide minerals.

Trace Metal Adsorption on Kaolinite. The constant capacitance model has been modified and extended to describe trace metal adsorption data on the clay mineral, kaolinite (*34*). In addition to the amphoteric surface hydroxyl group, SOH, Schindler et al. (*34*) postulated a second surface functional group, XH, which is weakly acidic and can undergo ion exchange with trace metal ions, M^{2+}, and cations, C^+, from the background electrolyte. Cations from the background electrolyte can react with the SOH functional group to form weak outer-sphere surface complexes. Thus, in addition to equations (2), (3), (4), and (5) the following surface complexation reactions are defined:

$$SOH + C^+ \rightleftharpoons SO^- -C^+ + H^+ \tag{24}$$

$$XH + C^+ \rightleftharpoons XC + H^+ \tag{25}$$

$$2XC + M^{2+} \rightleftharpoons X_2 M + 2C^+ \tag{26}$$

In addition to equations (8), (9), (10), and (11) the equilibrium constants for trace metal adsorption on kaolinite are:

$$K_{C+} = \frac{[SO^- -C^+][H^+]}{[SOH][C^+]} \tag{27}$$

$$K_{XC+} = \frac{[XC][H^+]}{[XH][C^+]} \tag{28}$$

$$K_{XM} = \frac{[X_2M][C^+]^2}{[XC]^2[M^{2+}]} \tag{29}$$

The fit of the constant capacitance model to adsorption of trace metal ions on hydrogen kaolinite is indicated in Figure 4. Despite the fact that values of the equilibrium constants, $K_{\pm}(\text{int})$, K_{C+}, K_{XC+}, and K_{XM}, were obtained using the computer program FITEQL (*23*), the assumption was made that ψ = zero. Schindler et al. (*34*) considered the model fit acceptable but suggested that systematic errors might be due to the use of the same capacitance density value for all ionic strengths and to the extension of the Davies equation up to ionic strengths of 1 M. Table IV provides values for

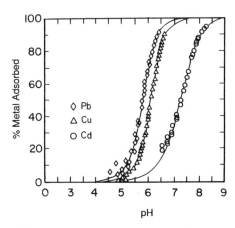

Figure 3. Fit of the constant capacitance model to trace metal adsorption on silicon oxide. Model results are represented by solid lines. Model parameters are provided in Table III. (Adapted from ref. 33.).

Table III. Values of trace metal surface complexation constants obtained using graphical methods for oxide minerals

Solid	Metal	Ionic Medium	$\log^c K^1_M$	$\log^c K^2_M{}^b$	Reference
γ-Al_2O_3	Pb^{2+}	0.1 M $NaClO_4$	-2.2	-8.1	(9)
Fe_3O_4	Co^{2+}	I = O	-2.44	-6.71	(35)
SiO_2(am)	Cu^{2+}	1 M $NaClO_4$	-5.52	-11.19	(33)
SiO_2(am)	Cd^{2+}	1 M $NaClO_4$	-6.09	-14.20	(33)
SiO_2(am)	Pb^{2+}	1 M $NaClO_4$	-5.09	-10.68	(33)
TiO_2, rutile	Co^{2+}	1 M $NaClO_4$	-4.30	-10.16	(16)
TiO_2, rutile	Cu^{2+}	1 M $NaClO_4$	-1.43	-5.04	(15)
TiO_2, rutile	Cd^{2+}	1 M $NaClO_4$	-3.32	-9.00	(15)
TiO_2, rutile	Pb^{2+}	1 M $NaClO_4$	0.44	-1.95	(15)

[a] For divalent metal ions $\log K^2_M(int) = \log^c K^2_M$.
SOURCE: Adapted from ref. (22).

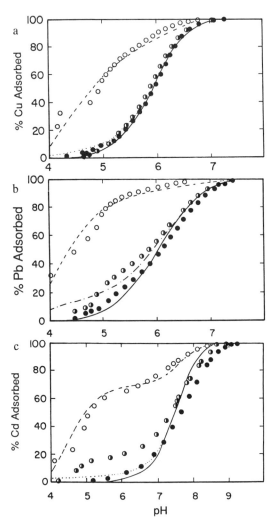

Figure 4. Fit of the constant capacitance model to trace metal adsorption on kaolinite: (a) copper (b) lead (c) cadmium. Model results are represented by dashed line (I = 0.01 M NaClO$_4$), dotted line (I = 0.1 M NaClO$_4$), and solid line (I = 1.0 M NaClO$_4$). Model parameters are provided in Table IV.
(Reproduced with permission from ref. 34. Copyright 1987.)

trace metal surface complexation constants obtained with the extended constant capacitance model for kaolinite.

Surface Precipitation Model. The surface precipitation model considers precipitation of ions on solid surfaces. Farley et al. (*18*) developed the surface precipitation model and incorporated it into the constant capacitance model. Sorption of ions is described by surface complex formation at low solution concentration and by surface precipitation as an ideal solid solution at high solution concentration. The composition of the solid solution varies continuously between the original solid and the pure precipitate containing the sorbing ion. Solid-phase activities are given by mole fractions. Reactions for surface precipitation of divalent trace metal ions onto a trivalent oxide mineral are (*18*):

Adsorption of M^{2+} onto $S(OH)_3(s)$:

$$\equiv SOH + M^{2+} + 2H_2O \rightleftharpoons S(OH)_{3(s)} + =MOH_2^+ + H^+ \qquad (30)$$

Precipitation of M^{2+}:

$$=MOH_2^+ + M^{2+} + 2H_2O \rightleftharpoons M(OH)_{2(s)} + =MOH_2^+ + 2H^+ \qquad (31)$$

Precipitation of S^{3+}:

$$\equiv SOH + S^{3+} + 3H_2O \rightleftharpoons S(OH)_{3(s)} + \equiv SOH + 3H^+ \qquad (32)$$

The equilibrium constants for the above reactions are: K_{adsM} for equation (30), $1/K_{spM}$ for equation (31), and $1/K_{spS}$ for equation (32). The ability of the constant capacitance model containing the surface precipitation model to describe lead sorption on amorphous iron hydroxide is presented in Figure 5. The model describes the experimental data well, although very few data points are available. Farley et al. (*18*) also applied the surface precipitation model containing the constant capacitance model to copper, cadmium, and zinc sorption on amorphous iron hydroxide. No additional applications of the model have been carried out.

Modeling of Trace Element Anion Adsorption Reactions on Oxides, Clay Minerals, and Soils

Adsorption envelopes are curves that characterize anion adsorption behavior as a function of solution pH. The constant capacitance model has been used to describe trace element anion adsorption envelopes on iron oxide (*26-28, 37*), aluminum oxide (*26, 28, 37, 38*), clay minerals (*26, 39-41*), and soils (*26, 40-43*). Intrinsic conditional equilibrium constants for trace element anions in the constant capacitance model were obtained using the computer programs MICROQL (*44*) and FITEQL (*23*). Figure 6 presents the ability of

Table IV. Values of trace metal surface complexation constants obtained
using computer optimization for kaolinite[a]

Metal	$logK_+(int)$	$logK_-(int)$	$log^cK^1_M$	$log^cK^2_M$	$logK_{C+}$	$logK_{XC+}$	$logK_{XM}$
Cu^{2+}	4.37	9.18	-2.50	-7.46	-9.84	-2.9	2.39
Cd^{2+}	4.37	9.18	-4.29	-10.4	-9.84	-2.9	2.27
Pb^{2+}	4.37	9.18	-2.45	-8.11	-9.84	-2.9	2.98

[a] Data from ref. (34).

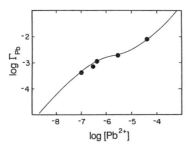

Figure 5. Fit of the constant capacitance model containing the surface precipitation model to a lead sorption isotherm on amorphous iron hydroxide at pH 4.5. Model results are represented by a solid line. $\Gamma_{Pb} = ([=PbOH] + [Pb(OH)_{2(s)}])/TOTFe$. $LogK_{adsPb} = 5.0$, $logK_{spPb} = 6.9$, $logK_{spFe} = 2.6$. Based on experimental data of Benjamin (36).

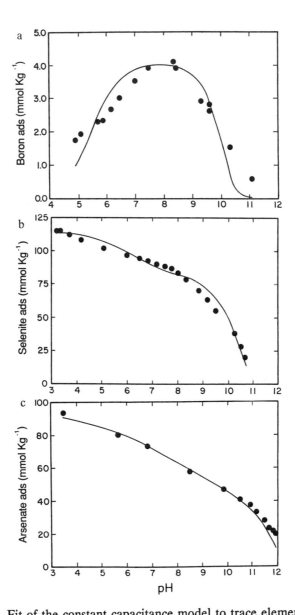

Figure 6. Fit of the constant capacitance model to trace element anion adsorption on goethite, α-FeOOH: (a) boron (b) selenite (c) arsenate. Model results are represented by solid lines. Model parameters are provided in Table V.
(Reproduced with permission from (a) ref. 37 (b) ref. 27 (c) ref. 28. Copyright (a,b) 1985, (c) 1986.)

the constant capacitance model to describe boron, selenite, and arsenate adsorption on goethite. The ability of the model to describe the adsorption data is good. Table V provides values for intrinsic trace element anion surface complexation constants obtained for various oxides.

Values of the boron surface complexation constants obtained using average values of $logK_\pm(int)$ from a literature compilation for oxide minerals (*24*) were not statistically significantly different for aluminum and iron oxide minerals (*37*).

Adsorption on Clay Minerals. Applications of the constant capacitance model to trace element anion adsorption envelopes on clay minerals have been carried out for boron (*39*), selenite (*40*), and arsenate (*41*). To describe boron, arsenate, and selenite adsorption on kaolinite and selenite adsorption on montmorillonite, $logK_\pm(int)$ values for the aluminol group were based on averages for a literature compilation of aluminum oxides. Figure 7 presents the ability of the constant capacitance model to describe boron, selenite, and arsenate adsorption on kaolinite. The model describes the general shapes of the adsorption envelopes. To describe boron and arsenate adsorption on montmorillonite and boron adsorption on illite, values of $logK_\pm(int)$ were optimized along with the trace element anion surface complexation constants (*39, 41*). Although the fit to anion adsorption was generally good, in some cases the optimized value of $logK_+(int)$ was larger than the optimized absolute value for $logK_-(int)$. This is a chemically unrealistic situation which potentially reduces the model application to a curve-fitting procedure. Table VI presents values for intrinsic anion surface complexation constants obtained for various clay minerals and soils.

Adsorption on Soils. The constant capacitance model was applied to heterogeneous soil systems in the study of Sposito et al. (*43*) of selenite adsorption on five alluvial soils and the study of Goldberg and Glaubig (*41*) of arsenate adsorption on a calcareous soil. These authors used average $logK_\pm(int)$ values obtained from a literature compilation for both aluminum and iron oxide minerals (*24*). The constant capacitance model is well able to describe arsenate adsorption on the calcareous soil as indicated by Figure 8. Sposito et al. (*43*) assumed that two types of sites in soil are selenite reactive. Monodentate selenite surface species are formed on one set of sites and bidentate selenite surface species are formed on another set of sites. The monodentate surface complexation constants were optimized by using adsorption data below pH 7, while the bidentate surface complexation constant was optimized using data above pH 7. Figure 9 presents the ability of the constant capacitance model to describe selenite adsorption on the Panoche soil. Using the intrinsic selenite surface complexation constants obtained for the Panoche soil, Sposito et al. (*43*) were able to qualitatively predict the selenite adsorption envelopes for the other four alluvial soils.

To describe boron adsorption on fourteen soils (*42*) and selenite

Table V. Values of intrinsic trace element anion surface complexation constants obtained using computer optimization for oxide minerals[a]

Solid	Anion	Ionic Medium	$\log K^1_A(\text{int})$	$\log K^2_A(\text{int})$	$\log K^3_A(\text{int})$	Reference
δ-Al$_2$O$_3$	B(OH)$_4^-$	0.1 M NaCl	5.13			(37)
δ-Al$_2$O$_3$	B(OH)$_4^-$	0.1 M NaCl	5.56			(37)
δ-Al$_2$O$_3$	B(OH)$_4^-$	0.1 M NaCl	2.87			(38)
activated alumina[b]	B(OH)$_4^-$	I = 0	5.09			(26)
α-Al(OH)$_3$[b]	AsO$_4^{3-}$	0.1 M NaCl	9.72	3.41	-3.58	(28)
α-Al(OH)$_3$[b]	SeO$_3^{2-}$	0.1 M NaCl	9.74	3.64		(26)
pseudoboehmite	B(OH)$_4^-$	0.1 M NaCl	5.09			(37)
Al(OH)$_3$(am)[b]	AsO$_4^{3-}$	0.01 M NaClO$_4$	9.89	3.32	-4.52	(28)
Al(OH)$_3$(am)[b]	AsO$_4^{3-}$	0.01 M NaClO$_4$	11.06	3.55	-3.19	(26)
Al(OH)$_3$(am)	B(OH)$_4^-$	0.1 M NaCl	5.92			(37)
α-FeOOH[b]	AsO$_4^{3-}$	0.1 M NaCl	10.10	5.80	-0.63	(28)
α-FeOOH[b]	AsO$_4^{3-}$	0.1 M NaCl	10.87	6.52	0.29	(28)
α-FeOOH[b]	SeO$_3^{2-}$	0.1 M NaCl	10.02	5.36		(27)
α-FeOOH[b]	SeO$_3^{2-}$	0.1 M NaCl	11.10	5.80		(27)
α-FeOOH	B(OH)$_4^-$	0.1 M NaCl	5.25			(37)
α-Fe$_2$O$_3$	B(OH)$_4^-$	0.1 M NaCl	4.88			(37)
Fe(OH)$_3$(am)	B(OH)$_4^-$	0.1 M NaCl	5.63			(37)

[a] Optimizations are based on $\log K_+(\text{int}) = 7.38$, $\log K_-(\text{int}) = -9.09$ for aluminum oxides and $\log K_+(\text{int}) = 7.31$, $\log K_-(\text{int}) = -8.80$ for iron oxides.
[b] Experimental data source provided in the reference.

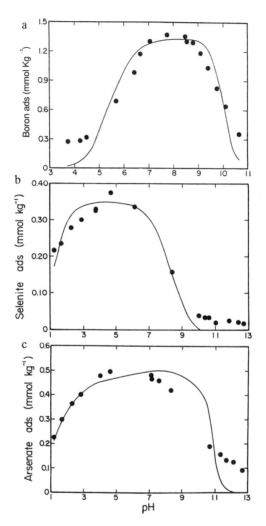

Figure 7. Fit of the constant capacitance model to trace element anion adsorption on KGa-2, poorly crystallized kaolinite: (a) boron (b) selenite (c) arsenate. Model results are represented by solid lines. $\text{LogK}_B(\text{int}) = 5.20$. All other model parameters are provided in Table VI.

(Reproduced with permission from (a) ref. 39 (b) ref 40 (c) ref 41. Copyright (a) 1986 (b,c) 1988.)

Table VI. Values of intrinsic trace element anion surface complexation constants obtained using computer optimization for clay minerals and soils[a]

Solid	Anion	Ionic Medium	logK^1_A(int)	logK^2_A(int)	logK^3_A(int)	Reference
Kaolinites[b]	B(OH)$_4^-$	0.1 M NaCl	5.28 ± 0.2			(39)
Kaolinite	AsO$_4^{3-}$	0.1 M NaCl	11.04	8.24	-3.21	(41)
Kaolinite	SeO$_3^{2-}$	0.1 M NaCl	11.07	3.34		(40)
Montmorillonites[c]	B(OH)$_4^-$	0.1 M NaCl	6.37 ± 1.3			(39)
Montmorillonite	SeO$_3^{2-}$	0.1 M NaCl	10.92	3.40		(40)
Illites[d]	B(OH)$_4^-$	0.1 M NaCl	5.39 ± 0.8			(39)
Soils[e]	B(OH)$_4^-$	0.01 M NaCl	5.48 ± 0.4			(42)
Panoche soil	SeO$_3^{2-}$	0.05 M NaCl	7.35[f]	0.85[f]		(43)
Imperial soil	AsO$_4^{3-}$	0.1 M NaCl	9.94	3.71	-4.78	(41)

[a] The work of Goldberg and co-workers was based on logK$_+$(int) = 7.38, logK$_-$(int) = -9.09 for kaolinites. All soils work was based on logK$_+$(int) = 7.35, logK$_-$(int) = -8.95 unless indicated otherwise.
[b] Average for four kaolinites.
[c] LogK$_+$(int) = 10.62 ± 1.6 and logK$_-$(int) = 10.46 ± 1.3 were also optimized. Average for three montmorillonites.
[d] LogK$_+$(int) = 9.30 ± 1.3 and logK$_-$(int) = 10.43 ± 0.5 were also optimized. Average for three illites.
[e] LogK$_+$(int) = 9.34 ± 0.8 and logK$_-$(int) = -10.64 ± 0.9 were also optimized. Average for fourteen soils.
[f] LogK$^1_{Se}$(int) = 20.05 defined for reaction equation (13) was also optimized.

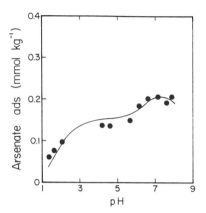

Figure 8. Fit of the constant capacitance model to arsenate adsorption on Imperial soil. Model results are represented by a solid line. Model parameters are provided in Table VI. (Adapted from ref. 41.)

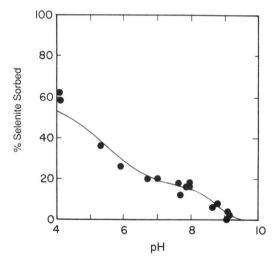

Figure 9. Fit of the constant capacitance model to selenite adsorption on Panoche soil. Model results are represented by a solid line. Model parameters are provided in Table VI.
(Reproduced with permission from ref. 43. Copyright 1988.)

adsorption on a soil (40), logK$_\pm$(int) were optimized along with the trace element anion surface complexation constants. The optimized values of logK$_-$ (int) for some of the soils were insignificantly small. This is another unrealistic situation which potentially reduces the application of the model to a curve-fitting procedure. Using an average set of intrinsic conditional surface complexation constants, obtained from all fourteen soils, the constant capacitance model was able to predict boron adsorption on most of the soil samples studied (42). The ability of the model to describe boron adsorption on soil samples from two depths is presented in Figure 10. The model describes the adsorption data well. Additional research into the application of the constant capacitance model for describing trace element anion adsorption on natural materials such as clay minerals and soils is needed.

Modeling of Competitive Trace Element Adsorption Reactions on Oxides

Metal-Metal Competition. The ability of the constant capacitance model to describe metal-metal competition has been investigated only preliminarily. The constant capacitance model containing the surface precipitation model was used to describe the competitive effect of copper on cadmium sorption by amorphous iron hydroxide (18). The model overpredicted the competitive effect for the adsorption data of Benjamin (36) obtained after four hours of reaction time. Using the results from one competitive adsorption data point after thirty hours of reaction time, Farley et al. (18) concluded that the constant capacitance model containing the surface precipitation model is capable of predicting competitive adsorption of metal ions when slow adsorption kinetics are considered. Clearly, much additional research is needed to substantiate the conclusion of Farley et al. (18) based on one data point.

Anion-Anion Competition. The ability of the constant capacitance model to predict competitive anion adsorption on goethite from solutions containing both phosphate and selenite using the intrinsic anion surface complexation constants obtained from single adsorbate systems has been evaluated (27). The model predicted competitive anion adsorption qualitatively, reproducing the shapes of the adsorption envelopes; however, phosphate adsorption was overestimated while adsorption of selenite was underestimated. Competitive anion adsorption of phosphate and arsenate on gibbsite (28) and goethite (26) and of phosphate and selenite on goethite (45) could be described by direct optimization of the surface complexation constants of the mixed-anion adsorption data. The fit of the constant capacitance model to the competitive anion adsorption data using the mixed-anion approach was much improved over that obtained by prediction from single anion systems. In the mixed-anion approach, the intrinsic anion surface complexation constants become conditional constants, dependent upon the anion surface composition. Goldberg and Traina (45) suggested that this composition dependence is the

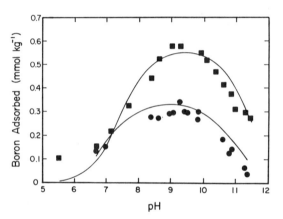

Figure 10. Fit of the constant capacitance model to boron adsorption on Fallbrook soil. Squares represent 0-25 cm sample: $logK_+(int) =$ 9.62, $logK_B(int) = 5.56$. Circles represent 25-51 cm sample: $logK_+(int) = 8.78$, $logK_B(int) = 5.14$. $LogK_(int) = -30.95$. Model results are represented by solid lines. (Based on experimental data from ref. 42).

result of surface site heterogeneity. Goldberg (*28*) found that arsenate and phosphate surface complexation constants obtained using the mixed-anion approach for one competitive gibbsite system could be used to predict competitive anion adsorption for other gibbsite systems containing different proportions of solution arsenate and phosphate (see Figure 11). The model predicts the competitive adsorption data well especially in the pH range 4.5 to 8.5.

Metal-Ligand Interactions. Ternary surface complexes can be formed upon adsorption of metal-ligand complexes, ML. A type A ternary surface complex, S-O-M-L, is formed when the metal-ligand complex attaches to the mineral surface via the metal ion (*46*). The constant capacitance model has been used to describe type A ternary surface complex formation on silicon oxide (*46, 47*) and titanium oxide (*16*). In this model application the following reactions are defined (*16*):

$$SOH + ML^{(m-1)+} \rightleftharpoons SOML^{(m-2)} + H^+ \tag{33}$$

$$2SOH + ML^{(m-1)+} \rightleftharpoons (SO)_2 ML^{(m-3)} + 2H^+ \tag{34}$$

The conditional equilibrium constant expressions for these reactions are:

$$K_{ML}^1 = \frac{[SOML^{(m-2)}][H^+]}{[SOH][ML^{(m-1)+}]} \tag{35}$$

$$K_{ML}^2 = \frac{[(SO)_2 ML^{(m-3)}][H^+]^2}{[SOH]^2 [ML^{(m-1)+}]} \tag{36}$$

Table VII provides values for type A ternary surface complexation constants. The ability of the constant capacitance model to describe copper adsorption on silicon oxide by considering a ternary surface complex is indicated in Figure 12 for ethylenediamine. The model describes the adsorption data very well.

Incorporation of the Constant Capacitance Model Into Computer Codes

The constant capacitance model has been incorporated into various chemical speciation models. The computer program MINTEQ (*48*) combines the mathematical framework of MINEQL (*49*) with the thermodynamic data base of WATEQ3 (*50*) and contains constant capacitance modeling. The constant capacitance model is incorporated into the computer speciation program SOILCHEM (*51*). The computer program HYDRAQL (*52*) was developed

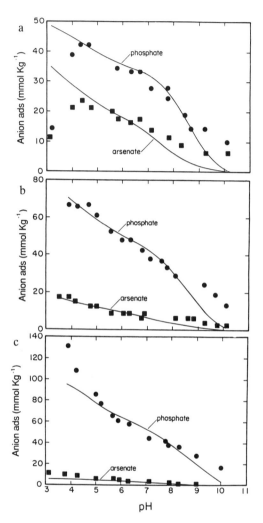

Figure 11. Fit of the constant capacitance model to competitive phosphate and arsenate adsorption on gibbsite (b). Prediction of competitive adsorption (a) and (c) using the surface complexation constants obtained from fit (b). Model results are represented by solid lines. $LogK^1_{As}(int) = 6.59$, $logK^2_{As}(int) = 0.96$, $logK^3_{As}(int) = -5.66$, $logK^1_P(int) = 7.69$, $logK^2_P(int) = 1.92$, $logK^3_P(int) = -4.68$. (Reproduced with permission from ref. 28. Copyright 1986.)

Table VII. Values of type A ternary surface complexation constants for oxide minerals

Solid	Metal	Ligand	Ionic Medium	$logK^1_{ML}$	$logK^2_{ML}$	Reference
SiO_2	Cu^{2+}	ethylenediamine	1 M $NaClO_4$	-5.22	-12.57	(47)
SiO_2	Cu^{2+}	glycine	1 M KNO_3	-5.64	-12.29	(46)
TiO_2, rutile	Co^{2+}	glycine	1 M $NaClO_4$	-4.40	-10.94	(16)

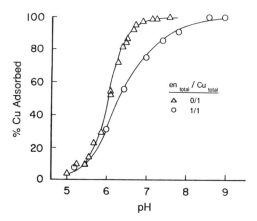

Figure 12. Fit of the constant capacitance model to copper adsorption on silicon oxide using a type A ternary copper-ethylenediamine surface complex. Model results are represented by solid lines. Model parameters are provided in Table VII. (Reproduced with permission from ref. 47. Copyright 1978.)

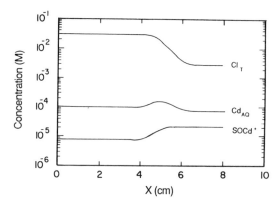

Figure 13. TRANQL simulation of cadmium transport in a solution of 1×10^{-4} M Cd_T containing 3×10^{-3} M Cl_T and 1×10^{-3} M Br_T. A step input increases Cl_T to 3×10^{-2} M and Br_T to 1×10^{-2} M. $LogK^1_{Cd}(int) = -7.0$, pH = 7.0. (Adapted from ref. 54.).

from the computer program MINEQL (*49*) and includes the constant capacitance model. Unlike the above mentioned computer programs, the programs MICROQL (*44*) and FITEQL (*23*) do not contain thermodynamic data files. Instead, the user enters the values of the equilibrium constants for the problem of interest.

Transport Models. A few studies have incorporated the constant capacitance model into transport models (*53, 54*). Jennings et al. (*53*) linked the constant capacitance model into a transport model and simulated competitive sorption of two hypothetical ligands at constant pH. The computer program TRANQL (*54*) links the program MICROQL (*44*) with the transport program ISOQUAD (Pinder, 1976, unpublished manuscript). The TRANQL program was used to simulate cadmium transport in the presence of chloride and bromide in a one-dimensional laboratory column. Figure 13 presents TRANQL simulations for cadmium transport where total cadmium concentration is held constant and the total concentrations of chloride and bromide experience a step increase. Validation of these predictions has not yet been carried out.

Acknowledgment

Gratitude is expressed to Dr. Samuel J. Traina, Department of Agronomy, Ohio State University for presenting this paper at the Emerging Technologies in Hazardous Waste Management III Symposium in Atlanta, GA, October 1991.

Literature Cited

(1) Sposito, G. *The surface chemistry of soils.* Oxford Univ. Press, Oxford, UK, 1984.
(2) Harter, R.D., Smith, G. In *Chemistry in the soil environment*; Dowdy, R.H., et al., Ed.; Soil Science Society of America. Madison, WI., 1981, pp. 167-181.
(3) Schindler, P.W., Gamsjäger, H. *Kolloid-Z. Z. Polymere,* 1972, *250,* 759-763.
(4) Stumm, W., Kummert, R., Sigg, L. *Croatica Chem. Acta,* 1980, *53,* 291-312.
(5) Schindler, P.W. In *Adsorption of inorganics and organic ligands at solid-liquid interfaces*; Anderson, M.A., Rubin, A.J., Ed.; Ann Arbor Science. Ann Arbor, MI., 1981, pp. 1-49.
(6) Westall, J.C. *Am. Chem. Soc. Symp. Ser.,* 1986, *323,* 54-78.
(7) Sposito, G. *J. Colloid Interface Sci.,* 1983, *91,* 329-340.
(8) Westall, J. *Am. Chem. Soc. Adv. Chem. Ser.,* 1980, *189,* 33-44.
(9) Hohl, H., Stumm, W. *J. Colloid Interface Sci.,* 1976, *55,* 281-288.
(10) Kummert, R., Stumm, W. *J. Colloid Interface Sci.,* 1980, *75,* 373-385.
(11) Pulfer, K., Schindler, P.W., Westall, J.C., Grauer, R. *J. Colloid Interface Sci.,* 1984, *101,* 554-564.

(12) Bleam, W.F., Pfeffer, P.E., Goldberg, S., Taylor, R.W., Dudley, R. *Langmuir*, 1991, *7*, 1702-1712.

(13) Sigg, L. *Untersuchungen über Protolyse und Komplexbildung mit zweiwertigen Kationen von Silikageloberflächen*, M.Sc. Thesis, University of Bern, Bern, Switzerland, 1973.

(14) Sigg, L.M. *Die Wechselwirkung von Anionen und schwachen Säuren mit αFeOOH (Goethit) in wässriger Lösung*. Ph.D. Thesis. Swiss Federal Institute of Technology, Zürich, Switzerland, 1979.

(15) Fürst, B. *Das Koordinationschemische Adsorptionsmodel: Oberflächenkomplexbildung von Cu(II), Cd(II) und Pb(II) an SiO₂ (Aerosil) und TiO₂ (Rutil)*. Ph.D. Thesis. University of Bern, Bern, Switzerland, 1976.

(16) Gisler, A. *Die Adsorption von Aminosäuren an Grenzflächen Oxid-Wasser*. Ph.D. Thesis. University of Bern, Bern, Switzerland, 1980.

(17) Regazzoni, A.E., Blesa, M.A., Maroto, A.J.G. *J. Colloid Interface Sci.*, 1983, *91*, 560-570.

(18) Farley, K.J., Dzombak, D.A., Morel, F.M.M. *J. Colloid Interface Sci.*, 1985, *106*, 226-242.

(19) Lövgren, L., Sjöberg, S., Schindler, P.W. *Geochim. Cosmochim. Acta*, 1990, *54*, 1301-1306.

(20) Schindler, P.W., Kamber, H.R. *Helv. Chim. Acta*, 1968, *51*, 1781-1786.

(21) Osaki, S., Miyoshi, T., Sugihara, S., Takashima, Y. *Sci. Total Environ.*, 1990, *99*, 105-114.

(22) Schindler, P.W., Stumm, W. In *Aquatic Surface Chemistry*; Stumm, W., Ed.; Wiley-Interscience, New York, NY., 1987, pp. 83-110.

(23) Westall, J.C. *FITEQL: A computer program for determination of equilibrium constants from experimental data*. Rep. 82-01. Department of Chemistry, Oregon State University, Corvallis, OR, 1982.

(24) Goldberg, S., Sposito, G. *Soil Sci. Soc. Am. J.*, 1984, *48*, 772-778.

(25) Westall, J., Hohl, H. *Adv. Colloid Interface Sci.*, 1980, *12*, 265-294.

(26) Goldberg, S. In ENVIROSOFT 86 Zanetti, P. Ed.; CML Publications, Ashurst, Southampton, UK., 1986, pp. 671-688.

(27) Goldberg, S. *Soil Sci. Soc. Am. J.*, 1985, *49*, 851-856.

(28) Goldberg, S. *Soil Sci. Soc. Am. J.*, 1986, *50*, 1154-1157.

(29) Goldberg, S. *J. Colloid Interface Sci.*, 1991, *145*, 1-9.

(30) Goldberg, S., Sposito, G. *Soil Sci. Soc. Am. J.*, 1984, *48*, 779-783.

(31) Benjamin, M.M., Leckie, J.O. In *Contaminants and sediments;* Baker, R.A., Ed.; Ann Arbor Science. Ann Arbor, MI., 1980, Vol. 2. pp. 305-322.

(32) Benjamin, M.M., Leckie, J.O. *J. Colloid Interface Sci.*, 1981, *79*, 209-221.

(33) Schindler, P.W., Fürst, B., Dick, R., Wolf, P.U. *J. Colloid Interface Sci.*, 1976, *55*, 469-475.

(34) Schindler, P.W., Liechti, P., Westall, J.C. *Netherlands J. Agric. Sci.*, 1987, *35*, 219-230.

(35) Tamura, H., Matijevic, E., Meites, L. *J. Colloid Interface Sci.*, 1983, *92*, 303-314.

(36) Benjamin, M.M. *Effects of competing metals and complexing ligands on*

trace metal adsorption at the oxide/solution interface. Ph.D. Thesis, Stanford University, Stanford, CA., 1978.

(37) Goldberg, S., Glaubig, R.A. *Soil Sci. Soc. Am. J.,* 1985, *49,* 1374-1379.

(38) Goldberg, S., Glaubig, R.A. *Soil Sci. Soc. Am. J.,* 1988, *52,* 87-91.

(39) Goldberg, S., Glaubig, R.A. *Soil Sci. Soc. Am. J.,* 1986, *50,* 1442-1448.

(40) Goldberg, S., Glaubig, R.A. *Soil Sci. Soc. Am. J.,* 1988, *52,* 954-958.

(41) Goldberg, S., Glaubig, R.A. *Soil Sci. Soc. Am. J.,* 1988, *52,* 1297-1300.

(42) Goldberg, S., Glaubig, R.A. *Soil Sci. Soc. Am. J.,* 1986, *50,* 1173-1176.

(43) Sposito, G., deWit, J.C.M., Neal, R.H. *Soil Sci. Soc. Am. J.* 1988, *52,* 947-950.

(44) Westall, J.C. *MICROQL. I. A chemical equilibrium program in BASIC. II. Computation of adsorption equilibria in BASIC.* Technical Report, Swiss Federal Institute of Technology, EAWAG, Dübendorf, Switzerland, 1979.

(45) Goldberg, S., Traina, S.J. *Soil Sci. Soc. Am. J.,* 1987, *51,* 929-932.

(46) Schindler, P.W. *Rev. in Mineralogy,* 1990, *23,* 281-307.

(47) Bourg, A.C.M., Schindler, P.W. *Chimia,* 1978 *32,* 166-168.

(48) Felmy, A.R., Girvin, D.C., Jenne, E.A. *MINTEQ: A computer program for calculating aqueous geochemical equilibria.* EPA-600/3-84-032. Office of Research and Development. U.S. EPA, Athens, GA., 1984.

(49) Westall, J.C., Zachary, J.L., Morel, F.M.M. *MINEQL: A computer program for the calculation of chemical equilibrium composition of aqueous systems.* Technical Note 18, Ralph M. Parsons Laboratory, Department of Civil Engineering, Massachusetts Institute of Technology, Cambridge, MA, 1976.

(50) Ball, J.W., Jenne, E.A., Cantrell, M.W. *WATEQ3: A geochemical model with uranium added.* USGS, Washington, D.C. Open File Report 81-1183, 1981.

(51) Sposito, G., Coves, J. *SOILCHEM: A computer program for the calculation of chemical speciation in soils.* Kearney Foundation of Soil Science, University of California, Riverside, CA., 1988.

(52) Papelis, C., Hayes, K.F., Leckie, J.O. *HYDRAQL: A program for the computation of chemical equilibrium composition of aqueous batch systems including surface-complexation modeling of ion adsorption at the oxide/solution interface.* Technical Report No. 306. Dept. of Civil Engineering, Stanford University, Stanford, CA., 1988.

(53) Jennings, A.A., Kirkner, D.J., Theis, T.L. *Water Resour. Res.,* 1982, *18,* 1089-1096.

(54) Cederberg, G.A., Street, R.L., Leckie, J.O. *Water Resour. Res.* 1985, *21,* 1095-1104.

RECEIVED October 6, 1992

Chapter 15

Modeling Brackish Solution Influences on the Chemical and Physical Behavior of Temperate Region Soils

M. Marsi[1] and V. P. Evangelou[2]

[1]College of Agriculture, University of Sriwijaya, Jl. Padang Selasa No. 524, Bukit Besar, 30139, Palembang, Indonesia
[2]Department of Agronomy, N−122 Agricultural Science Center−North, University of Kentucky, Lexington, KY 40546−0091

Oil well brine discharges on to soils of mixed mineralogy, commonly found in the eastern U.S., have become an environmental issue. Because of this, there is need to understand sodicity of soils that have developed under a temperate climate. In this study, two Kentucky soils of mixed mineralogy (Pembroke; fine-silty, mixed, mesic, Mollic Paleudalf and Uniontown; fine-silty, mixed mesic, Typic Hapludalf) were chosen for modeling purposes to evaluate the influence of ionic strength (I), sodium adsorption ratio (SAR) and pH on the Vanselow exchange coefficient (K_v), and dispersion/saturated hydraulic conductivity relationships. The results showed that both soils exhibited at least two classes of exchange sites but the Na^+-Ca^{2+} exchange behavior of the Pembroke soil was relatively independent of pH and ionic strength. Soil dispersion was shown to increase as electrolyte concentration was decreased and sodium adsorption ratio (SAR) and pH were increased. Furthermore, dispersion of the Uniontown soil was more dependent on electrolyte concentration, and SAR changes, but less dependent on pH changes than that of the Pembroke soil. The soil components shown to regulate saturated hydraulic conductivity were: 1) clay dispersion and 2) clay swelling. The first mechanism was associated with Pembroke, a soil with relative low expansion potential, while the second mechanism was associated with Uniontown, a soil with relative high expansion potential.

Need for oil from the developing and industrialized world has put significantly higher demand on oil production. Often, this higher demand for oil is met through "stripper wells." Such oil wells are producing a large quantity of brine,

0097−6156/93/0518−0308$09.75/0

an environmental pollutant, which is discharged onto agricultural lands and/or into natural water supplies. For example, in the state of Kentucky, it has been estimated that more than 100,000 gallons of brine per day are discharged onto land and surface waters. These brines contain approximately 0.5 mol L^{-1} sodium chloride. Similar brine problems exist in many other southeastern and northeastern states (1).

Information is available on the behavior and influence of sodium chloride on soils and on soil-solution suspensions (2, 3). However, most of this information pertains to soils of the arid west which are often alkaline and consist mostly of 2:1 clay minerals. In the temperate regions of the U.S., soils are often acid and their mineralogy is highly mixed (1:1 plus 2:1 clay minerals) and the 2:1 minerals are highly interlayered (4, 5). Information is needed on these soils regarding their chemical and physical behavior with respect to sodium loads.

This study will evaluate Na^+-Ca^{2+} exchange reactions and dispersion/saturated hydraulic conductivity relationships of soils of a temperate climatic region with differing mineralogy which are likely to receive discharges of oil well brines.

Theoretical and experimental considerations.

Na - Ca exchange. A binary exchange reaction at equilibrium involving Na^+ and Ca^{2+} on a soil system can be written as (5)

$$1/2\ Ex_2Ca + Na^+ \underset{\longleftarrow}{\overset{\longrightarrow}{}} ExNa + 1/2\ Ca^{2+} \qquad [1]$$

where Ex denotes an exchanger phase taken to have a charge of negative one (-1) and Na^+ and Ca^{2+} denote solution species. A criterion of chemical-reaction equilibrium is (6)

$$\Sigma v_i\ \mu_i = 0 \qquad [2]$$

where

v_i stoichiometric coefficient in chemical reaction for species i, and
μ_i chemical potential for species i.

The chemical potential μ_i of species i in solution is identical to the partial molar Gibbs energy G_i and at constant T:

$$d\mu_i = dG_i = RTd\ln f_i \qquad [3]$$

relates these quantities to the fugacity (f) in solution. Integration of Eq. [3] from the standard state of species i to a state of species i in solution gives:

$$\mu_i - G_i^\circ \ = \ RT\ln\frac{f_i}{f_i^\circ} \tag{4}$$

where
  G_i° molar Gibbs energy for species i,
  R gas constant and
  T system temperature.

The ratio f_i/f_i° is defined as the activity α_i in solution. For a gas, the standard state f_i° is the ideal-gas state of pure i at a pressure of 1 bar (or 1 atm). Thus for gas phase reactions $\alpha_i = f_i/f_i^\circ = f_i$. For solids and liquids the usual standard state is the pure solid or liquid at 1 bar (or 1 atm) and the system temperature.

From the preceding equations and definitions:

$$\mu_i = G_i^\circ + RT\ln\alpha_i \tag{5}$$

and at thermodynamic equilibrium for a chemical reaction:

$$\Sigma v_i(G_i^\circ + RT\ln\alpha_i) = 0 \tag{6}$$

from which it follows that:

$$\Pi(\alpha_i)v_i = \exp\frac{-\Sigma v_i G_i^\circ}{RT} = K_{eq} \tag{7}$$

where Π signifies the product over all species i in the chemical reaction and K_{eq} is the equilibrium constant for the reaction. Also,

$$-RT\ln K_{eq} = \Sigma v_i^\circ \, G^\circ = \Delta G^\circ \tag{8}$$

The pure component Gibbs energy, G_i°, is a property of pure species i in its standard state and fixed pressure. It depends only on temperature. It follows from Eq. [8] that K_{eq} is also only a function of temperature and ΔG° is the standard Gibbs energy change of reaction.

For any reaction at equilibrium:

$$K_{eq} = \Pi(\alpha_i{}^\circ)\nu_i = \Pi\left(\frac{f_i}{f_i{}^\circ}\right)^{\nu_i} \tag{9}$$

and it follows that the activities α_i are not completely defined without also defining the pure component reference states $f_i{}^\circ$ and $G_i{}^\circ$.

From the above, the thermodynamic exchange equilibrium constant K_{eq} for reaction [1] at room temperature (22° C) and one atmosphere pressure is represented by:

$$K_{eq} = \frac{\alpha_{Ca}^{1/2}\ \alpha_{ExNa}}{\alpha_{Na}\ \alpha_{Ex_2Ca}^{1/2}} \tag{10}$$

where

α_{Ca}, α_{Na} activity of solution phase Na^+ or Ca^{2+} and
$\alpha_{ExNa}, \alpha_{ExCa}$ activity of exchange phase Na^+ or Ca^{2+}.

Activity α_i, is defined by the equation

$$\alpha_i = \gamma_i c_i \tag{11}$$

where

γ_i = activity coefficient of species i and
c_i = concentration of species i.

In order to define solution phase α_i, its value is set to 1, hence $\gamma_i = 1$, when solution ionic strength (I) approaches zero. For mixed electrolyte solutions when $I > 0$, the single ion activity concept introduced by Davies (7) is employed to estimate γ_i. The equations for estimating γ_i for mixed electrolyte solutions are given in the Materials and Methods section.

The activity component of the adsorbed or solid phase is defined by employing the mole fraction concept (X_i) introduced by Vanselow (8). For a heterovalent binary exchange reaction such as Na^+-Ca^{2+}, assuming that the system obeys ideal solid-solution theory (9), the activity term (α_{Exi}) is defined by (8)

$$\alpha_{ExNa} \approx X_{Na} = \frac{ExNa}{ExNa + Ex_2Ca} \tag{12}$$

and

$$^{\alpha}Ex_2Ca \approx X_{Ca} = \frac{Ex_2Ca}{ExNa + Ex_2Ca} \qquad [13]$$

where

X_{Na}, X_{Ca} denote mole fraction of Na^+ or Ca^{2+} and Ex denotes the exchange phase with a valence of -1.

For a system where ideal solid-solution behavior is not obeyed

$$\alpha_{Exi} = f_i X_i \qquad [14]$$

where

f_i denotes adsorbed ion activity coefficient.

In the mole fraction concept the sum of exchangeable Na^+ (ExNa) and exchangeable Ca^{2+} (Ex$_2$Ca) is expressed in moles per kilogram soil. Equivalent fractions (E_i) for Na^+ and Ca^{2+} are estimated by:

$$E_{Na} = \frac{ExNa}{ExNa + 2Ex_2Ca} \qquad [15]$$

and

$$E_{Ca} = \frac{2Ex_2Ca}{ExNa + 2Ex_2Ca} . \qquad [16]$$

Exchangeable sodium percentage (ESP) is estimated by simply multiplying E_{Na} of Eq. [15] times one hundred. The cation exchange capacity (CEC) of soil is taken to be:

$$CEC = ExNa + 2Ex_2Ca \qquad [17]$$

In order to justify Eq. [17] it is assumed that any other cations such as exchangeable K^+ and/or H^+ are present in negligible quantities and do not interfere with Na^+-Ca^{2+} exchange, or H^+ is tightly bound to the solid surface giving rise only to pH-dependent charge (10). The absence of a third cation (ternary system) from the solution-exchange system in controlled laboratory experiments is of importance because it simplifies the calculations for estimating adsorbed ion activity coefficients. For more information on two cation exchange (binary) versus ternary cation exchange see Elprince et al.11), Sposito (12) and Sposito and LeVesque (13).

Based on the above concepts, an equilibrium exchange expression for reaction [1] can be given as

$$K_v = \frac{X_{Na} \, \alpha_{Ca}^{1/2}}{X_{Ca}^{1/2} \, \alpha_{Na}} \qquad [18]$$

where $\alpha_{Na}/\alpha_{Ca}^{1/2}$ is known as the sodium adsorption ratio (SAR) and K_v is the Vanselow exchange selectivity coefficient. Commonly, the magnitude of K_v is taken to represent relative affinity of Na^+ with respect to Ca^{2+} by the clay surface (13, 14). When K_v equals 1 at a given level of exchangeable Na^+, the exchanger at that level of Na load shows no preference for either Na^+ or Ca^{2+}. On the other hand, a $K_v > 1$ at any given level of exchangeable Na^+ signifies exchanger preference for Na^+ and a $K_v < 1$ at any given level of exchangeable Na^+ signifies preference for Ca^{2+}.

A K_v that is variable with respect to ESP magnitude can be transformed to the thermodynamic exchange constant (K_{eq}) as follows:

$$K_{eq} = K_v \, \frac{f_{Na}}{f_{Ca}^{1/2}} \qquad [19]$$

where

f_{Na}, f_{Ca} denote adsorbed ion activity coefficient for Na^+ or Ca^{2+}. Argersinger et al. (15) noted that any variation in K_v with respect to exchange phase composition is followed by a variation in the solid phase activity coefficients (f_i) such that

$$d\ln K_v = -d\ln f_{Na} + 1/2 d\ln f_{Ca} \qquad [20]$$

Furthermore, any variation in f_{Na} must be compensated for by a variation in f_{Ca}. This produces equation (16)

$$X_{Na} d\ln f_{Na} + X_{Ca} d\ln f_{Ca} = 0 \qquad (21)$$

The solutions of Eqs. [20] and [21] with respect to f_{Na} and f_{Ca} are

$$1/2 d\ln f_{Ca} = -E_{Na} d\ln K_v \qquad [22]$$

and

$$d\ln f_{Na} = (1-E_{Na}) d\ln K_v \qquad [23]$$

where E_{Na} is the equivalent charge fraction of adsorbed Na^+ [Sposito, 1984]. Integration of Eqs. [22] and [23] generates two equations which give values for lnf_{Na} and lnf_{Ca} at any value of E_{Na}. These equations are:

$$\ln f_{Na} = (1-E_{Na}) \ln K_v - \int_{E_{Na}}^{1} \ln K_v \, d E_{Na} \qquad [24]$$

and

$$1/2 \ln f_{Ca} = -E_{Na} \ln K_v + \int_{0}^{E_{Na}} \ln K_v \, d E_{Na} \qquad [25]$$

For a detailed discussion on Eqs. [24] and [25] refer to Evangelou and Phillips (1989).

Exchangeable Na/saturated hydraulic conductivity.

The presence of exchangeable Na^+ significantly decreases soil permeability (19, 20, 21). The mechanism(s) responsible for decreasing soil permeability in the presence of Na^+ can be demonstrated by looking into the components controlling water or soil solution movement potential under saturating conditions. This potential is described by Darcy's equation

$$V = KH \qquad [26]$$

where

  V = Flow velocity per unit cross-sectional area of the flow bed,
  H = total potential gradient and
  K = soil saturated hydraulic conductivity.

Soil saturated hydraulic conductivity is described by (22)

$$K = kg/n \qquad [27]$$

where

  k = permeability of the soil,
  g = gravitational constant and
  n = kinematic viscosity or the ratio of solution viscosity over the fluid density.

For soil systems contaminated with brackish solutions, kinematic viscosity is not significantly affected (22) thus, the components controlling water flow velocity are the hydraulic gradient (H) and soil permeability (k). The latter component (k) is influenced by clay dispersion, clay migration and clay swelling.

The theory of colloidal stability developed by Derjaguin and Landau (23) and Verwey and Overbeek (24) (DLVO theory) generally accounts for the influences of ion valence and concentration on clay dispersion and/or clay swelling. According to the DLVO theory the long range repulsive potential resulting from diffuse double layers (DDLs) of like charged colloids retards the coagulation or flocculation rate of clay colloids. Colloid stability (maximum dispersion) depends on maximum repulsive electrical potential between planar colloidal surfaces. This maximum repulsive electrical potential is controlled by surface electrical potential and solution electrolyte concentration (25). The surface electrical potential is controlled by pH, assuming that the colloids involved exhibit pH dependent charge. Suarez et al. (26) was able to demonstrate soil dispersion as a function of pH and relate this relationship to soil saturated hydraulic conductivity.

In addition to the above components (surface electrical potential, ion valence, electrolyte concentration) controlling colloidal flocculation or stability (3, 25, 27) additional components are also involved. These additional components include relative proportion of monavalent to divalent cations in the bulk solution (3) type of cations (cations residing in region II versus region III of the TLM (5, 28), shape of particles and initial particle concentration in suspension (29), type of clay minerals present and relative proportion of clay minerals (30). Observations on the influence of various clay minerals and their relative proportion on dispersion/flocculation behavior suggest that certain physico-chemical interactions between various colloids change their dispersive behavior. The mechanism(s) for this apparent behavior is (are) not understood at the present time. Based on these observations, soils of mixed mineralogy and with various proportions of different clay minerals are expected to have unique dispersive properties. Deterioration of soil physical properties influencing soil permeability is accelerated directly or indirectly by the presence of high Na^+ on the soil's exchange complex and the electrolyte composition and concentration of the soil solution (19, 31, 32, 33, 34). To improve soil physical properties of Na-affected soils, Ca^{2+} is usually added to replace Na^+ on the exchange sites. Calcium reduces clay swelling and enhances clay flocculation (35).

Additional components influencing the effect of Na^+ on saturated hydraulic conductivity of soil include clay content, soil bulk density, Fe and Al oxide content, organic matter content, salt concentration and Na^+ to Ca^{2+} ratio (36, 37, 38, 39, 40, 41, 42). McNeal and Coleman (31) stated that each soil has a unique saturated hydraulic conductivity response threshold because of its unique properties.

Materials and methods.

Soil sample collection and preparation.

The soils involved in this study were the Pembroke soil (fine-silty, mixed, mesic, Mollic Paleudalf) from Hardin County, Kentucky and the Uniontown (fine-silty, mixed, mesic, Typic Hapludalf) from Union County, Kentucky. Both soils are described in detail by USDA (43), and USDA (44). Soil sampling procedures, preparation and characterization are described in detail by Marsi and Evangelou (5). Some of the characterization procedures will be repeated for the purpose of convenience
     The soils were analyzed for pH and acidity employing techniques described by Thomas (45). Exchangeable Ca, Mg, K, Na, and CEC were determined by $BaCl_2$ extraction (46) and 1 $\underline{M}$ ammonium acetate (Chapman, 1965). Particle size analysis was measured by using the pipette method described by Gee and Bauder (48). Organic carbon content was determined by the dry combustion method (49). The < 2 μm clay fraction was prepared for x-ray analysis using the procedure of Jackson (50).

Exchange.

Subsamples of Pembroke and Uniontown soils were adjusted to pH ranges of 4.2 to 4.4, 6.1 to 6.3, and 7.5 to 7.7. The pH adjustments were carried out by either calcium hydroxide $[Ca(OH)_2]$ saturated solution or 0.2 mol L^{-1} hydrochloric acid (HCl) over approximately 1 month period. Details are given in Marsi and Evangelou (5).
     The Na^+-Ca^{2+} exchange was carried out employing duplicate 5-gram soil samples weighed into preweighed 50-ml test tubes. To each of these test tubes was added 25 ml of a solution composed of various Na^+ to Ca^{2+} ratios with a constant chloride concentration at 1 mol L^{-1}. After shaking for 3 hours, the test tubes were centrifuged at 2000 rpm for 10 min and the clear supernatants were decanted. To these test tubes was added a 25-ml solution of the corresponding Na^+ to Ca^{2+} ratio at three different chloride concentrations (5 mmol L^{-1}, 50 mmol L^{-1}, or 200 mmol L^{-1}). The test tubes were shaken for 3 hours, centrifuged at 10000 or 2000 rpm for 10 min, depending on the Na^+ to Ca^{2+} ratio and ionic strength, and the clear supernatants were again decanted. This last treatment was repeated two more times. At the third wash, the test tubes were shaken for 24 hrs and the supernatants were collected for Ca^{2+} and Na^+ determinations in the solution phase. The test tubes were again weighed to quantify the entrapped solution. The exchangeable Na^+ and Ca^{2+} were displaced with 25 ml of 0.1 mol L^{-1} $BaCl_2$ (46).
     Vanselow exchange selectivity coefficients (K_v) were calculated with Eq. [18]. Single ion activity (α) for Na^+ and Ca^{2+} employed in Eq. [18] were estimated by the equation:

$$\alpha_{Na,Ca} = C_{Na,Ca}\,\gamma_{Na,Ca} \qquad [28]$$

where $C_{Na,Ca}$ denotes effective molality of ionic species Na^+ or Ca^{2+} and $\gamma_{Na,Ca}$ dentoes the single ion activity coefficient of ionic species Na^+ or Ca^{2+}. The values for $\gamma_{Na,Ca}$ were estimated by employing the effective ionic strength (I) equation:

$$I = 1/2 \sum_{\beta=1}^{n} C_{\beta} Z^2_{\beta} \qquad [29]$$

where Z denotes charge of the ionic species β, (β denotes all ionic species in solution) and the Davies equation (7)

$$Log\, \gamma_{Na,Ca} = -0502\, Z^2_{Na,Ca}\, [I^{1/2}/(I + I^{1/2})] + 0.1505\, Z^2_{Na,Ca}(I) \qquad [30]$$

Adsorbed-ion activity coefficients (f_i) for Na^+ and Ca^{2+} based on the Argersinger et al. (15) convention were calculated with Eqs. [24] and [25], respectively.

Dispersion/hydraulic conductivity.

Subsamples of the soil were loaded with either Ca^{2+} or Na^+. This Ca^{2+} or Na-loading process was carried out by weighing 400 g of soil into 1 L plastic containers. To each of the plastic containers, 800 ml of 1 mol_c L^{-1} $CaCl_2$ or NaCl was added. The soil suspension was shaken for 3 hours and centrifuged at 2000 rpm for 10 minutes. The clear supernatants were then decanted. This process was repeated three more times. At the last salt wash the suspensions were shaken for 24 hours. The soil was then washed repeatedly with distilled water until solutions were free of chloride according to the $AgNO_3$ test. The soil was then air dried, ground and passed through a 2-mm sieve.

Subsamples of the soil sample in the Ca- or Na- form were adjusted to three pH values. These pH values were in the range of 4.2 to 4.4, 6.1 to 6.3 and 7.5 to 7.7. High pH adjustments were carried out as previously described and details are given in Marsi and Evangelou (5).

Clay dispersion measurements were carried out employing the Imhoff cone test (5, 51). The procedure for the actual measurement of suspended solids by the Imhoff cone was as follows: One gram of pH-adjusted soil with a given ESP was weighed and placed in a 50-ml centrifuge tube. About 25 ml of the salt solution with the corresponding SAR (determined from the ESP-SAR data) was added. The soil suspension was shaken for 1 hour in order to disperse the soil. At the end of 1 hour of shaking the soil suspension was transferred into an Imhoff cone and filled with solution with corresponding SAR up to the 1 L mark. The soil suspension in the Imhoff cone was shaken for 1 minute and allowed to settle 1 hour. The suspension was then carefully siphoned into a 1-L volumetric flask and passed through a 0.22 μm GS millipore filter. The clay collected on top of the filter was dried and weighed. A clay dispersion index (DI) was calculated by dividing the amount of dispersed clay by the amount of clay found in 1 g of soil.

Laboratory saturated hydraulic conductivity measurements were made using a plexiglass permeameter. Detailed description of the permeameter is given in Marsi and Evangelou (52).

In order to attain various ESP values, given quantities of Ca-loaded soil were mixed with given quantities of Na-loaded soil. Furthermore, to this soil 40 percent, on a weight basis, of 20 to 30-mesh Ottawa sand was added and mixed thoroughly in order to maintain water flux under clay dispersive conditions.

Each sample of soil-sand mixture was spread out on a flat surface, moistened and stirred vigorously with a stirring rod in an effort to create aggregates and to attain a uniform mixture of sand and soil and was left to air dry. The air dry soil-sand mixture was introduced to the permeameter with enough compaction to attain a bulk density of 1.35 cm^{-3}.

The soil-sand mixture in the permeameter was slowly saturated from the bottom up by passing a solution that corresponded to the pre-established ESP at the given pH and chloride concentration. A constant hydraulic head of 3 cm was maintained throughout the experiment. The soil columns were leached until they reached a constant saturated hydraulic conductivity before solution flow measurements were recorded. Saturated hydraulic conductivity (K) was calculated employing Darcy's equation (52, 53, 54, 55).

Results and discussion.

Exchange.

The physical properties as well as some chemical properties for the Pembroke and Uniontown soils were reported in Marsi and Evangelou (5). Important differences between these two soils relative to this study were the clay content and the mineralogy. The Pembroke soil was dominated by kaolinite and to a lesser extent by mica and vermiculite, and the Uniontown soil was dominated by vermiculite and to a lesser extent by mica, kaolinite and hydroxy-interlayered vermiculite. Another important difference was the much greater iron and clay content of the Pembroke soil.

The data in Tables 1 and 2 show the mean value of the summation of exchangeable Na$^+$ (ExNa) and exchangeable Ca^{2+} (ExCa) as a function of pH and chloride concentration of the Pembroke and the Uniontown soils, respectively. The plus or minus value associated with each mean value represents the difference in metal adsorption when one of the metals (Na$^+$ or Ca^{2+}) on the exchange phase approaches zero. Therefore, for any mean value plus the deviation from the mean the sum signifies the effective charge (EC$_g$) of the soil when the latter is loaded with Ca^{2+}, and for any mean value minus the deviation from the mean the difference signifies the EC$_g$ of the soil when the latter is loaded with Na$^+$. The data in Tables 1 and 2 clearly demonstrate that the EC$_g$ of these two soils is highly ionic strength dependent, specific ion dependent, and to a lesser degree pH dependent.

Exchangeable sodium percentage versus SAR plots for the Uniontown soil are shown in Figs. 1 and 2. These two figures demonstrate that the ESP-

Table 1. The mean-sum (M) of exchangeable Na^+ and exchangeable Ca^{2+} of the Pembroke soil as a function of pH and chloride concentration (Reproduced with permission from ref. 5. Copyright 1991.)

Chloride	pH		
	4.3	6.1	7.5
mmol L^{-1}	- cmol$_c$kg^{-1} - - - - - - - - - - - - - - - - - - -		
5	7.44$\pm$0.20	8.76$\pm$0.31	9.69$\pm$0.36
50	11.21$\pm$1.4	12.69$\pm$1.31	13.59$\pm$1.21
200	27.69$\pm$3.29	27.74$\pm$5.54	28.92$\pm$5.40

Note: M + S = Effective charge (EC$_g$) of Ca-loaded soil

M - S = Effective charge (EC$_g$) of Na-loaded soil

S = deviation from the average.

Table 2. The mean-sum (M) of exchangeable Na^+ and exchangeable Ca^{2+} of the Uniontown soil as a function of pH and chloride concentration (Reproduced with permission from ref. 5. Copyright 1991.)

Chloride	pH		
	4.3	6.3	7.7
mmol L^{-1}	- cmol$_c$kg^{-1} - - - - - - - - - - - - - - - - - - -		
5	8.94$\pm$0.36	9.54$\pm$0.52	10.47$\pm$0.64
50	11.23$\pm$1.37	13.36$\pm$1.64	14.08$\pm$1.52
200	20.79$\pm$2.10	23.17$\pm$2.83	25.43$\pm$2.90

Note: M + S = Effective charge (EC$_g$) of Ca-loaded soil

M - S = Effective charge (EC$_g$) of Na-loaded soil

S = deviation from the average.

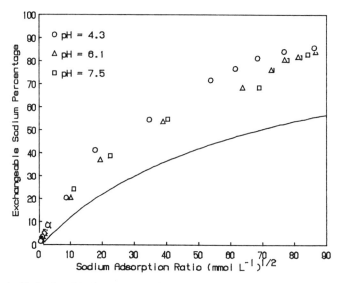

Figure 1. Relationship between exchangeable sodium percentage (ESP) and sodium adsorption ratio (SAR) at a chloride concentration of 5 mmol L^{-1} of the Pembroke soil at three pH values (the solid line without data represents most salt affected soils in the western USA).
(Reproduced with permission from ref. 5. Copyright 1991.)

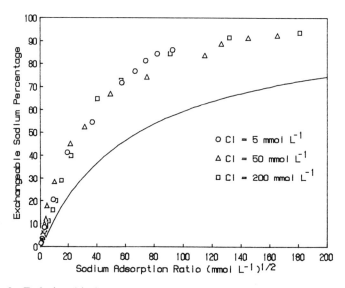

Figure 2. Relationship between exchangeable sodium percentage (ESP) and sodium adsorption ratio (SAR) at three concentrations of chloride of the Pembroke silt loam soil at pH 4.3 (the solid line without data represents most salt affected soils in the western USA).
(Reproduced with permission from ref. 5. Copyright 1991.)

SAR relationship is independent of pH and ionic strength. Furthermore, these data show that the Pembroke soil exhibits a higher affinity for Na^+ than salt affected soils commonly found in the western USA. The data in Fig. 3 and 4 demonstrate that there are at least two classes exchange sites with respect to Na^+-Ca^{2+} exchange on the Pembroke soil. These data also point out that at low ESP values (ESP < 20) the soil exhibits a higher affinity for Na^+ than at high soil ESP values. However, at ESP > 20, the magnitude of K_v remains constant, approximately 1 ($\Delta G° = 0$) which suggest no ion preference (17). These observations are consistent with the information presented by van Bladel et al. (56), Levy and Hillel (57), and Sposito and Mattigod (58).

Exchangeable sodium percentage versus SAR plots for the Uniontown soil are shown in Figs. 5 and 6. These data demonstrate that pH and ionic strength have a strong influence on the ESP-SAR relationship. These data also show that the Uniontown soil exhibits a higher affinity for Na^+ than salt affected soils commonly found in the Western USA. The K_v data in Fig. 7 show that as pH increases so does K_v; consequently, Na^+ is preferred by the solid phase. Furthermore, as ESP increases K_v also increases. According to Stumm and Bilinski (59) deprotonating clay edge surfaces have greater affinity for a monovalent cation than a divalent cation because the former (monovalent cation) requires much less free energy to desolvate and thus come closer to the adsorbing surface. However, according to the data shown in Fig. 8 as ionic strength increases, K_v decreases. Therefore, under high ionic strength the soil appears to prefer Ca^{2+}. This suggests that under high ionic strength divalent cations are most likely to carry out surface soil deprotonation.

A number of other researchers have carried out various studies involving binary heterovalent exchange on various clay minerals. For example, Sposito and Mattigod (58) showed that for the exchange reactions of Na^+ with trace metal cations (Cd^{2+}, Co^{2+}, Cu^{2+}, Ni^{2+} and Zn^{2+}) on Camp Berteau montmorilonite, K_v was constant and independent of exchanger composition up to an equivalent fraction of trace metal cations of 0.70. However, Van Bladel et al. (56) studied Na^+-Ca^{2+} exchange reaction on the same kind of clay mineral and found that there is a more pronounced selectivity of clay for Ca^{2+} exchange on montmorilontic soils.

Based on the above Na^+-Ca^+ exchange studies and on the data presented in this paper the magnitude of K_v is variable. In general, it can be said that the selectivity coefficient (K_v) for a binary exchange reaction depends primarily on the ionic strength, pH and ESP.

Surface charge behavior with respect to heterovalent binary systems, such as Na^+-Ca^{2+} exchange, has been evaluated based on the double layer model by Shainberg et al. (14). These authors separated clay surfaces into two major classes: a) internal surfaces and b) external surfaces. Internal surfaces showed high preference for Ca^{2+} due to their high electrical potential. When ionic strength in the bulk solution was increased and the double layers were suppressed, internal surfaces showed increasing preference for Na^+. Also, because of electrical potential suppression the authors concluded that internal surfaces were expected to increase preference for Na^+ as ESP was decreased.

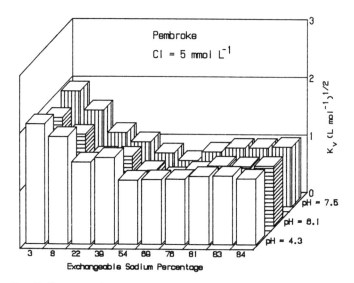

Figure 3.   Influence of exchangeable sodium percentage (ESP) on the Vanselow exchange coefficient of the Pembroke soil at a chloride concentration of 5 mmol L^{-1} and at three pH values.
(Data used with permission from ref. 5. Copyright 1991.)

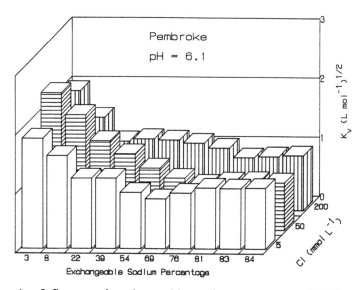

Figure 4.   Influence of exchangeable sodium percentage (ESP) on the Vanselow exchange coefficient of the Pembroke soil at pH 6.1 and at three chloride concentrations.
(Data used with permission from ref. 5. Copyright 1991.)

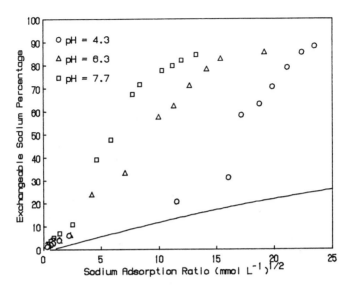

Figure 5. Relationship between exchangeable sodium percentage (ESP) and sodium adsorption ratio (SAR) at a chloride concentration of 5 mmol L^{-1} of the Uniontown soil at three pH values (the solid line without data represents most salt affected soils in western USA).

(Reproduced with permission from ref. 5. Copyright 1991.)

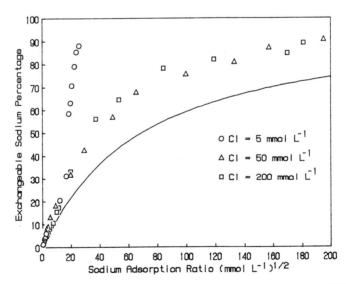

Figure 6. Relationship between exchangeable sodium percentage (ESP) and sodium adsorption ratio (SAR) at three concentrations of chloride of the Uniontown silt loam soil at pH 4.3 (the solid line without data represents most salt affected soils in the western USA).

(Reproduced with permission from ref. 5. Copyright 1991.)

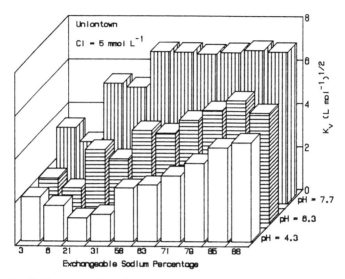

Figure 7.　Influence of exchangeable sodium percentage (ESP) on the Vanselow exchange coefficient of the Uniontown soil at a chloride concentration of 5 mmol L^{-1} and at three pH values.
(Data used with permission from ref. 5. Copyright 1991.)

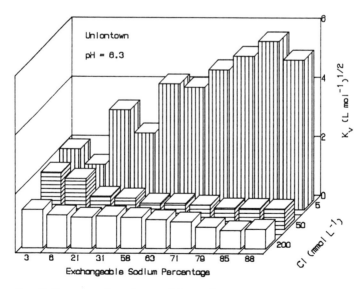

Figure 8.　Influence of exchangeable sodium percentage (ESP) on the Vanselow exchange coefficient of the Uniontown soil at pH 6.3 and at three chloride concentrations.
(Data used with permission from ref. 5. Copyright 1991.)

Based on the double layer model, external clay surface areas because of their low electrical potential showed preference for the Na^+ instead of Ca^{2+}. However, when ionic strength in the bulk solution was increased, the external surfaces showed increasing preference for Ca^{2+}. Finally, when Ca^{2+} surface coverage was increased, external surfaces were expected to show an increasing preference for Na^+.

Experimental evidence on Na^+-Ca^{2+} exchange for internal and external surfaces does not always agree with the conclusions drawn based on the double layer model. Shainberg et al. (14) argued that the reason for this discrepancy was that the assumptions made in employing the double layer model may not always be valid. These assumptions included uniformly charged surfaces, ions as point charges and nonspecific interactions of the ions with the charged surfaces. For example, their experimental data on Na^+-Ca^{2+} exchange showed that the affinity of illite for Ca^{2+} decreased when exchangeable Ca^{2+} was increased, however, the reverse was true for montmorilonite.

Soils of temperate regions in the U.S. cannot be classified as soils dominated by either internal or external surface areas. Their mineralogy is highly mixed. Because of mixed mineralogy, Na^+-Ca^{2+} exchange behavior on these soils would be dictated by the specific interactions between the various clay minerals (30).

Lack of influence of pH and ionic strength on the K_v in the case of the Pembroke soil (Figs. 3 and 4) could be related to a number of processes that could take place on a clay surface as pH and/or ionic strength was increased. For example, Pratt et al. (1962) have demonstrated on a number of soils that when pH was decreased, the exchange selectivity coefficient of Na^+-Ca^{2+} exchange was also increased. This increase in the K_v signified an increase in affinity of the Na^+ by the clay surface due to a decrease in surface charge density (14). Their data supported this conclusion. Additionally, Shainberg et al. (14) showed that for Na^+-Ca^{2+} exchange, when ionic strength was increased affinity for the Na^+ by the illitic surface was also increased. The latter observation would depend on whether one deals with an external surface or an internal surface. For example, if the case of an external surface an increase in ionic strength could increase affinity for Ca^{2+}, while in the case of an internal surface an increase in ionic strength could decrease affinity for Ca^{2+}. Considering that the Pembroke soil is composed of external and internal surfaces, a cancelling effect on the magnitude of K_v due to an increase in ionic strength could be obtained. Variations in soil pH may not have any influence on the magnitude of K_v for several reasons. One reason could be that the soil surface charge density remains unaffected by pH. A second reason maybe the new charge sites generated due to pH increase do not differ in cation affinity fror the charge sites present at lower pH values.

Marsi and Evangelou (5) postulated that the location of a cation with respect to a charged surface has meaning with respect to clay dispersion. They justified this postulation by linking absorbed-ion activity coefficients (f_i), in a physical way to the triple layer model (TLM) (17). Based on the TLM, a negatively charged surface consists of three compartments or regions: I. the inner-sphere region; II. the outer-sphere region; and III. the diffuse ion swarm region. The inner-sphere region represents completely dehydrated cations. It is assumed that such an adsorption mechanism is not representative for Na^+ or Ca^{2+}. These cations most likely occupy regions II and III (Sposito, 1984). Region II is characterized by water tightly bound to the adsorbed cations, and anions are excluded. Region III is characterized by water tightly bound to the cations as well as water loosely bound to the cations, and anions are partially excluded. For Na^+-Ca^{2+} exchange, when E_{Na} approaches zero, f_{Ca} equals unity; and, when Ca-equivalent fraction on the exchange approaches zero, f_{Na} also equals unity (Evangelou, et al., 1989). Therefore, an f_i ($i = Na^+$ or Ca^{2+}) of 1 is taken to represent an outer-sphere complex. A f_i value greater than 1 is taken to signify that cation i is located in the diffuse layer while a f_i value less than 1 is taken to signify that the cation is specifically interacting with the charged surface.

Adsorbed-ion activity coefficients (f_i) are taken to reflect changes in the fugacity of an ion held by a clay surface. Fugacity is defined as the degree of freedom an ion has to leave the adsorbed state relative to a standard of maximum freedom set at unity. Thus, for an $f > 1$ it can be assumed that the adsorbed ion has nearly escaped the influence of the electric field of the charged surface, while for an $f < 1$, it can be assumed that the adsorbed ion is strongly influenced by the electric field of the charged surface. The former case signifies an expanded double layer while the latter case signifies a compressed double layer.

The data in Fig. 9 show that in the case of the Pembroke soil Ca^{2+} was tightly bound to the charged surface ($f_{Ca} < 1$). This binding strength was increased as ESP increased. Also at low ESP values, f_{Na} was less than one which signified that the Na ion was specifically interacting with the surface. Furthermore, when ESP was increased, f_{Na} increased and became approximately 1. Ionic strength also appeared to have influence on the magnitude of f_{Ca}. The data showed that f_{Ca} at 200 mmol L^{-1} chloride concentration was large than f_{Ca} at 5 and 50 mmol L^{-1} Cl. This could be because the Pembroke soil is dominated by external adsorption sites (14).

The findings shown in Fig. 9 representing the Pembroke soil are not in full agreement with the findings shown in Fig. 10 representing the Uniontown soil. This is especially true for f_{Na}. The data in Fig. 11 show that f_{Na} at 5 mmol L^{-1} Cl is greater than 1 up to an ESP of approximately 60. At 50 and 200 mmol L^{-1} Cl, f_{Na} remains near 1 for the entire exchange isotherm. The value for f_{Ca}, on the other hand, decreases below 1 as ESP increases for the 50 and 200 mmol L^{-1} Cl but f_{Ca} increases as ESP increases at the 5 mmol L^{-1} Cl. Therefore, the largest difference in the magnitude of f_{Na} between the Pembroke and the Uniontown soils appears to be at low ESP and at low ionic strength.

The Uniontown soil appears to exhibit high Na^+ fugacity at low ESP and low ionic strength, while the Pembroke soil for the same ESP and ionic strength exhibits low Na^+ fugacity. These differences could be attributed to the mineralogical differences of these two soils. The Pembroke soil, because of high kaolinite content is dominated by external surface area; therefore, it is

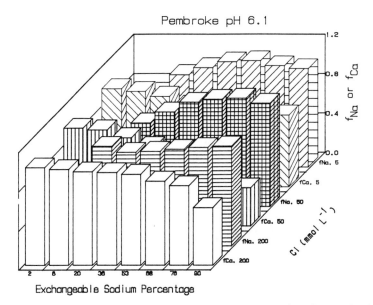

Figure 9. Influence of exchangeable sodium percentage (ESP) on adsorbed-ion activity coefficients of the Pembroke soil at pH 6.1 and at three chloride concentrations. (Data used with permission from ref. 5. Copyright 1991.)

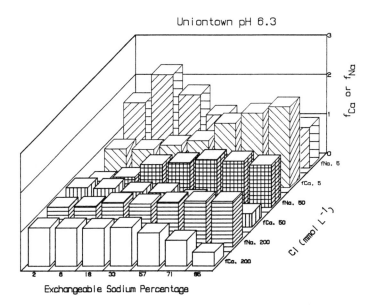

Figure 10. Influence of exchangeable sodium percentage (ESP) on adsorbed-ion activity coefficients of the Uniontown soil at pH 6.3 and at three chloride concentrations. (Data used with permission from ref. 5. Copyright 1991.)

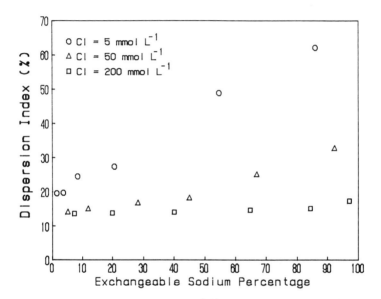

Figure 11.   Influence of exchangeable soidum percentage (ESP) on the dispersion index (DI) of the Pembroke soil at pH 4.3 with three chloride concentrations. (Data used with permission from ref. 69. Copyright 1991.)

expected to exhibit high specificity for Na$^+$ (14). On the other hand, the Uniontown soil, because of high vermiculite content (large internal surface area), is expected to exhibit low specificity for Na$^+$, especially at low ionic strength (14).

Imhoff Cone Test.

The data in Figs 11 and 12 show that at low ionic strength the potential of the two soils to undergo dispersion was related to ESP. When ionic strength was adjusted to 200 mmol L^{-1}, no apparent effect of ESP on soil dispersion was observed. The repulsive force, due to a contracted double layer was decreased when ionic strength was increased (60).

Suppression of repulsive double-layer forces can be accomplished by decreasing pH. A decrease in pH is associated with an increase in positive charge (61, 62, 63). As the colloids approach the pH of zero point of net charge (pH at which positive surface potential equals negative surface potential), edge-to-face interactions increase, leading to clay flocculation (23, 27, 64, 65). The data in Fig. 13 and 14 show that, even at pH 4.3, both soils exhibit dispersion at low ionic strength. This observation suggests that at pH 4.3 both soils exhibit net surface negative charge.

The data in Fig. 13 show there was a relatively small increase effect on clay dispersion of the Pembroke soil at the three pH values tested. This was not surprising since there was a very small pH effect on the surface charge (CEC) of the soil. Data in Marsi and Evangelou (5; Table 4) show that CEC (ExNa + 2Ex$_2$Ca) of the Pembroke soil at 5 mmol L^{-1} chloride increased from 7.5 cmol$_c$ kg^{-1} when pH increased from 4.3 to 7.5.

Dispersion data on Uniontown sample (Fig. 15) show that the dispersion behavior of this soil was relatively independent of pH when chloride concentration was set at 5 mmol L^{-1}. This could have been caused by the fact that the effective charge (ExNa + 2Ex$_2$Ca) of the Uniontown soil increased slightly (8.9 to 10.5 cmol$_c$ kg^{-1}) when pH was increased from 4.3 to 7.7 (5).

In summary the Uniontown soil was more sensitive to dispersion under decreasing electrolyte concentration and increasing ESP but less sensitive to pH changes than the Pembroke soil. The data also showed that for any given electrolyte concentration, pH and ESP, the dispersion index of the Uniontown soil was always greater than that of the Pembroke soil. This appeared to be in agreement with the interpretation assigned earlier to adsorbed ion activity coefficients. The magnitude of adsorbed-ion activity coefficient (f_{Na}) for the Uniontown soil at low ESP was greater than one. Considering that $f_{Na} > 1$ is taken to signify that Na$^+$ "resides" in the diffuse layer of the triple-layer model, one would expect that the Uniontown soil to be highly dispersive. On the other hand, the magnitude of f_{Na} of the Pembroke soil at low ESP is less than 1. Assuming this signifies that Na$^+$ forms outer-sphere complexes with the clay surfaces, the Pembroke soil would be expected to be less dispersive than the Uniontown soil.

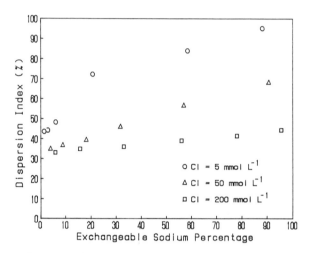

Figure 12. Influence of exchangeable sodium percentage (ESP) on the dispersion index (DI) of the Pembroke soil equilibrated with a solution of 5 mmol L^{-1} chloride concentration at three chloride concentrations.
(Data used with permission from ref. 69. Copyright 1991.)

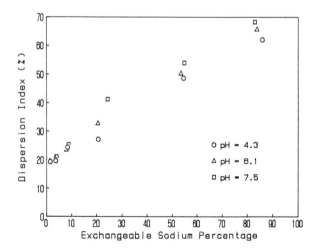

Figure 13. Influence of exchangeable sodium percentage (ESP) on the dispersion index (DI) of the Pembroke soil equilibrated with a solution of 5 mmol L^{-1} chloride concentration at three pH values.
(Data used with permission from ref. 69. Copyright 1991.)

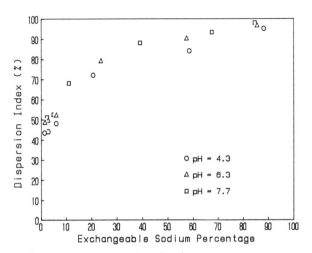

Figure 14. Influence of exchangeable sodium percentage (ESP) on the dispersion index (DI) of the Uniontown soil equilibrated with a solution of 5 mmol L^{-1} chloride concentration at three pH values. (Data used with permission from ref. 69. Copyright 1991.)

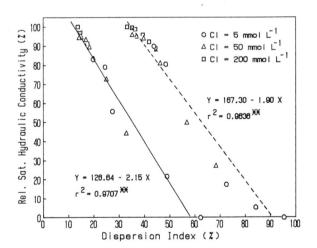

Figure 15. Relationship between relative saturated hydraulic conductivity (RHSC) and dispersion index (DI) of the Pembroke (____) and Uniontown (-----) soils at pH 4.3 with three chloride concentrations. (Reproduced with permission from ref. 69. Copyright 1991.)

Imhoff Cone-Saturated Hydraulic Conductivity.

In order to predict soil saturated hydraulic conductivity of such salt affected soils relative saturated hydraulic conductivity (RSHC) values were correlated with Imhoff cone results expressed as a dispersion index (DI). These data, shown graphically in Fig. 15 through Fig. 19 and summarized in Tables 3 and 4, demonstrate the following points: a) relative saturated hydraulic conductivity was related to clay dispersion (DI), b) the relationship between RSHC and DI was dependent on ionic strength and pH, and c) the two soils exhibited different RSHC-DI relationships.

The data in Fig. 15 showed that at pH 4.3 the RSHC-DI relationship was independent of chloride concentration; but, at pH of approximately 7.5 the soils (Fig. 16 a, b) exhibited two unique RSHC-DI relationships. The first RSHC-DI relationship belonged to the 50 and 200 mmol L^{-1} chloride systems, and the second belonged to the 5 mmol L^{-1} chloride system.

The data in Fig. 17a revealed that the Pembroke soil at 5 mmol L^{-1} chloride solution showed two unique RHSC-DI relationships. One occurred at pH 4.3 while the other occurred at pH 6.1 and 7.5. The Uniontown soil (Fig. 17b) showed a unique RSHC-DI relationship for each of the pH values tested. When the chloride concentration was raised to 200 mmol L^{-1}, the RSHC-DI relationship became independent of pH in both soils (Fig. 18).

Generally, the data in Figs. 15 through 19 demonstrated that the slope of the RSHC-DI relationship of the Pembroke soil was greater than the Uniontown soil. This suggested that the saturated hydraulic conductivity of the Uniontown soil was less affected by change in DI than was the Pembroke soil. Moreover, these data also showed that in order to attain the same relative suppression in saturated hydraulic conductivity on both soils, a greater DI was needed for the Uniontown soil than the Pembroke soil. This was probably due to soil texture. The Pembroke soil contained 59 percent clay while the Uniontown contained only 28 percent clay. Hamid and Mustafa (33) reported that RSHC-DI relationships are highly affected by soil texture as well as pore size distribution.

Figure 16a,b shows that for each of the two soils there was a unique RSHC-DI relationship at the 5 mmol L^{-1} chloride concentration. More importantly, at this chloride concentration in comparison to the two higher chloride concentrations a lower DI was needed to suppress significantly soil saturated hydraulic conductivity. This suggested that at the lower salt concentration, clay swelling was also implicated in reducing saturated hydraulic conductivity (19, 31, 32, 66, 67).

Clay swelling effects due to pH were also implicated in this study. When pH was increased, a smaller DI imposed a large suppression in saturated hydraulic conductivity. It is postulated that the increased pH increased clay swelling potential. This is likely because of the removal of Al-OH polymers from the interlayer at the higher pH values. The presence of Al-OH polymers at the lower pH values may limit interlayer swelling (68).

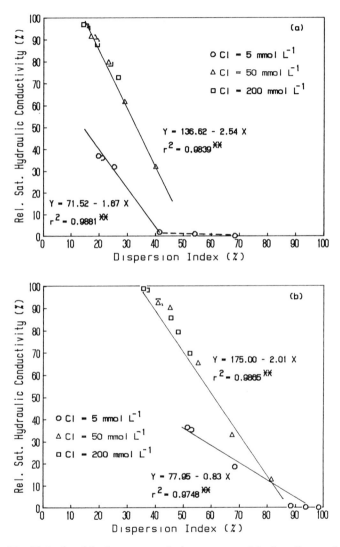

Figure 16. Relationship between relative saturated hydraulic conductivity (RSHC) and dispersion index (DI) of Pembroke soil at pH 7.5 (a) and Uniontown soil at pH 7.7 (b) with three chloride concentrations. (Reproduced with permission from reference 69; Fig. 7.)

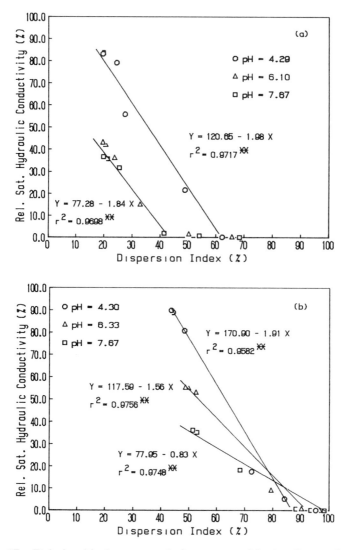

Figure 17. Relationship between relative saturated hydraulic conductivity (RSHC) and dispersion index (DI) of the Pembroke (a) and Uniontown (b) soils equilibrated with solutions of 5 mmol L^{-1} chloride concentration at three pH values. (Reproduced with permission from reference 69; Fig. 8.)

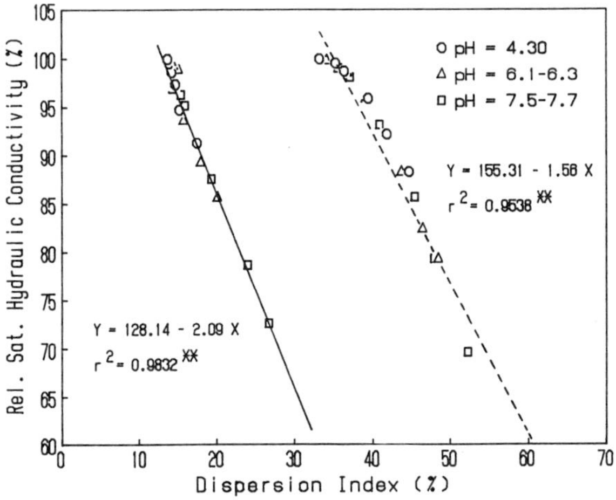

Figure 18. Relationship between relative saturated hydraulic conductivity (RSHC) and dispersion index (DI) of the Pembroke (_____) and Uniontown (-----) Uniontown soils equilibrated with solutions of 200 mmol L^{-1} chloride concentration at three pH values.
(Reproduced with permission from ref. 69. Copyright 1991.)

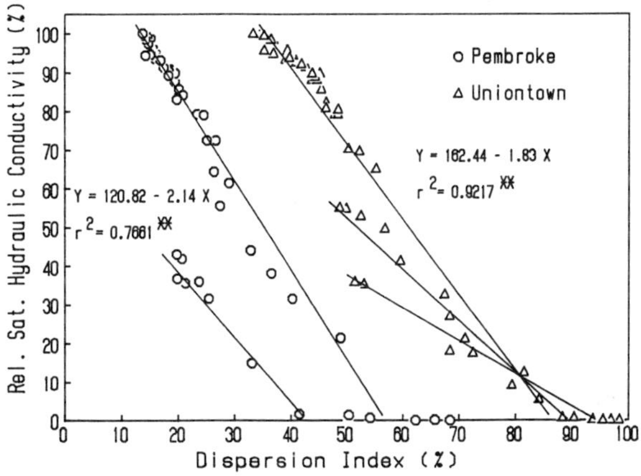

Figure 19. Relationship between relative saturated hydraulic conductivity (RSHC) and dispersion index (DI) of the Pembroke and Uniontown soils equilibrated with three chloride concentration at three pH values.
(Reproduced with permission from ref. 69. Copyright 1991.)

Table 3. Regression equations and coefficient correlations relating relative saturated hydraulic conductivity (RSHC) to dispersion index (DI) for the Pembroke soil under various pH and solution chloride concentration treatments

pH	Cl (mmol L^{-1})	Regression equation	df	r^2
4.29	5	Y = 120.65 - 1.98 X	4	0.9717**
4.29	50	Y = 136.80 - 2.73 X	4	0.9718**
4.29	200	Y = 131.38 - 2.33 X	4	0.9614**
4.29	All Cl	Y = 126.30 - 2.15 X	16	0.9707**
6.10	5	Y = 58.42 - 1.00 X	4	0.8754**
6.10	50	Y = 141.27 - 2.85 X	4	0.9952**
6.10	200	Y = 134.20 - 2.45 X	4	0.9559**
6.10	All Cl	Y = 118.51 - 2.17 X	16	0.7184**
7.48	5	Y = 50.85 - 0.86 X	4	0.8352**
7.48	50	Y = 140.35 - 2.70 X	4	0.9944**
7.48	200	Y = 126.72 - 2.02 X	4	0.9969**
7.48	All Cl	Y = 115.33 - 2.02 X	16	0.6888**
All pH	5	Y = 78.56 - 1.31 X	16	0.6295**
All pH	50	Y = 138.08 - 2.70 X	16	0.9805**
All pH	200	Y = 128.14 - 2.09 X	16	0.9832**
All pH	All Cl	Y = 120.82 - 2.14 X	52	0.7661**

** denotes significantly different at P < 0.01
SOURCE: Reproduced with permission from ref. 69. Copyright 1991.

Table 4. Regression equations and coefficient correlations relating relative saturated hydraulic conductivity (RSHC) to dispersion index (DI) for the Uniontown soil under various pH and solution chloride concentration treatments

pH	Cl (mmol L^{-1})	Regression equation	df	r^2
4.30	5	Y = 170.90 - 1.91 X	4	0.9582**
4.30	50	Y = 176.84 - 2.19 X	4	0.9836**
4.30	200	Y = 136.49 - 1.06 X	4	0.9579**
4.30	All Cl	Y = 167.30 - 1.90 X	16	0.9636**
6.33	5	Y = 117.59 - 1.56 X	4	0.9756**
6.33	50	Y = 191.05 - 2.42 X	4	0.9860**
6.33	200	Y = 153.67 - 1.52 X	4	0.9743**
6.33	All Cl	Y = 161.00 - 1.83 X	16	0.9270**
7.67	5	Y = 77.95 - 0.83 X	4	0.9748**
7.67	50	Y = 179.65 - 2.09 X	4	0.9881**
7.67	200	Y = 163.13 - 1.75 X	4	0.9781**
7.67	All Cl	Y = 157.93 - 1.76 X	16	0.8785**
All pH	5	Y = 131.14 - 1.44 X	16	0.8453**
All pH	50	Y = 177.86 - 2.14 X	16	0.9683**
All pH	200	Y = 155.31 - 1.56 X	16	0.9538**
All pH	All Cl	Y = 162.44 - 1.83 X	52	0.9217**

** denotes significantly different at P < 0.01
SOURCE: Reproduced with permission from ref. 69. Copyright 1991.

Conclusions.

This investigation has shown that the Pembroke and Uniontown soils exhibited some unique Na^+-Ca^{2+} exchange characteristics. In the case of the Pembroke soil, Na^+-Ca^{2+} exchange behavior was independent of pH and ionic strength and ideal solid-solution behavior appeared to be obeyed in the ESP range of approximately 20 to 100. Furthermore, this soil showed no preference for either Na^+ or Ca^{2+}. In contrast, the Uniontown soil exhibited Na^+-Ca^{2+} exchange dependence on pH and ionic strength and displayed a strong preference for Na^+, but ideal solid-solution behavior was only displayed at the highest ionic strength.

Surface behavior analysis of these two soils by estimating absorbed-ion activity coefficients and relating them to the most probably residence of the ions with respect to the triple layer model (inner-sphere, outer-sphere, and diffuse layer) showed that for a significant ESP range (0 to 20) in the Uniontown soil, Na^+ resided in the diffuse layer. Adsorbed-ion activity coefficients at low ionic strength appeared to be greater than 1 up to an ESP of approximately 60. On the other hand, adsorbed-ion activity coefficient data for the Pembroke soil suggested that Na^+ formed mostly outer-sphere complexes.

The dispersion index data clearly pointed out that the dispersive behavior of the Pembroke and Uniontown soils were affected by electrolyte concentration, solution composition, and to a lesser extent, by pH. In general, clay dispersion increased as electrolyte concentration decreased and SAR and pH increased. The Uniontown soil dispersed more than the Pembroke soil as salt concentration decreased and ESP increased. No pH influence on clay dispersion was apparent in the Uniontown soil. These observations appear to be consistent with the observations and conclusions drawn from adsorbed-ion activity coefficient data.

The overall study demonstrates that soil dispersion could be employed for predicting soil saturated hydraulic conductivity (RSHC). However, the relationship between RSHC and soil dispersion is not universal; it is unique to a particular soil under certain conditions. These conditions include soil mineralogy, soil texture, soil pH, ionic strength and solution composition. The mechanisms influencing the RSHC-DI relationship are clay dispersion potential and clay swelling potential. Indirect evidence implicate both of the above mechanisms. Clay dispersion appears to be implicated at a wide range of pH values, wide range of solution composition and low ionic strength. On the other hand, clay swelling appears to be implicated at high pH and low ionic strength.

References.

1.   Brine Disposal in Northeast Ohio. A report on the problem and recommendation for action. *Report of Northeast Ohio Brine Disposal Task Force, Northeast Ohio Area wide Coordinating Agency,* May 1984.

2.   U.S. Government Printing Office, Washington, DC. *Diagnosis and Improvement of saline and alkali soils.* U.S. Salinity Laboratory Staff, USDA Agric. Handbook No. 60, 1954, 160 pp.

3.   Shainberg, I.; Letey, J.; "Response of soils to sodic and saline conditions";. *Hilgardia* 1984, 52:1-57.

4.   Lumbanraja, J; Evangelou, V.P.; "Acidification and liming influence on surface charge behavior of Kentucky subsoils"; *Soil Sci. Soc. Am. J.* 1991, 54:26-34.

5.   Marsi, M.; Evangelou, V.P.; "Chemical and physical behavior of two Kentucky soils: I. sodium-calcium exchange"; *J. Env. Sci. & Health*, 1991a, A26:1147-1176.

6.   Smith, J.M.; Van Ness, H.C.; *Introduction to Chemical Engineering Thermodynamics, fourth edition*; McGraw-Hill Book Company, New York, NY, 1987.

7.   Davies, C.W.; *Ion association*; Butterworth, Washington, D.C., 1962, 190 pp.

8.   Vanselow, A.P.; "Equilibria of the base exchange reaction of bentonite, permutites, soil colloids, and zeolites"; *Soil Sci.* 1932, 33:95-113.

9.   Evangelou, V.P.; Phillips, R.E.; "Sensitivity analysis on the comparison between the Gapon and Vanselow exchange coefficients"; *Soil Sci. Soc. Am. J.*, 1987, 51:1473-1479.

10.   Sposito, G.; *The thermodynamics of soil solutions*; Clarendon Press: Oxford, UK, 1981.

11.   Elprince, A.M.; Vanselow, A.P.; Sposito, G.; "Heterovalent, ternary cation exchange equilibria: NH_4^+-Ba^{2+}-La^{3+} exchange on montmorilonite"; *Soil Sci. Soc. Am. J.* 1980, 44:964-969.

12.   Sposito, G.; "Thermodynamics of the soil solution"; *Soil Physical Chemistry*; CRC Press: Boca Raton, FL; 1986.

13.   Sposito, G.; LeVesque, C.S.; "Sodium-calcium-magnesium exchange on silver Hill Illite"; *Soil Sci. Soc. Am. J.* 1985, 49:1153-1159.

14.   Shainberg, I.; Oster, J.D.; Wood, J.D.; "Na/Ca exchange in montmorilonite and illite suspension"; *Soil Sci. Soc. Am. J.* 1980, 44:960-964.

15.   Argersinger, Jr., W.J.; Davidson, A.W.; Bonner, O.D.; "Thermodynamics and ion exchange phenomena. *Trans. Kansas Acad. Sci.* 1950, 53:404-410.

16.   Guggenheim, E.A.; *Thermodynamics*; North-Holland, Amsterdam, 1967, 389 pp.

17.   Spositio, G.; *The surface chemistry of soils*; Oxford Univ. Press: New York, NY. 1984.

18.   Evangelou, V.P.; Phillips, R.E.; "Theoretical and experimental interrelationship of thermodynamics exchange parameters obtained by the Argersinger and Gaines and Thomas conventions; *Soil Sci.* 1989, 148:311-321.

19.   Quirk, J.P.; Scholfield, R.K.; "The effect of electrolyte concentration on soil permeability"; *J. Soil Sci.* 1955, 6:163-178.

20.   Reeve, R.C.; Bower, C.A.; "Use of high-salt water as a flocculant and source of divalent cations for reclaiming sodic soils"; *Soil Sci.* 1960, 90:139-144.

21.   Park, C.S.; O'Connor, G.A.; "Salinity effect on hydraulic conductivity of soils"; *Soil Sci.* 1980, 130:167-174.

22. Lagerwerff, J.V.; Nakayama, F.S.; Frere, M.H.; "Hydraulic conductivity related to porosity and swelling of soil"; *Soil Sci. Soc. Am. Proc.* 1969, 33:3-11.

23. Derjaguin, B.V.; Landau, L.; "A theory of the stability of strongly charged hyophobic soils and the coalescence of strongly charged particles in electrolytic solutions"; *Acta Physiochim. URSS* 1941, 14:633-662.

24. Verwey, E.J.W.; Overbeek, J. Th. G.; *Theory of the stability of lyophobic colloids*; Elsevier Publ. Co.: Amsterdam, 1948.

25. Evangelou, V.P.; "Influence of water chemistry on suspended soils in coal-mine sedimentation ponds"; *J. Environ. Qual.* 1990, 19:428-434.

26. Suarez, D.L.; Rhoades, J.D.; Lavado, R.; Grieve, C.M.; "Effect of pH on saturated hydraulic conductivity and soil dispersion"; *Soil Sci. Soc. Am. J.* 1984, 48:50-55.

27. Schofield, R.K.; Samson, H.R.; "Flocculation of kaolinite due to the attraction of oppositely charged crystal faces"; *Discuss. Faraday Soc.* 1954, 18:135-145.

28. Hesterberg, D.; Page, A.L.; "Critical coagulation concentrations of sodium and potassium illite as affected by pH"; *Soil Sci. soc. Am. J.* 1990, 54:735-739.

29. Oster, J.D.; Shainberg, I.; Wood, J.D.; "Flocculation value and gel structure of sodium/calcium montmorilonite and illite suspension"; *Soil Sci. Soc. Am. J.* 1980, 44:955-959.

30. Arora, H.S.; Coleman, N.T.; "The influence of electrolyte concentration on flocculation of clay suspensions"; *Soil Sci.* 1979, 127:134-139.

31. McNeal, B.L.; Coleman, N.T.; "Effect of solution composition on soil hydraulic conductivity"; *Soil Sci. Soc. Am. Proc.* 1966, 30:308-312

32. Rowell, D.L.; Payne, D.; Ahmad, N.; "The effect of the concentration and movement of solution on the swelling, dispersion, and movement of clay in saline and alkali soils"; *J. Soil Sci.* 1969, 20:176-188.

33. Hamid, K.S.; Mustafa, M.A.; "Dispersion as an index of relative hydraulic conductivity in salt affected soils of Sudan"; *Geoderma* 1975, 14:107-114.

34. Yousaf, M.; Ali, O.M.; Rhoades, J.D.; "Clay dispersion and hydraulic conductivity of some salt affected arid land soils"; *Soil Sci. Soc. Am. J.* 1987, 51:905-907.

35. Greacen, E.L.; "Swelling forces in straining clays"; *Nature* 1959, 184:1695-1697.

36. Keren, R.; Singer, M.J.; "Effect of low electrolyte concentration on hydraulic conductivity of sodium/calcium montmorilonite-sand system"; *Soil Sci. Soc. Am. J.* 1988, 52:368-373.

37. Gardner, W.R.; Mayhugh, M.S.; Goertzen, J.O.; Bower, C.A.; "Effect of electrolyte concentration and exchangeable sodium percentage on diffusivity of water in soils"; *Soil Sci.* 1959, 88:270-274.

38. Shainberg, I.; Caiserman, A.; "Studies on Na/Ca montmorillonite systems. 2. The hydraulic conductivity. *Soil Sci.* 1971, 111:276-281.

39. Cass, A.; Sumner, M.E.; "Soil pore structural stability and irrigation water quality: I. Empirical sodium stability model"; *Soil Sci. Soc. Am. J.* 1982, 46:503-506.

40. McNeal, B.L.; Layfield, D.A.; Norvell, W.A.; Rhoades, J.D.; "Factors influencing hydraulic conductivity of soils in the presence of mixed salt solutions"; *Soil Sci. Soc. Am. Proc.* 1968, 32:187-190.

41. Wada, K.; Beppu, Y.; "Effects of Aluminum treatments on permeability and cation status of a smectite clay"; *Soil Sci. Soc. Am. J.* 1989, 53:402-406.

42. Frenkel, H.; Goertzen, J.O.; Rhoades, J.D.; "The effect of clay type and content, exchangeable sodium percentage, and electrolyte concentration on clay dispersion and soil hydraulic conductivity"; *Soil Sci. Soc. Am. J.* 1978, 42:32-39.

43. USDA, Soil Conservation Service, in cooperation with the Kentucky Agricultural Experiment Station; *Soil survey of Hardin and Larue counties, Kentucky,* 1976.

44. USDA, Soil Conservation Service, in cooperation with the Kentucky Agricultural Experiment station; *Soil Survey of Union and Webster counties, Kentucky,* 1981.

45. Thomas, G.W.; "Exchangeable cations"; In *Methods of soil analysis Part 2. Agronomy*; L.A. Page, Editor; 1982, 9:159-165.

46. Hendershot, W.H.; Duquette, M.; "A simple barium method for determining cation exchange capacity and exchangeable cations" *Soil Sci. Soc. Am. J.* 1986, 50:605-608.

47. Chapman, H.D.; "Cation exchange capacity" In *Methods of soil analysis*; C.A. Black et al., Editor; Methods of soil analysis. Part 1. *Agron.* 1965, 9:891-901.

48. Gee, G.W.; Bauder, J.W.; "Particle size analysis" In A. Klute, Editor; Methods of soil analysis. Part 1" *Agron.* 1982, 9:539-579.

49. Nelson, D.W.; Sommers, L.E.; "Total carbon, organic carbon, and organic matter"; In *Methods of soil analysis, Part 2. Agron.*; L.A. Page, et al.; Editor; 1982, 9:539-579.

50. Jackson, M.L.; *Soil chemical analysis - advanced course, 2nd edition*; Publisher: M.L. Jackson; Madison, WI, 1975, pages 169-249.

51. Barfield, B.J.; Warner, R.C.; Haan, C.T.; *Applied hydrology and sedimentology for disturbed areas.* Publisher: Oklahoma Technical Press, Stillwater, OK, 1981, 603 pp.

52. Marsi, M.; Evangelou, V.P.; "Chemical and physical behavior of tow Kentucky soils: II. Saturated hydraulic conductivity-exchangeable sodium relationship"; *J. Env. Sci. & Health,* 1991b, A26:1177-1194.

53. Hillel, D.; *Fundamentals of soil physics*; Academic Press, Inc., Orlando, FL, 1980, 413 pp.

54. Baver, L.D.; Gardner, W.H.; Gardner, W.R.; *Soil physics, 4th edition*; John Wiley & Sons, Inc., New York, NY, 1972, 498 pp.

55. Hands, R.J.; Ashcroft, G.L.; *Applied soil physics.   Soil water and temperature applications*, Springer-Verlag, Berlin, 1980, 159 pp.

56. Van Bladel, R.; Gavria, G.; Laudelout, H.; "A comparison of the thermodynamics, double-layer theory, empirical studies of the Na-Ca exchange equilibria in clay water system"; In *Proc. International Clay Conference*, 1972, pages 385-398.

57. Levy, R.; Hillel, D.; "Thermodynamic equilibrium constants of sodium-calcium exchange in some Israel soils. *Soil Sci.* 1968, 106:393-398.

58. Sposito, G.; Mattigod, S.V.; "Ideal behavior in Na^+-trace metal cation exchange on Camp Berteau montmorilonite"; *Clays and Clay Minerals*, 1979, 27:125-128.

59. Stumm, W.; Bilinski, H.; "Trace metals in natural waters: difficulties of interpretation arising from our ignorance on their speciation"; *Adv. Water Pollut. Res.* 1973, 6:39-49.

60. Luklema, J.; "Fundamentals of electrical double layers in colloidal systems"; In J.W. Goodwin; Editor; Colloidal dispersions; *Royal Soc. Chem. Spec. Pub. No. 43*; Burlington House, London, 1982, pages 47-70.

61. Rand, B.; Melton, I.E.; "Particle interactions in aqueous kaolinite suspensions. I. Effect of pH and electrolyte upon the mode of particle interaction in homoionic sodium kaolinite suspensions"; *J Colloid Interface Sci.* 1977, 60:308-321.

62. Shawney, B.L.; Norrish, K.; "pH dependent cation exchange capacity minerals and soils of tropical regions"; *Soil Sci.* 1971, 112:213-215.

63. Gillman, G.P.; "Using Variable charge characteristic to understand the exchange cation status of oxic soils"; *Aust. J. Soil Res.* 1984, 22:71-80.

64. Van Olphen; *An introduction to clay colloid chemistry, 2nd edition*; Publisher: John Wiley & Sons, New York, 1977, 318 pp.

65. Tama, K.; El-Swify, S.A.; "Charge, colloidal, and structural stability interrelationships for oxidic soils"; In *Modification of soil structure*, W.W. Emerson, et al., Editors; Publisher: John Wiley & Sons, Ltd., 1978, pages 40-49.

66. McNeal, B.L., Norvell, W.A.; Coleman, N.T.; "Effect of solution composition on the swelling of extracted soil clays"; *Soil Sci. Soc. Am. Proc.* 1966, 30:313-315.

67. Keren, R.; Shainberg, I.; Klein, Eva; "Settling and flocculation value of sodium-montmorilonite particles in aqueous media"; *Soil Sci. Soc. Am. J.* 1988, 52:76-80.

68. Barnhisel, R.I.; "Chlorites and hydroxy interlayered vermiculite and smectite"; In *Minerals in soil environments*; J.B. Dixon, et al. Editor; Soil Sci. Soc. Am., Madison, WI, 1977, pages 331-356.

69. M. Marsi; V.P. Evangelou; "Chemical and physical behavior of two Kentucky soils: III. Saturated hydraulic conductivity--Imhoff cone test relationships"; *J. Env. Sci. & Health*, 1991c, A26:1195-1115.

RECEIVED November 12, 1992

Chapter 16

Decomposition of Perchloroethylene and Polychlorinated Biphenyls with Fenton's Reagent

C. Sato[1], S. W. Leung[2], H. Bell[1], W. A. Burkett[3], and R. J. Watts[4]

[1]Department of Civil and Environmental Engineering, Polytechnic University, Six Metrotech Center, Brooklyn, NY 11201
[2]College of Engineering, Idaho State University, Pocatello, ID 83209
[3]Astoria Chemical Laboratory, Consolidated Edison Company of New York, 20th Avenue & 31st Street, Astoria, NY 11105
[4]Department of Civil and Environmental Engineering, Washington State University, Pullman, WA 99164

Many of the hazardous waste sites identified by U.S. EPA are contaminated with chlorinated organics such as perchloroethylene (PCE) and polychlorinated biphenyls (PCBs). Since these compounds are toxic to indigenous microorganisms, bioremediation is considered impractical at highly contaminated sites, and a chemical treatment method is often recommended. Fenton's reagent has recently drawn considerable attention because of its ability to decompose a variety of organic compounds. This study has been undertaken to examine the efficacy of Fenton's reagent to decompose PCE and PCBs adsorbed on sand. Results have shown that oxidation by Fenton's reagent can be an effective method to remediate PCE- and PCBs-contaminated soils. Over 90% of PCE was decomposed in 2 h, and over 70% of PCBs were degraded in 3 h. PCE followed the first-order decomposition kinetics with a constant value of 1.65/h, while the PCB degradation appeared to be zero-order with a constant value of 37.6 mg PCB/kg sand/hr at pH 3. Possible byproducts during the oxidation reaction were also examined.

Tetrachloroethylene (perchloroethylene, PCE) has been widely used as a reagent for dry cleaning, metal degreasing, and as an industrial solvent. PCBs were extensively used (before 1976) in dielectric fluids, plasticizers, cutting oils, and other purposes throughout industries because of their chemical and thermal stability, low or nonflammability, and good electrical insulating properties. These properties are responsible for PCB's extreme persistence in the environment. Both PCE and PCBs are known to be highly toxic, mutagenic and/or carcinogenic in experimental animals and have a potential health risk to humans.

A number of physical, chemical and biological processes have been proposed to treat soils contaminated with the hazardous waste, and some of them are already in use. These processes include nucleophilic substitution reaction, supercritical water oxidation, ultraviolet radiation, ozonation, thermal, and biodegradation processes. In this paper, the terms "degradation" and "decomposition" are used interchangeably. They are defined as the molecular degradation of an organic substance, measured by disappearance of a parent compound resulting from physical, chemical or biological reactions. The term "mineralization" is used according to Rochkind et al. (1) who defined it as the transformation of an organic molecule to its inorganic component parts with release of halide, CO_2 and/or methane.

Examples of the treatment technologies developed are the APEG-Plus treatment system, Light Activated Reduction of Chemicals (LARC) process, Ultrox system, Modar supercritical water oxidation (SCW) system, Advanced Electric Reactor (AER), and Bio-Clean process. The APEG-Plus system chemically decomposes chlorinated organics by removing chlorine atoms from the molecules. This process has been employed in many in-field pilot tests and small-scale clean up projects involving dioxins and PCBs. The LARC process involves a photochemical reductive reaction and dehalogenates various chlorinated compounds extracted from soil. The Ultrox process using $UV/O_3/H_2O_2$ has been accepted in the Superfund Innovative Technology Evaluation (SITE) program and tested at a hazardous waste site. The SCW process involves thermal destruction of hydrogen bonds. A bench-scale test of the Modar SCW process indicated complete destruction of chemically stable materials. Thermal treatment technologies include incineration (rotary kiln, fluidized bed), infrared thermal treatment, wet air oxidation, vitrification and pyrolysis. AER is a thermal treatment process which is permitted by EPA for the destruction of chlorinated organics. The principal products in this process are H_2, Cl_2, HCl, elemental carbon, and a granular free-flowing solid-derived waste. The Bio-Clean process is a biological treatment process that has been well demonstrated for pentachlorophenol (PCP), but it needs more comprehensive testing to bring it to a pilot stage for PCBs. More detailed information on these processes can be obtained from EPA's Alternative Treatment Technology Information Center (ATTIC).

Chemical treatment of chlorinated aromatics in soil is not only limited to polyethylene glycol based reagent, supercritical water, or ozone oxidation. Hydrogen peroxide has been known to provide the potential to degrade a variety of organic compounds. Enhanced oxidative power of hydrogen peroxide can be obtained when it is applied in conjunction with ferrous iron, due to increased generation of hydroxyl radicals. Hydroxyl radical is a highly potent oxidizing species (with oxidation potential only second to a fluorine atom) capable of reacting with most organic substances at near diffusion-controlled rates (2). The mixture of hydrogen peroxide and ferrous iron is known as Fenton's reagent.

The Fenton's reagent process involves oxidation of ferrous iron with hydrogen peroxide to generate hydroxyl radicals (*3, 4*):

$$H_2O_2 \;+\; Fe^{+2} \;\xrightarrow{k}\; \cdot OH \;+\; OH^- \;+\; Fe^{+3} \tag{1}$$

where k is the second-order rate constant and has a value of 76 L/mol-sec (*3*). There may be numerous reactions involved with Fenton's reagent, depending on the nature of reacting substrates. A reasonable pathway for the decomposition of substrate during the reaction may be written as (*5, 6*):

$$RH \;+\; \cdot OH \;\xrightarrow{}\; H_2O \;+\; R\cdot \tag{2}$$

$$R\cdot \;+\; O_2 \;\xrightarrow{}\; ROO\cdot \;\xrightarrow{}\; R' \;+\; \cdot O_2H \tag{3}$$

or

$$R\cdot \;+\; Fe^{+3} \;\xrightarrow{}\; Fe^{+2} \;+\; R'' \;+\; product(s) \tag{4}$$

where R′ and R″ are oxidized organic molecules, and R′, R″ and product(s) can be intermediate species generated from the oxidation reaction. The expected ultimate endproducts are carbon dioxide, water and chloride.

In recent years, the Fenton's oxidation process has drawn considerable attention in respect to chemical waste treatment applications. Fenton's reagent were used to treat wastewater containing sodium dodecylbenzenesulfonate (*7*), p-toluenesulfonic acid and p-nitrophenol (*8*), and azo dyes (*9*). Barbeni et al. (*10*) used Fenton's reagent to study oxidation of chlorophenols in water, including 2-chlorophenol, 3-chlorophenol, 4-chlorophenol, 3,4-dichlorophenol, and 2,4,5-trichlorophenol. They observed enhanced degradation of the chlorophenols with increased ferrous iron concentration. They further reported that some chlorinated aliphatic species (possible intermediate compounds) were expected to be formed because the rate of disappearance of the chlorophenols and the rate of formation of chloride ion was different. Based on a mass balance in their system involving the measurement of residual chlorophenol, chloride and total organic carbon, they concluded that the chlorophenols were mineralized. Murphy et al. (*11*) reported that Fenton-like reagent (H_2O_2/Fe^{3+}) oxidized a formaldehyde waste. Recently, Sedlak and Andren (*12*) studied a pathway of 2-chlorobiphenyl degradation. They observed considerable decomposition of 2-chlorobiphenyl and Aroclor 1242. Fenton's reagent has also been investigated for the treatment of organic contaminants in soils. Fenton's reagent successfully oxidized pentachlorophenol adsorbed on silica sand and natural soil (*13*).

The objectives of this study are: (1) to investigate the effectiveness of Fenton's reagent for the treatment of PCE- and PCB-contaminated sands, and (2) to examine the degradation kinetics of PCE and PCBs in soils. Intermediate species generated in the oxidation process and a kinetic mechanism for the PCE degradation are proposed. This paper presents the results of the first of many stages of the planned experiments to achieve the goal, i.e., to develop an in situ or on site remediation technology that can destroy the chlorinated contaminants at much faster, cheaper and safer than the currently available processes.

Experimental Section

Materials. Perchloroethylene (99%, Aldrich) and hydrogen peroxide (35% technical grade, Precision Laboratories) were purchased. Hydrogen peroxide was diluted to 7 to 10% before use. All other chemicals used were of reagent grade or HPLC grade. Aroclor 1260 was selected as a model contaminant because it was a widely used additive to transformer oil. PCB's QA standards (5,000 ug PCBs/g transformer oil) were used in this study. Commercial sand of 40-100 mesh and 80-100 mesh were used for the PCE and PCB experiments, respectively. Double deionized water (Barnstead Nanopure II) was used in all the sample and reagent preparations.

Experimental Procedures. All sample preparations and Fenton's reagent reactions and were performed at room temperature. The reactions were carried out with known quantity of silica sand (SiO_2) in a 40-mL borosilicate glass vial. To 3.5 g of PCE-contaminated silica sand (1000 mg/kg), 17.5 mL of hydrogen peroxide was added, following a $FeSO_4$ addition and pH adjustment. Since PCE has a relatively low boiling point (121 °C), the sample vials were kept at −5 °C for 2 hours before and after spiking with PCE to prevent loss due to evaporation. The PCE recovery was over 90%. The concentrations of H_2O_2 and $FeSO_4$ were approximately 7% by weight and 5 mM, respectively. Initially, pH of the sample was adjusted to 3 by an addition of dilute NaOH. The reacting vials were closed with Teflon coated septa and kept at 20±1 °C on a constant speed orbital shaker. At each measurement period, the Fenton's reactions were abruptly deactivated (quenched) by an addition of small amount of concentrated sulfuric acid. PCE and other hydrophobic organic products were extracted with 3 mL of ethylacetate for analysis.

The oxidation reaction of PCBs with Fenton's reagent was carried out with 2.5 g of PCB-contaminated soil in a 40 mL borosilicate glass. To the sample, $FeSO_4$ and dilute sulfuric acid were added to obtain 5 mM iron and pH 3. After the pH adjustment, 15 mL of 10% H_2O_2 were added to initiate the reaction. At the designated time, the reaction was stopped by an addition of 1 mL of concentrated sulfuric acid to the vials. The ability of concentrated sulfuric acid to quench the PCB oxidation was assumed based upon the results obtained by Watts et al. (13, 14) in monitoring the PCP degradation in acidified samples. No PCB degradation was documented for the PCB treatment with concentrated sulfuric acid (15). The sand slurries were extracted with 15 to 25 mL of hexane by shaking for 20 minutes.

Before injection of the extract into a gas chromatograph, an aliquot of the hexane extract was scrubbed with concentrated sulfuric acid to create a separate and clear layer of solvent containing PCBs. This additional step is critical because, without this scrubbing, the resolution of the chromatogram is very poor and thus erroneous data would result. The recovery after this treatment was found to be excellent (*15*).

Analysis. PCE and organic byproducts were analyzed by a Hewlett-Packard 5890 gas chromatograph (GC) with a flame-ionization detector (FID) and a 30 m x 0.32 mm (i.d.) Supelco Nukol capillary column, or a 3'x1/8" 60/80 Carbopack B/1% SP-1000 glass column; nitrogen was used as carrier gas. Carbon dioxide evolved from the mineralization of PCE as a result of the Fenton's oxidation was measured by the sealed ample method with an OI Corporation Model 700 Total Organic Carbon (TOC) Analyzer and equipped with a purging and sealing unit. High purity nitrogen (>99.99%) was used as purging gas. Intermediate species were identified by comparing the gas chromatogram with known references, or by a Finnigan 4023 gas chromatograph-mass spectrometer (GC/MS) with electron impact and chemical ionization capability and an INCOS data system.

The concentrations of residual PCBs were determined by GC (Perkin-Elmer 8500) equipped with an electron capture detector (ECD) and a 1.8 m x 3.5 mm (i.d.) glass column containing 4% OV-225 chromosorb HP 80/100 mesh. GC was operated under the following conditions: oven temperature (isothermal) 225 °C, injector temperature 350 °C, detector temperature 375 °C, carrier gas Argon/Methane 95:5, flow rate 35 mL/min.

An HP 3354 LAS computer was used for data reduction. The computer is a real-time on-line system with instrument and autosampler control and is compatible with any 1-volt analog instrument output. Calculation and Aroclor selection were done as follows. The computer searched for 5 key peaks for each Aroclor of interest. It also calculated the concentration for each Aroclor based on total peak area. Each standard peak was calibrated at total concentration, all peaks were averaged, and the relative standard deviation (RSD) was calculated. An error factor was calculated using both RSD and total peak-area. Aroclor with the lowest error factor was chosen. The computer then looked at the %RSD. If it was equal or less than 40%, the mean value was reported. If it was above 40%, the total-area value was reported.

Chloride measurement was accomplished with a solid state Fisher chloride ion electrode paired with a double junction reference electrode, or using a Dionex 2110i ion chromatograph according to the manufacture's instruction manual. Hydrogen peroxide was quantified by iodometric titration with sodium thiosulfate (*16*). The literature "Prudent Practices for Handling Hazardous Chemicals in Laboratories" (*17*) was consulted for laboratory safety.

Results and Discussion

One of the primary objectives of this study was to determine whether Fenton's reagent can mineralize PCE. Figure 1 shows the concentration of PCE remained, TOC removed (CO_2 generated) and Cl^- formed, and Figure

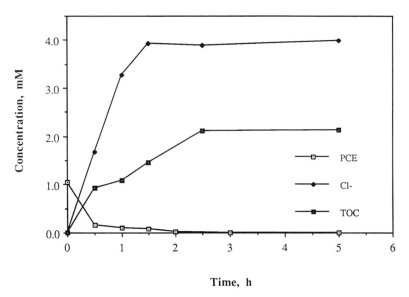

Figure 1. PCE degradation and corresponding TOC (CO_2) and chloride
generated upon exposure to Fenton's Reagent.

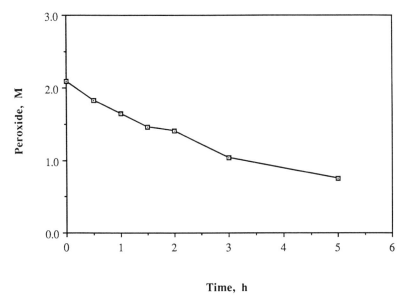

Figure 2.   Hydrogen peroxide consumption corresponding to PCE
degradation.

2 presents the corresponding concentration of hydrogen peroxide. As it is seen in Figure 1, about 80% of the initial 1,000 mg/kg of PCE was degraded in the first half hour, and PCE was totally degraded after 3 h. The chloride formed lagged behind the PCE decomposition indicating that the disappearance of PCE does not necessarily imply dechlorination and that daughter products of the Fenton's reaction could also contain chlorine. As supported by the TOC measurements, mineralization of PCE occurred after about 3 h of reaction. The TOC removal that lagged behind the Cl^- formation suggests that intermediate product (before it was evolved into CO_2) did not contain chlorine. During the course of reaction, hydrogen peroxide was steadily consumed as shown in Figure 2.

An attempt was made to identify intermediate species evolved in the decomposition process. Gas chromatographic analyses (using Carbopack column) indicated two daughter peaks in aqueous solution several minutes after the initiation of reactions. The peaks were only detectable at pH near 3, and were undetectable above pH 6. Size of the peaks reached maximum at about 30 min of the reaction and slowly disappeared in about 3 h. The intermediate compounds appeared to be unstable and remained reactive after the hydroxyl radical was quenched. (Note that quenching would not totally stop the Fenton's reactions.) The first peak (predominant peak) and second peak (relatively small) reached maxima within the first hour of reaction. The first peak was appeared to be dichloroacetic acid (DCAA). The normalized concentration of DCAA in 5 h of reaction is shown in Figure 3. A second peak compound has not been identified but it is not trichloroacetic acid. The compound appeared to have a similar concentration profiles as DCAA. This observation suggests that a second peak compound would be in pseudo-steady state with the DCAA, or a minor product generated by a parallel mechanism. Ethylacetate extract of the reacting solutions was also analyzed by GC/MS. Results showed no significant detectable species other than small fragments of low molecular organic acids and chlorinated compounds.

The formation of DCAA was not observed by Ollis et al. (*18*) in their study of photoassisted/TiO_2 catalyzed degradation of PCE. They proposed the formation of CO_2 and phosgen-like species in decarboxylation of dichloroacetyl acid. Dichloroacetylchloride (DCAC) was reported as a possible intermediate compound in homogeneous gas phase reaction of PCE with hydroxyl radicals. Pruden and Ollis (*19*) observed generation of dichloroacetaldehyde in oxidation of trichloroethylene with hydroxyl radical. However, our experimental data suggest that PCE reacts with a hydroxyl radical generated from Fenton's reagent producing intermediate species similar to the gas phase reaction (*20*):

$$PCE + \cdot OH \dashrightarrow DCAC$$

$$DCAC + H_2O \xrightarrow{\text{fast}} DCAA + HCl$$

Since PCE decomposes rapidly with Fenton's reagent producing no significant intermediates other than DCAA, and the fact that hydroxyl radicals react readily with simple organic compounds (2, 11, 21, 22), it is likely that hydroxyl radicals continue to attack DCAA and that the rate limiting step is the decomposition of DCAA. Possible decomposition mechanism is written as:

$$PCE + OH \xrightarrow{k_1} DCAC + Cl \cdot \tag{5}$$

$$DCAC + H_2O \xrightarrow{fast} DCAA + HCl \tag{6}$$

$$DCAA + 2 \cdot OH \xrightarrow{slow} 2HCl + CO_2 + HCOOH \tag{7}$$

$$HCOOH + 2 \cdot OH \xrightarrow{fast} 2H_2O + CO_2 \tag{8}$$

It has been demonstrated that HCOOH reacted with Fenton's reagent to form CO_2 and H_2O (11), thus reactions 7 and 8 are the reasonable steps to occur and consistent with our experimental results. The Cl· generated in reaction 5 could combine with ·OH to form HOCl.

The rate equation for reaction 5 may be given by

$$-d[PCE]/dt = k_1 [OH\cdot] [PCE]$$

We observed, however, that the kinetics of PCE degradation was first order as shown in Figure 4. Thus, the degradation of PCE can be expressed as:

$$-d[PCE]/dt = k_1 [OH\cdot] [PCE] = k_{obs1} [PCE]$$

where k_{obs1} is the observed (pseudo-first-order) rate constant, and determined to be 1.65/h (R = 0.95). The concentration of OH· appeared to be constant throughout the experiment and given by:

$$[OH\cdot] = k_{obs1}/k_1$$

Presuming the H_2O_2 decomposition follows reaction 1, then the rate equation may be given by:

$$-d[H_2O_2]/dt = k [H_2O_2] [Fe^{+2}]$$

Similarly, as it is shown in Figure 4, the decomposition of H_2O_2 followed the first-order kinetics with the observed pseudo-first order rate constant, k_{obs2}:

$$-d[H_2O_2]/dt = k[H_2O_2] [Fe^{+2}] = k_{obs2} [H_2O_2]$$

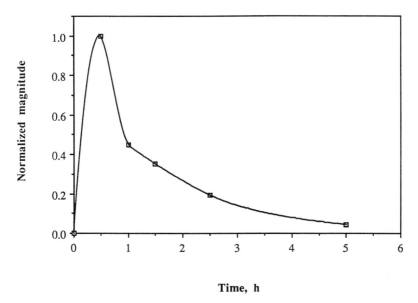

Figure 3. Normalized concentration profile of dichloroacetic acid (DCAA) with time.

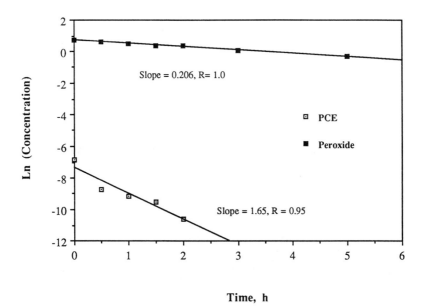

Figure 4. First-order plot of PCE degradation and hydrogen peroxide consumption.

$$[Fe^{+2}] = k_{obs2}/k$$

Thus, the concentration of Fe^{+2} is constant and calculated to be 7.53 x 10^{-4} mM. The value of k_{obs2}, was determined to be 0.206/h (R=1.0).

The calculated $[Fe^{+2}]$ indicated that most of the iron remained as Fe^{+3} in the degradation process, although the calculation did not take into consideration that H_2O_2 can be catalyzed by Fe^{+2} to release oxygen (23). During the experiment, we observed a heavy brownish formation that is characteristic of Fe^{+3}.

The PCB study focused on immediate response of PCBs (Aroclor 1260) to Fenton's reagent during a 3 h-exposure period. The concentrations of PCBs and chloride over the course of the Fenton's reaction are shown in Figure 5. The computer identified PCBs as Aroclor 1260 in all the samples. The measured initial concentration of PCBs in the surrogate soil was 152 mg/kg. The extent of PCB degradation was much greater at low pH. These findings were similar to those reported by Watts et al. (13, 14) in their study with pentachlorophenol, and Sedlak and Andren (24) with chlorobenzenes. Our data showed that approximately 77% of PCBs was degraded in 3 hours at pH 3 and only 13% at neutral pH.

An attempt was made to determine the rates at which PCBs were decomposed at pH 3 and pH 7. The PCB degradation rate appeared to be zero-order within 3 hours of treatment, although numerous reactions would occur simultaneously, and sequentially leading to complex reactions. The data were fitted by a least-square linear regression analysis. The rate constants obtained from the slopes were 37.6 ppm/hr (ug PCBs/g.hr) for pH 3 and 4.5 ppm/hr for pH 7. The correlation coefficients are 0.997 and 0.965 for pH 3 and pH 7, respectively. The higher PCB degradation rate at the low pH may be explained by examining the hydrogen peroxide reactions given above. The hydroxyl radical generation can be accelerated at low pH due to increased iron solubility and increased availability of ferrous iron for reaction 1. The low pH is favored for the reduction of Fe(III) to Fe(II) which is necessary to generate hydroxyl radicals. The formation of hydroxyl radical is enhanced at low pH, whereas the oxygen evolution is the predominant route of the hydrogen peroxide decomposition at neutral pH. The increased availability of hydroxyl radical led to the increased rate of PCB degradation.

This treatment measure is expected to oxidize PCBs to intermediate byproducts and chloride. As seen in Figure 5, the chloride concentration increased as the PCB degradation progressed. The statistical analysis yielded correlation coefficients of 1.00 and 0.91 at pH 3 and pH 7, respectively. The dechlorination rate appeared to be zero-order, which was consistent with the PCB data. The rates at which chloride was released are 16.3 and 1.6 ppm (ug Cl^-/g soil) at pH 3 and pH 7, respectively.

Figure 6 shows the relationship between the Cl^- and PCB concentrations. A linear regression of the PCBs and chloride data yielded correlation coefficients of 1.00 and 0.98 at pH 3 and pH 7, respectively. The slopes and intercepts for the both lines (pH 3 and pH 7) are similar. The slopes indicate the ratio of chloride released to PCB degraded, which are

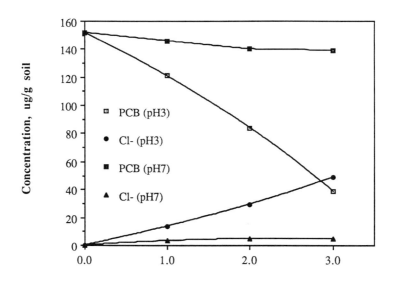

Figure 5. Concentrations of PCBs and chloride upon exposure to Fenton's reagent.

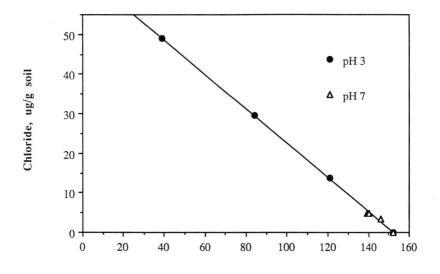

Figure 6. Chloride released vs. PCB remained in sand.

0.43 (ug Cl⁻/ug PCB) at pH 3 and 0.38 at pH 7. The intercepts indicate the amount of chloride accumulated after the completion of PCB degradation, which are 65.9 ppm (ug Cl⁻/g soil) at pH 3 and 57.4 ppm at pH 7. The theoretical chloride recovery is about 0.6 ug Cl⁻/ug PCB. The recovery of chloride in this experiment was 67% of the expected value. Based on this recovery value, it appeared that decomposition reaction was incomplete and that other species of organic chlorides still existed. An extended reaction time may be required for total mineralization. Possible intermediate compounds postulated in the literature are hydroxychlorobiphenyls. Sedlak and Andren (*12*), in their study on Fenton's oxidation of 2-chlorobiphenyl in aqueous solution, suggested 5-hydroxy-2-chlorobiphenyl as one of the intermediate byproducts. They also reported that dechlorination reactions are insignificant relative to hydroxylation reactions in the initial attack of ·OHs on chlorinated aromatic compounds.

PCB-contaminated soils are generally classified into three groups based on levels of contamination: (i) greater than 500 ppm as PCB material; (ii) 50 to 500 ppm as PCB-contaminated material; (iii) less than 50 ppm as waste material. In terms of disposal, PCB material is subject to severe restrictions, PCB-contaminated material requires moderate restrictions, and soils contaminated with less than 50 ppm can be disposed as waste material. Therefore, there is an incentive for PCB waste generators to develop methods to reduce the PCB level to the range where the wastes can be disposed under less severe restrictions. In this study, sand contaminated with PCBs in transformer oil was treated with Fenton's reagent. Since the contaminated soil with the level of 150 ppm resulted in the residual concentrations less than 50 ppm at pH 3, the treated soil should be reclassified and subject to less severe restrictions for its disposal.

Clearly, much work needs to be done to evaluate and improve the proposed method. The needs include: (1) the determination of optimum conditions such as concentrations of H_2O_2 and Fe(II) and pH, (2) the evaluation of degradability of other PCBs such as Aroclor 1016, 1242, 1254, (3) the examination of the effect of different soil types, and (4) the identification of intermediate byproducts by GC/MS. Especially, soils containing minerals (e.g., Mn) and organics (e.g., humic substances) have various effects on the process which may consume more peroxide or contribute to slow rate of degradation by reducing ·OH availability to target contaminants. Determination of intermediate byproducts is also important because byproducts can have an impact on the environment and human health.

Conclusions

The laboratory investigations and data analysis provided the following conclusions.

— Total mineralization of 1,000 mg/kg of PCE occurred in 3 h.

— The PCE decomposition followed first-order kinetics with a rate constant of 1.65/hr.

— Dichloroacetic acid (DCAA) appeared to be the only significant intermediate product during the PCE decomposition, and the formation of DCAA was believed to be the rate limiting step toward the PCE mineralization.

— Over 70% of PCBs (Aroclor 1260) was degraded in 3 h at pH 3.

— The degradation of PCBs (the initial concentration of 150 mg/kg) was zero-order in a 3-h of treatment, and provided the decay constants of 37.6 ppm/h at pH 3 and 4.5 ppm/hr at pH 7.

— The Fenton's oxidation can be an effective method to remediate PCE- or PCB-contaminated soils.

Literature Cited

1. Rochkind, M.L.; Blackburn, J.W.; Sayler, G.S.; "Introduction"; *Microbial Decomposition of Chlorinated Aromatic Compounds*; Hazardous Waste Engineering Research Laboratory, Cincinnati, 1986; EPA/600/2-86/090, 1-3.
2. Hoigne, J.; Faust, B.C.; Haag, W.R.; Scully, Jr. F.E.; Zepp, R.G.; "Aquatic humic substances as sources and sinks of photochemically produced transient reactants"; In *Aquatic Humic Substances*; American Chemical Society, Washington, D.C., 1989; 363-381.
3. Walling, C.; "Fenton's reagent revised"; *Acc. Chem. Res.* **1975**, *8*, 125-131.
4. Haber, H.; Weiss, J.; "The catalytic decomposition of hydrogen peroxide by iron salts" *Proceedings of the Royal Soc.*, London, **1934**, *147A*, 332-351.
5. Swallow, A.; "Reactions of free radicals produced from organic compounds in aqueous solution by means of radiation"; *Progress in Reaction Kinetics* **1978**, *9*, 195-365.
6. Cooper, W.; Herr, F.; "Introduction and Overview"; In *Photochemistry of Environmental Aquatic Systems*; Zika, R.G.; Cooper, W.J.; ACS Symposium Series 327; American Chemical Society, Washington, D.C., 1987, 1-8.
7. Sato, S.; Kobayashi, T.; Sumi, Y.; "Removal of sodium dodecylbenzene sulfonate with Fenton's reagent"; *Yukagaku* **1975**, *24*, 863-868.
8. Feuerstein, W.; Gilbert, E.; Eberle, S.H.; "Model experiments for the oxidation of aromatic compounds by hydrogen peroxide in wastewater treatment"; *Vom Wasser* **1981**, *56*, 35-54.
9. Kitao, T.; Kiso, Y.; Yahashi, R.; "Studies on the mechanism of decolorization with Fenton's reagent"; *Mizu Shori Gijutsu* **1982**, *23*, 1019-1026.
10. Barbeni, M.; Minero, C.; Pelizzetti, E.; Borgarello, E.; Serpone, N.; "Chemical degradation of chlorophenols with Fenton's reagent"; *Chemosphere* **1987**, *16*, 2225-2237.

11. Murphy, P.; Murphy, W.J.; Boegli, M.; Price, K.; Moody, C.D.; "A Fenton-like reaction to neutralize formaldehyde waste solutions"; *Environ. Sci. Technol.* **1989**, *23*, 166-169.
12. Sedlak, D.L.; Andren, A.W.; "Aqueous-phase oxidation of polychlorinated biphenyls by hydroxyl radicals"; *Environ. Sci. Technol.* **1991**, *25*, 1419-1427.
13. Watts, R.J.; Rauch, P.A.; Leung, S.W.; Udell, M.D.; "Treatment of pentachlorophenol-contaminated soils using Fenton's reagent"; *Hazardous Waste and Hazardous Materials;* **1990**, *7*, 333-345.
14. Watts, R.J.; Udell, M.D.; Leung, S.W.; "Treatment of contaminated soils using catalyzed hydrogen peroxide"; In *Chemical Oxidation, Technologies for the Nineties*; Eckenfelder, W.W.; Bowers, A.R.; Roth, J.A.; Techmonic Press, Lancaster, PA. 1991, 37-50.
15. Lerman, S.I.; Gordon, H.; Hendricks, J.P.; "Analysis of PCBs in transformer oils"; *American Laboratory* **1982**, *2*, 176-182.
16. Masschelein, W.; Denis, M.; Ledent, R.; "Spectrophotometric determination of residual hydrogen peroxide"; *Water and Sewage Works* **1977**, *124*, 69-72.
17. *Prudent Practices for Handling Hazardous Chemicals in Laboratories*; Committee on Hazardous Substances in the Laboratory; Assembly of Mathematical and Physical Sciences; National Research Council. National Academy Press; Washington, D.C. 1981.
18. Ollis, D.F.; Hsiao, C.Y.; Budiman, L.; Lee, C.L.; "Heterogeneous photoassisted catalysis: Conversion of perchloroethylene, dichloroethane, chloroacetic acids, and chlorobenzenes"; *J. Catalysis* **1984**, *88*, 89-96.
19. Pruden, A.L.; Ollis, D.F.; "Photoassisted heterogeneous catalysis: the degradation of trichloroethylene in water"; *J. Catalysis* **1983**, *82*, 404-417.
20. Leung, S.W.; Wattas, R.J.; Miller, G.C.; "Degradation of perchloroethylene by Fenton's reagent: speciation and pass way"; *J. Envir. Quality,* (in press).
21. Cooper, W.J.; Zika, R.G.; Petasne, R.G.; Ficher, A.M.; "Sunlight-induced photochemistry of humic substances in natural waters: Major reactive species"; In *Aquatic Humic Substances*; Suffet, I.; McCarty, P.; American Chemical Society, Washington, D.C., 1989, 333-362.
22. Walling, C.; Amarnath K.; "Oxidation of mandelic acid by Fenton's reagent"; *J. Am. Chem. Soc.* **1982**, *104*, 1185-1189.
23. Sung, W.; Morgan, J.J.; "Kinetics and product of ferrous iron oxygenation in aqueous systems"; *Environ. Sci. Technol.* **1980**, *14*, 561-568.
24. Sedlak, D.L.; Andren, A.W.; "Oxidation of chlorobenzene with Fenton's Reagent"; *Environ. Sci. Technol.* **1991**, *25*, 777-782.

RECEIVED September 3, 1992

Treatment of Volatile
Compounds

Chapter 17

Catalytic Oxidation of Trichloroethylene and Methylene Chloride

Henry Shaw[1], Yi Wang[1], Tia-Chiang Yu[1], and Anthony E. Cerkanowicz[2]

[1]Department of Chemical Engineering, Chemistry, and Environmental Science and [2]Department of Mechanical and Industrial Engineering, New Jersey Institute of Technology, Newark, NJ 07102

Two noble metal catalysts (Pt and PdO) and a transition metal oxide (MnO_2) were evaluated for their ability to oxidize two chlorinated compounds, viz., methylene chloride (DCM) and trichloroethylene (TCE). A 1.5% Pt on γ-alumina catalyst either on a cordierite monolith with 62 channels/cm^2, or as a powder, completely oxidizes 200 ppm TCE in air to CO_2, HCl and Cl_2 at 773 K and space velocities of 30,000 v/v/hr. DCM is similarly oxidized at space velocities of 17,000 v/v/hr and 823 K. A 4% PdO catalyst in a similar system is just as effective, but at 100 K higher temperature. However, highly chlorinated products are produced with PdO. The MnO_2 catalyst favored more chlorinated compounds than combustion products. The presence of hydrogen sources such as water or methane reduces the production of more chlorinated compounds.

Hydrocarbons contaminated with halogen compounds are emitted from many industrial processes. These compounds are often found in trace amounts and are best disposed of by incineration. One such example involves chlorinated hydrocarbons used commercially as stripping and dry cleaning solvents, refrigerants, transformer fluids, etc. These materials can become toxic wastes for which cost effective and environmentally sound methods of disposal are being sought. Incineration provides an option which can be applied to a wide range of such wastes. Thermal incineration requires high temperatures, with concurrent high fuel costs and the potential for formation of acid gases such as NO_x. Frequently, more highly chlorinated, and hence, more toxic products than the starting materials are formed. Use of a catalytic approach results in lower temperatures, less toxic products, and greater flexibility when compared to homogeneous thermal processes.

Subbanna, et al., (1) studied various catalytic materials and their activities and selectivities towards incinerating polychlorinated biphenyls (PCB). The results show significant performance differences between catalysts. For example, Cr_2O_3 converted 69 % of the PCB's at 873 K, but the yield of oxides of carbon was only 50 %. An automotive catalytic converter, composed of Pt and Pd, destroyed 87 % of the PCB's with the yield of carbon oxides about 76 %. Clearly, a substantial amount of oxychlorination occurs over these catalysts. An explanation

0097–6156/93/0518–0358$06.50/0

for this observation could be offered by appealing to a Mars-van Krevelen mechanism (2) based on chlorides rather than oxides. This mechanism is consistent with the following two steps:

1.   A reaction takes place between the catalytic oxide or chloride, and the hydrocarbon. The hydrocarbon is oxidized and the surface oxide or chloride is reduced.

2.   The reduced oxide reacts with O_2 or with the chlorocarbon, returning to its initial state as an oxide or becoming a chloride. The surface component directly responsible for the oxidation is generally assumed to be the O^{2-} ion, or Cl^- ion.

The Mars-van Krevelen mechanism led Sachtler and de Boer (3) to postulate that the tendency of an oxide to donate its oxygen should be of major importance in determining whether it is a selective oxidation catalyst. If reduction of the oxide is easy (i.e., if the enthalpy of dissociation is small), then O can easily be donated to a molecule from the gas phase. Under these conditions, the catalyst is expected to be active and nonselective. On the other hand, if it is difficult to dissociate O_2 because the metal-oxygen bond is strong, the oxide is expected to have low catalytic activity. In the intermediate range, oxides might be moderately active and selective.

The temperature at which 50 percent of the reactant is converted into CO_2 + CO at a particular set of flow conditions was considered by Simons, et al., (4) to be a characteristic measure of the activity of a catalyst. This measure allows a convenient comparison of the various metal oxide catalysts. Their data show that the temperature for 50 percent conversion to CO_2 + CO increases in roughly linear fashion with the heat of reaction Qo, defined as:

$$MO_n \longrightarrow MO_{n-1} + 1/2O_2 - Qo \qquad (1)$$

The corresponding equation for chlorination is:

$$MCl_n \longrightarrow MCl_{n-1} + 1/2Cl_2 - Qo \qquad (2)$$

A comparison of equation 1 for oxidation with chlorination in equation 2 is provided in Table I.

Table I. Bond Energies of Catalytic Materials

	Equation 1 $M-O$, kcal/mol		Equation 2 $M-Cl$, kcal/mol
Co_3O_4	38	Co_2Cl_5	28
Cr_2O_3	NA	$CrCl_3$	40
CuO	34	$CuCl_2$	17
Fe_2O_3	54	$FeCl_3$	15
MnO_2	17	$MnCl_4$	NA
NiO	58	$NiCl_2$	38
PdO	20	$PdCl_2$	23
SnO_2	70	$SnCl_4$	23
TiO_2	69	$TiCl_4$	51
V_2O_5	29	VCl_5	29
ZnO	83	$ZnCl_2$	50

NA = Not Available

According to this mechanism, one would expect MnO_2, PdO, and V_2O_5 to be among the best metal oxides for destruction of hydrocarbons and possibly chlorocarbons. But in the case of chlorocarbons, it may be possible that the stepwise mechanism for donating electrons to oxygen is short-circuited by the presence of chlorine. Under such circumstances, active metal chlorides would be produced on the surface. It is important to note that Cu and Fe catalysts, which have particularly good oxychlorination activity are among those having the lowest metal chloride bond energies.

In support of this viewpoint, Ramanthan and Spivey (5) report not being able to close the chlorine balance in the oxidation of 1,1-dichloroethane, and they speculate on the formation of chlorine monoxide or dioxide. It may be possible, however, that the chlorine combines with the catalytic metal, and depending on the bond strength, is irreversibly absorbed or follows a Mars-van Krevelen type mechanism.

Although the above mentioned references illustrate the potential feasibility as well as some of the theoretical problems of the catalytic approach to acceptable chlorocarbon destruction, the literature falls short of demonstrating commercial feasibility. Additionally, means of extending efficient operation of catalytic combustors to very low contaminant concentration need to be explored. This research addresses some of the key issues necessary to move this technology closer to commercial application.

Experimental

The noble metal catalysts used in this research were provided by Engelhard Corporation and the MnO_2 was purchased from Strem Co. The properties of the catalysts used in these studies are summarized in Table II. They include both powdered catalysts and catalysts supported on cordierite mononoliths. The noble metal catalysts were deposited with an alumina washcoat on either 62 cells per cm^2 (cpsc) or 31 cpsc, which are equivalent to 400 cells per square inch (cpsi) or 200 cpsi in a cordierite honeycomb.

Table II. Catalysts tested

	MnO_2	Pt	PdO
Manufacturer	Strem	Engelhard	Engelhard
Powder			
BET, m^2/g	112		
Apparent diameter, μm	7.1	200	200
Bulk density, g/cm^3	1.06	0.36	0.36
Theoretical density	5.03	4.0	4.0
Melting point K	808	~2,273	~2,273
Monolith			
Support		Cordierite	Cordierite
Washcoat		γ-Alumina	γ-Alumnia
Metal content, % (wt)		1.5	4
Structure, cells/cm^2		62	62

To obtain high space velocities, high temperature resistant cement was used to block a number of cells, thus reducing catalyst volume. Only 25 square cells were allowed to remain in the 2.54 cm in diameter monolith. The length of the catalyst monolith was 7.62 cm; therefore, the actual volume of modified catalyst was:

$$V_c = 7.62*(25/400)*2.54^2 = 3.0726 \text{ cm}^3$$

Different space velocities were obtained during the experiments by using different flow rates. Thus, using total flow rate of 1530 cm³/min corrected to standard conditions of 273 K and one atmosphere pressure, the space velocity for the 7.62 cm catalyst is calculated as follow:

Space Velocity = total flow rate/catalyst volume

= (1530 cm³/min*60min/hr)/3.0726 cm³
= 30,000 v/v/hr

Residence times for the kinetic studies were estimated by inverting the space velocity after correcting the flow rate to the operating temperatures and pressures.

The experiments were conducted in a laboratory-scale tubular reactor system shown in Figure 1. This system consists of a 2.5 cm inside diameter quartz tube reactor residing in a vertical three zone controlled furnace containing known volume of catalyst. The middle zone was designed to maintain a flat temperature profile over the length of the catalyst monolith. Space velocity was varied by changing catalyst volume or gas flow rate.

A glass U-tube containing the chlorocarbon feed in liquid form was placed in an ice-bath and part of the air feed was bubbled through the U-tube, becoming saturated with the chlorocarbon at room temperature. The chlorocarbon containing air stream was mixed with the rest of the air before entering the reactor. The flow rates of inlet gases were measured with four calibrated Cole Parmer rotameters.

The reactor temperature was monitored with two 0.16 cm Chromel-Alumel thermocouples which were placed in the centerline of the inlet and outlet to the catalyst. Since the measured temperatures were kept below 873 K, no corrections were made for radiation.

The HCl concentrations were determined by absorbing it in a flask containing a measured volume of a standardized solution of silver nitrate. The amount of HCl was measured using a Cl⁻ specific ion electrode to indicate the point when the HCl just exceeded the $AgNO_3$ concentration. This measurement provided an average HCl effluent rate over the period needed to reach the end-point. The flask was then immediately replaced with a fresh flask containing a measured volume of the standard $AgNO_3$ solution. The procedure was repeated until the end of the experiment. The HCl measurements were checked periodically with a Drager colorimetric tube and found to be in excellent agreement. Chlorine was measured using a Drager tube. The colorimetric method of analysis was reported to have a variability of about 10%.

The gases were purchased from the Liquid Carbonic Co. and used directly from cylinders. Air was of zero air research grade purity, with less than 5 ppm H_2O and less than 1 ppm hydrocarbons. The purity of the chlorocarbons used was reported to be 99+ percent by the supplier.

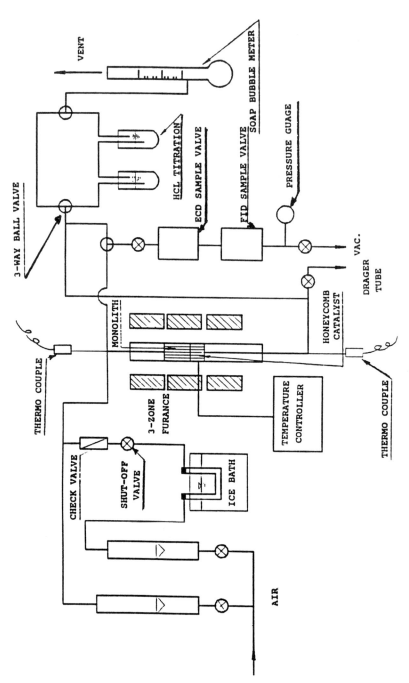

Figure 1. Schematic of catalytic oxidation system.

The concentrations of feed and product compounds were measured using two Hewlett Packard 5890 gas chromatographs. One has flame ionization (FID) and thermal conductivity (TCD) and the other has electron capture (ECD) and flame photometric (FPD) detectors.

The CO_2, CH_4 and CO were separated on a 1/8 inch in diameter by 6 feet long stainless steel column packed with 80/100 mesh Porapak Q and hydrogenated at 623 K over Ni-catalyst system to CH_4 before detection by flame ionization. Thus, the CO, CH_4 and CO_2 peaks are detected as CH_4, but identified based on retention time. The chlorinated hydrocarbons were separated on a 1/8 inch in diameter by 10 feet long stainless steel column packed with 80/100 mesh Chromosorb GAW and detected by electron capture.

Both carbon dioxide and carbon monoxide concentrations were calibrated with purchased standard gas mixtures. The chlorinated hydrocarbon concentrations were calibrated with the pure compounds by liquid injection. Periodic GC/MS analyses were made to verify that no other chlorocarbons were produced. Hewlett Packard 3396A integrators were used as both recorders and integrators.

Figure 2 shows typical peak resolution and retention time for the chlorinated products from TCE oxidation. Similar excellent peak separation was obtained for the products from DCM oxidation.

Results

The results presented in this section are subdivided into aging studies conducted with 1.5% Pt/γ-alumina on 62 cpsc cordierite and 40 to 60 ppm (v) TCE in air. Kinetic parameters were derived for this system. The effect of hydrogen containing additives H_2O and CH_4 on HCl, Cl_2 and C_2Cl_4 yield were evaluated. Results from an identical system containing 4% Pd instead of Pt are presented. It should be emphasized that the PdO results were obtained at much lower space velocities and higher TCE concentrations. A transition metal oxide catalyst, MnO_2, was evaluated as a powder at similar conditions to those used with the two noble metals on monoliths, except the chlorocarbon considered was DCM. DCM oxidation with the noble metals monoliths and powders were also conducted, but the details are omitted here for brevity. However, the results from these experiments are included here.

Aging Experiments. Experiments were run to determine the effect of TCE oxidation over Pt catalyst activity for 100 hours. To avoid losing Pt sites because of vaporization, the maximum temperature was limited to 723 K. At this temperature, over 90% conversion of TCE can be achieved. As a frame of reference, a fresh catalyst was used to oxidize 40 ppm trichloroethylene at 30,000 v/v/hr space velocity over the temperature range 423 to 723 K. After this run, the catalyst bed was cooled down to room temperature. Then the furnace was heated directly to 723 K, and 50 ppm C_2HCl_3 was fed at 30,000 v/v/hr space velocity and this condition was maintained for 25 hours. This was followed by a cool-down to room temperature and a characterization run to compare the change in activity of the catalyst (deactivation) to the activity of the fresh catalyst. The 25 hours aging experiments were repeated 3 more times with the same catalyst at the same operating conditions as the initial 25 hour run. Interspersed between the 25 hour runs were 423 to 723 K activity checks.

The effect of temperature on conversion and product distribution for the fresh catalyst is plotted in Figure 3. Since the experiment was conducted at high space velocity 30,000 v/v/hr, the light-off temperature of 573 K is significantly higher than would be expected at a more usual operating space velocity of 8,000

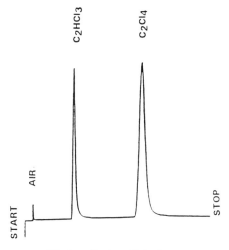

Electron Capture Detection

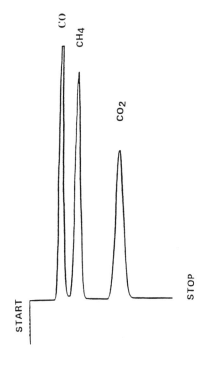

Flame Ionization Detection
after Hydrogenation

Figure 2. Typical gas chromatography peak resolution of chlorinated compounds with electron capture detection and carbon compounds, after hydrogenation over a nickel catalyst, with flame ionization detection.

v/v/hr. Figure 4 shows that in the temperature range 523 to 603 K, conversion increases slowly with increasing temperature. As temperature increases further to 723 K, about 95% conversion is obtained. The main products are carbon dioxide, hydrogen chloride and chlorine. Trace amounts of carbon monoxide and perchloroethylene are also produced. Almost 90% selectivity to CO_2 is obtained, but only 22% selectivity to HCl is achieved. As temperature increases above light off, very low concentrations of CO are found. At 623 K, the amount of CO reaches a maximum and then decrease with increasing temperature. The distribution of C_2Cl_4 follows a similar trend to CO. At 673 K, the amount of C_2Cl_4 reaches maximum and then decreases.

After each of the 4 aging runs of 25 hours at 723 K, tests were conducted at same operating conditions that were used with the fresh catalyst to measure any change in activity. The results show that in all cases, there were very slight changes in conversion with temperature, and the catalyst was not obviously deactivated. The trend of product distribution for all cases is the same. The comparison of conversion versus temperature curves at different aging time is shown in Figure 4.

The chemical kinetics of the catalytic oxidation of trichloroethylene were determined at conversions of less than 30%. Space velocities were varied at each temperature to allow evaluation at different residence times. Figure 5 shows the linear relationship between log retention of TCE and residence time. The slope of each line is the first order rate constant at that temperature. The linearity and zero intercept on the ordinate is consistent with first order kinetics. A plot of the logarithm of the rate constant k versus the inverse of temperature for the fresh catalyst is shown in Figure 6. Thus, the Arrhenius parameters for this catalytic reaction are E = 19.4 kcal/mol and A = $2.23*10^7 sec^{-1}$.

Thermodynamics indicate that since trichloroethylene contains only one hydrogen atom, it will dissociate at high temperatures to yield equal molar concentrations of HCl and Cl_2. To improve the selectivity to HCl, a hydrogen-rich fuel such as methane or water vapor could be added to the feed. With methane as the additive, the selectivity to HCl was improved to more than 80% at 723 K, and the selectivity to HCl increased gradually with increasing temperature. Methane also inhibited perchloroethylene production. In the case of water as the additive, the results were even better, viz., 100% selectivity to HCl was reached at 723 K. Also, selectivity increases with increasing temperature, and water inhibits perchloroethylene production. A comparison of conversion versus temperature curves for the cases of no additive, water, and methane is provided in Figure 7. The results indicate that methane accelerates C_2HCl_3 oxidation, while water slightly inhibits this reaction.

Catalytic Oxidation of TCE over 4% PdO/γ-Alumina/Monolith. The product distribution from the catalytic oxidation of TCE as a function of temperature is given in Figure 8 at 6,000 v/v/hr. The products at 823 K are CO_2, C_2Cl_4, HCl, Cl_2 and C_2HCl_3. The total mass of the only chlorinated hydrocarbon product C_2Cl_4 peaks at 5% of initial concentration of TCE. Chlorine gas is found in comparable concentrations to HCl in these experiments because TCE contains more chlorine atoms than hydrogen atoms. The overall stoichiometry of the catalytic oxidation of TCE is represented by the following global reaction:

$$C_2HCl_3 + 2O_2 \longrightarrow 2CO_2 + HCl + Cl_2 \tag{3}$$

As described above, experiments to inhibit the formation of chlorine gas and enhance the selectivity to hydrogen chloride in this system were conducted

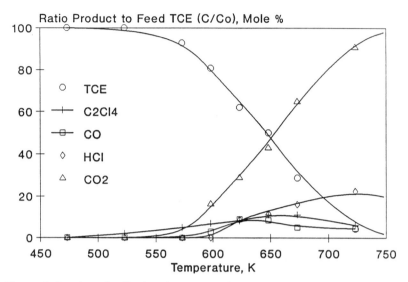

Figure 3. Product distribution from the Oxidation of TCE with air over fresh 1.5% Pt/γ-Al$_2$O$_3$ on a cordierite monolith as a function of temperature. The feed concentration of TCE was 40 ppm(v) and the space velocity 30,000 v/v/hr.

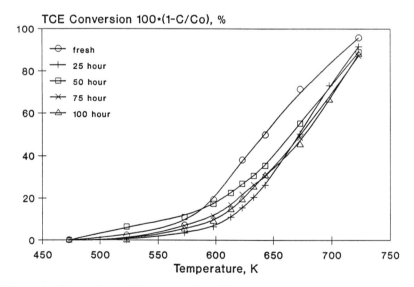

Figure 4. Comparison of the conversion of TCE with air over aged 1.5% Pt/γ-Al$_2$O$_3$ catalyst as a function of temperature. The feed concentration of TCE was 40 to 60 ppm(v) and the space velocity 30,000 v/v/hr.

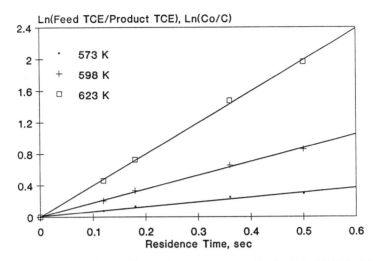

Figure 5. Determination of first order rate constant for fresh 1.5% Pt/γ-Al$_2$O$_3$ catalyst at TCE conversions of less than 30%. The feed concentration of TCE was 40 ppm(v) and the space velocity 30,000 v/v/hr.

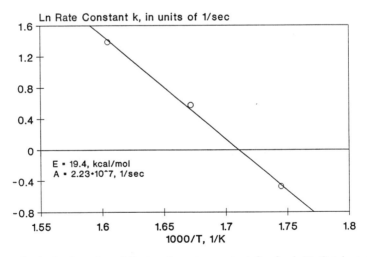

Figure 6. Arrhenius plot of first order rate constant for fresh Pt Catalyst. The feed concentration of TCE was 40 ppm(v) and the space velocity 30,000 v/v/hr.

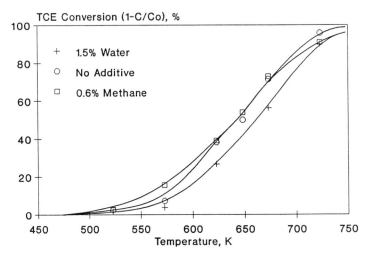

Figure 7. Effect of hydrogen containing additives on TCE conversion over fresh Pt catalyst. TCE feed concentrations were 50 ppm(v) with no additive, 240 ppm(v) with methane, and 236 ppm(v) with water.

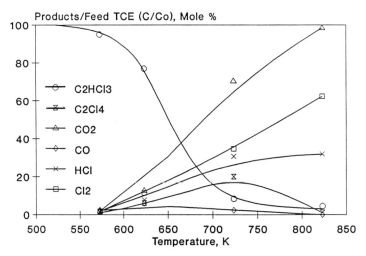

Figure 8. Product distribution from the oxidation of TCE with air over fresh 4% $PdO/\gamma-Al_2O_3$ on cordierite as a function of temperature. The feed concentration of TCE was 200 ppm(v) and the space velocity 6,000 v/v/hr.

with methane and water. The effect of 0.5% methane on TCE oxidation is shown in Figure 9. An additional benefit is that the oxidation of CH_4 produces heat which increases the rate of TCE oxidation. It was found that the concentration of hydrogen chloride increases, and the concentration of chlorine is reduced substantially with increasing temperature. This observation is consistent with equilibrium calculations. Formation of byproduct C_2Cl_4 is also significantly reduced. However, the rate of CO formation at low temperatures increases.

When 1.5% water is added to the feed, the Cl_2 normally produced from the oxidation of TCE is converted to HCl. Water can dissociate on the oxygen precovered surface of the catalyst which provides hydrogen atoms to react with chlorinated hydrocarbons at low temperatures (6). Alternatively, Narayanan and Greene (7) proposed a mechanism that includes surface gas phase reactions of Cl atoms with water to give HCl and OH. The product distribution from the catalytic oxidation of TCE with the addition of 1.5% water is shown in Figure 10. Activity is not affected by the addition of this relatively low quantity of water, and the rate of C_2Cl_4 byproduct formation is reduced drastically. No significant amount of CO is detected at low temperatures.

Kinetics of TCE Destruction with PdO. In these experiments, air was employed as an oxidant and O_2 was in large stoichiometric excess over 200 ppm TCE. In order to obtain kinetic parameters, the measured rates of oxidation of TCE in air were correlated first using an empirical power law of the form:

$$-r_{TCE} = k_1[C_2HCl_3]^a[O_2]^b \tag{4}$$

where, $[C_2HCl_3]$ and $[O_2]$ are the concentrations of C_2HCl_3 and O_2, respectively. The reaction order, with respect to O_2, was studied by changing the concentration of oxygen. It was found that reducing the oxygen concentration by a factor of 4 has no marked effect on the rate of oxidation of C_2HCl_3. Therefore, it was concluded that the rate of C_2HCl_3 oxidation was zero order with respect to oxygen within the experimental conditions studied. The rate reaction can be written:

$$-r_{TCE} = k_2[C_2HCl_3]^a \tag{5}$$

where, $k_2 = k_1[O_2]^b$

The order of reaction with respect to C_2HCl_3 was estimated by a least square regression of equation 5. The slope from the log-log relationship of measured reaction rate versus mean concentration of C_2HCl_3 is 0.97. Therefore, this reaction can be assumed to be first order.

Figure 11 is a plot of the logarithm of the rate constant versus 1/T. The Arrhenius activation energy, E, is estimated at 34 kcal/mole and the pre-exponential factor, A, is $7.1*10^{11}$ sec^{-1}.

DCM Oxidation Over the Noble Metal Catalysts. Four series of experiments were run to determine if any special differences exist between the noble metals with a γ-alumina washcoat as a powder, or on a monolith. These experiments were run using the standard techniques of varying the reactant concentration in the feed, while keeping space velocity constant. The results are presented in Figure 12 for the two catalysts on 62 cpsc monoliths and Figure 13 for the catalyst in powder form. It is interesting to note that the activation energy for Pt is 20 kcal/mol regardless of whether it is used as a powder or monolith. This activation energy is equal to that obtained for TCE oxidation with Pt on the monolith. The results for

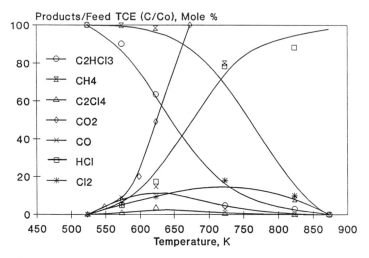

Figure 9. Effect 0.5% CH$_4$ on the product distribution from TCE oxidation over fresh PdO catalyst. The TCE feed concentration was 200 ppm(v) and the space velocity 6,000 v/v/hr.

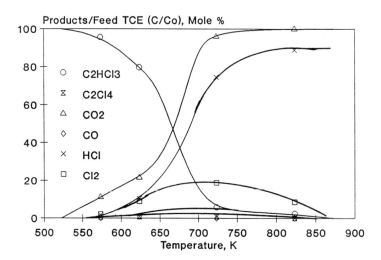

Figure 10. Effect of 1.5% water on the product distribution from TCE oxidation over fresh PdO catalyst. The TCE feed concentration was 200 ppm(v) and the space velocity 6,000 v/v/hr.

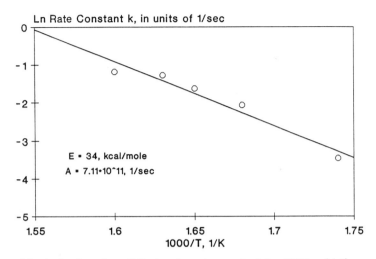

Figure 11. Arrhenius plot of first order rate constant for TCE oxidation over 4% PdO/γ-Al$_2$O$_3$ catalyst on a cordierite monolith.

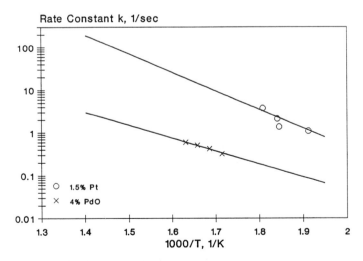

Figure 12. Comparison of the Arrhenius relationship for the first order rate constants for oxidation of DCM over 1.5% Pt, vs., 4% PdO, both with γ-Al$_2$O$_3$ washcoats and 62 cells per square centimeter monolith.

PdO are similarly equivalent at 13 kcal/mol for the powder and monolith, but much lower than the 34 kcal/mol obtained in TCE oxidation.

DCM Oxidation Over MnO_2. Figure 14 and 15 compare DCM conversion for a feed containing no methane with one containing 2.33% methane, respectively. Conversion of CH_2Cl_2 in both systems was not significantly different in the temperature range of 513-625 K. Since no reaction was observed when running only methane and air through the catalyst bed in the 450-750 K temperature range, CH_4 is probably not involved in the decomposition of CH_2Cl_2. However, from Figure 16, it is clear that the product distribution of chlorine compounds with CH_4 present has changed compared to Figure 17 with no methane present. CH_3Cl is detected at about 575 K. In the temperature range of 625-675 K, CH_3Cl and $CHCl_3$ increase, whereas CCl_4 decreases. At temperatures over 675 K, the concentrations of CH_3Cl, $CHCl_3$ and CCl_4 become constant, which indicates that the reaction is controlled by mass transfer. It suggests that the global reaction of methane with carbon tetrachloride may take place on the catalyst surface, forming products as follows:

$$CH_4 + CCl_4 \longrightarrow CH_3Cl + CHCl_3 \tag{6}$$

The product CH_3Cl is only found when CH_4 is present, adding evidence to this reaction. On the other hand, carbon dioxide, water and hydrogen chloride product distributions do not change when methane is present. It appears that methane acts as a source of hydrogen. This is in contrast with chemical equilibrium predictions that show water increases in the system with methane because CH_4 oxidizes to CO_2 and H_2O.

First order kinetics also fit the results in this case very well. Table III summarizes the final results of the kinetics analysis.

Table III. Chemical Kinetics Data

System	A 1,000/sec	E kcal/mol
CH_2Cl_2+air	9.6($\pm$0.6)	12.7($\pm$0.2)
CH_2Cl_2+CH_4+air	18.8($\pm$0.8)	11.4($\pm$0.9)

Discussion

The overall stoichiometry of oxidation of trichloroethylene at 1 atm and at the temperature range used in this study may be represented by the following global reaction:

$$C_2HCl_3 + 2O_2 \longrightarrow 2CO_2 + HCl + Cl_2 \tag{7}$$

For complete oxidation, each mole of C_2HCl_3 yields two moles of CO_2. Neglecting the change in total volume of gas due to the large excess of the air, the CO_2 concentration should be twice the amount of inlet concentration of C_2HCl_3, if all carbon atoms are oxidized to CO_2.

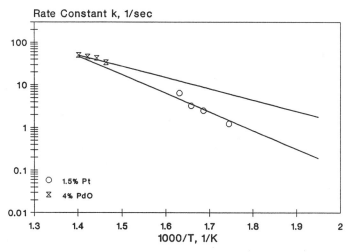

Figure 13. Comparison of the Arrhenius relationship for the first order rate constants for oxidation of DCM over 1.5% Pt, vs. 4% PdO, both over γ-Al₂O₃ powder and no monolith.

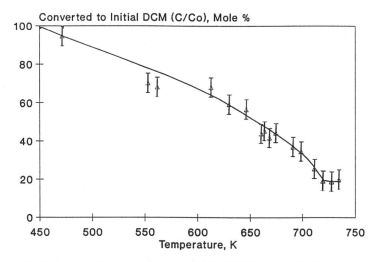

Figure 14. Retention (inverse of conversion) of 100 ppm(v) DCM over MnO₂ powder catalyst at 12,000 v/v/hr as a function of temperature.

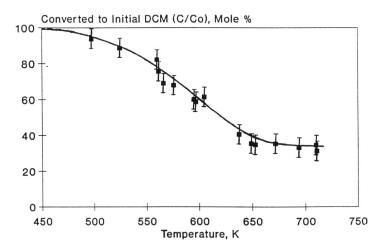

Figure 15. Retention of 100 ppm DCM over MnO_2 powder catalyst at 12,000 v/v/hr with the addition of 2.3% methane as a function of temperature.

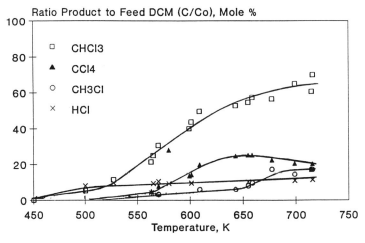

Figure 16. Distribution of chlorinated products as a function of temperature from the oxidation of DCM in the presence of 2.3% methane.

The formation of CO at low temperature usually occurs in homogeneous combustion followed by burnout to CO_2. Bose (8) invoked a two-step oxidation reaction of chlorinated hydrocarbons:

$$C_2HCl_3 + O_2 \longrightarrow 2CO + HCl + Cl_2 \qquad (8)$$

$$CO + 1/2O_2 \longrightarrow CO_2 \qquad (9)$$

The concentration of CO decreases with increasing temperature because reaction 9 is favored at high temperatures.

The formation of C_2Cl_4 can be explained by the catalytic promotion of the following free radical reactions (9), or as indicated in the introduction by a Mars-van Krevlen mechanism:

$$C_2HCl_3 \longrightarrow C_2HCl_2 + Cl \qquad (10)$$

$$C_2HCl_3 + Cl \longrightarrow C_2HCl_4 \qquad (11)$$

$$C_2HCl_4 + C_2HCl_3 \longrightarrow C_2HCl_5 + C_2HCl_2 \qquad (12)$$

$$C_2HCl_5 \longrightarrow C_2Cl_4 + HCl \qquad (13)$$

The Pt aging experiments were conducted to help identify catalyst deactivation rate parameters. Aging experiments were conducted over 4 periods of 25 hours each. Activity checks were made before and after each aging run.

As shown in Figure 4, only the fresh catalyst was obviously deactivated. In the other three cases, there were only slight changes in conversion with temperature, and the catalyst was not significantly further deactivated. Huang and Pfefferle (10) addressed this phenomena. Their research shows that the deactivation of a platinum/γ-alumina catalyst by chlorinated hydrocarbons is reversible for hydrocarbon oxidation. Similar results are seen in the petroleum industry where reforming catalysts are regenerated with Cl_2, air and steam injection. In our research, we believe each catalyst achieves a steady state between deactivation and regeneration since all ingredients for both effects are present in the feed. This could explain why the latter three conversion curves essentially overlapped, taking normal experimental errors into account.

The only exception is the case with fresh Pt catalyst. In this case, the fresh catalyst was obviously deactivated. The reason for this is because for most catalysts the activity declines sharply at first due to redispersion and irreversible poisoning, and then reaches a state where the catalyst activity decreases much more slowly with time (11).

Activation energy, E, represents the energetics of the rate determining step on the surface of the catalyst, and the pre-exponential factor, A, represents catalyst specific characteristics. Consequently, deactivation can be correlated with changes in the A factor.

The activation energies of the five different experiments were averaged, and this E_{av} was used in the Arrhenius equation to estimate a new A factor for each activity check. At the same temperature, the five modified A factors were calculated using 5 measured rate constants. The plot of modified A factors versus aging times is shown in Figure 18. The modified A factors decreased sharply over the first 25 hours and seemed to level out over the rest of the aging run. Since the mechanism for deactivation is not known, further investigation is necessary. Also,

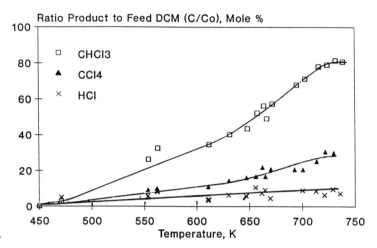

Figure 17. Distribution of chlorinated products as a function of temperature from the oxidation of DCM with air and no hydrogen containing additives.

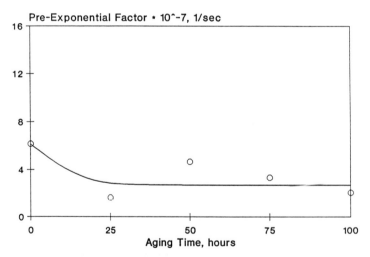

Figure 18. Variation of Arrhenius pre-exponential factor with catalyst age for an average activation energy of 20 kcal/mol. Data in the temperature range of 523 to 643 K and conversion of less than 30% were used at space velocity of 30,000 v/v/hr.

as a consequence of the scatter of the data, these experiments need to be replicated to clearly establish the degree of catalyst deactivation.

The kinetics for the oxidation of TCE and DCM over the three catalysts tested in this study are summarized in Table IV. It is very interesting that the activation energy for Pt destruction of the chlorocarbons does not appear to depend on the chlorocarbon, nor whether the Pt/γ-alumina is used as a powder or deposited on a monolith. The kinetics associated with PdO, on the other hand, are strongly dependent on the chlorocarbon but are not affected by the state of the catalyst. Finally, the kinetics for oxidation of DCM over MnO_2 appear similar to PdO kinetics, but the products are more highly chlorinated compounds ($CHCl_3$ and CCl_4) rather than CO_2, H_2O and HCl.

Table IV. Comparison of Results

	MnO_2	4% PdO	1.5% Pt
Trichloroethylene (TCE)			
Concentration, ppm		200	50
Conversion, %		97.1	99.9+
Temperature, K		823	823
Space Velocity, v/v/hr		6,000	30,000
Arrhenius Function			
(A, E)1/sec, kcal		7.1×10^{11}, 34	2.2×10^7, 20
Dichloromethane (DCM)			
Concentration, ppm	100	520	200
Conversion, %	80+	99+	99+
Temperature, K	720	773	823
Space Velocity, v/v/hr	12,000	6,000	17,000
Arrhenius Function			
(A, E)1/sec, k/cal			
Monolith	----	4.9×10^4, 14	2.4×10^8, 20
Powder	9.6×10^4, 13	2.6×10^5, 12	6.4×10^7, 20

Conclusions

As a consequence of the research, the following conclusions were reached:
- Pt is the best oxidation catalyst we tested for TCE destruction,
  - PdO is also good, and
  - MnO_2 is not suitable.
- Major products from TCE oxidation with noble metal catalysts are HCl, Cl_2, and CO_2. Trace quantities of CO and C_2Cl_4 are found at low temperatures (600 - 700 K).
- Major products from DCM oxidation with noble metals are HCl and CO_2. Trace quantities of CO and $CHCl_3$ are found.
- The effluent from MnO_2 promoted oxidation of DCM at 700 K consists of 19% CO_2 and CO, 36% H_2O, 13% unreacted DCM, 23% chloroform, 7% carbon tetrachloride, and 2% HCl.

- Aging a Pt catalyst for 100 hours at 723 K and 30,000 v/v/hr had no substantial effect on conversion and product distribution, after an initial drop in activity.
- Water or methane additives increase HCl formation, and reduce production of chlorine and more highly chlorinated chlorocarbons.
- Transition metals and oxides, such as manganese, iron, and copper catalysts tend to promote oxychlorination reactions and may not be suitable for catalytic oxidation of chlorocarbons.
- Kinetics show that the oxidation of DCM or TCE occur with an activation energy of 20 kcal/mol with the Pt catalyst on a monolith or powder. However, PdO on a monolith or powder destroys TCE with an activation energy of 34 kcal/mol and DCM with an activation energy 13 kcal/mol. The MnO powder catalyst converts DCM with an activation energy of 13 kcal/mol.

Acknowledgments

The authors thank the NSF Industry/University Cooperative Research Center for Hazardous Substance Management for support of the research. The Engelhard Corporation kindly donated the noble metal catalysts described in this paper. The Altamira Cooperation helped the researchers obtain the instrumentation to characterize the catalysts. The authors are particularly grateful to Dr. Robert J. Farrauto for his encouragement and advice.

Literature Cited

1. Subbanna, P.; Greene, H.; Desai, F. "Catalytic Oxidation of Poly-chlorinated Biphenyls in a Monolithic Reactor System," *Environ. Sci. Technol.* **1988**, *vol. 22*, pp. *557-561.*
2. Mars, P.; van Krevelen, D. W. "Oxidation Carried Out by Means of Vanadium Oxide Catalysts," *Chem. Eng. Sci. Suppl.* **1954**, *vol. 3*, pp. *41-59.*
3. Sachtler, W. M. H.; de Boer, N. H. "Catalytic Oxidation of Propylene to Acrolein," Proc. 3d Int. Cong. Catal.; North-Holland: Amsterdam, **1965**; pp. *252-265.*
4. Simons, Th. G.J.; Verheyen, E. J., Batist, P. A.; Schuit, G.C.A. "Oxidation on Metal Oxide Surfaces," *Adv. Chem. Ser.* **1968**, *vol. 76(II),* pp. *261-275.*
5. Ramanathan, K.; Spivey, J. J. "Catalytic Oxidation of 1,1-Dichloroethane," *Combust. Sci. and Tech.* **1989**, *vol. 63*, pp. *247-255.*
6. Heras, J. M.; Viscido, L. "The Behavior of Water on Metal Surface," *Cata. Rev. & Sci. Eng.* **1988**, *vol. 24*, pp. *233-309.*
7. Narayanan, S.; Greene, H. L. "Deactivation by H_2S of Cr_2O_3 Emission Control Catalyst for Chlorinated VOC Destruction," presented at the 83rd Annual AWMA Meeting Pittsburgh, PA, June 24-29, 1990.
8. Bose, D.; Senkan, S. M. "On the Combustion of Chlorinated Hydrocarbon: Trichloroethylene," *Combustion Sci. & Tech.* **1983**, *vol. 35*, pp. *187-202.*
9. Senkan, S. M.; Chang W. D.; Karra, S. B. "A Detailed Mechanism for the High Temperature Oxidation of C_2HCl_3," *Combustion Sci. & Tech.* **1986**, *vol. 49*, pp. *107-121*, and Senkan, S. M.; Weldon, J. "Catalytic

Oxidation of CH_3Cl by Cr_2O_3," *Combustion Sci. & Tech.* **1986**, *vol. 47*, pp. *229-237*.

10. Huang, Stephen L.; Pfefferle, L. D. "Methyl Chloride and Methylene Chloride Incineration in a Catalytically Stabilized Thermal Combustor," *Eviron. Sci. & Tech.* **1989**, *vol. 23*, pp. *1085-1091*.

11. Hughes, R. *Deactivation of Catalysts*; Academic Press, Inc.: London, **1984**.

RECEIVED September 3, 1992

Chapter 18

Advanced Ultraviolet Flash Lamps for the Destruction of Organic Contaminants in Air

Paul G. Blystone[1], Mark D. Johnson[1], Werner R. Haag[1], and Paul F. Daley[2]

[1]**Purus Inc., 2713 North First Street, San Jose, CA 95134**
[2]**Lawrence Livermore National Laboratory, Livermore, CA 94550**

This chapter describes a new process for photo-oxidation of volatile organic compounds (VOCs) in air using an advanced ultraviolet source, a xenon flashlamp. The flashlamps have greater output at 200 - 250 nm than medium-pressure mercury lamps at the same power and therefore cause much more rapid direct photolysis of VOCs. The observation of quantum efficiencies greater than unity indicate the involvement of chain reactions for trichloroethene (TCE), perchloroethene (PCE), 1,1-dichloroethene (DCE), chloroform, and methylene chloride. A full scale air emission control system (>95% removal) for TCE has been constructed, capable of continuous treatment of at least 300 scfm with a residence time of about 2 seconds. Further treatment of photo-oxidation products is likely to be needed for chloroethenes.

Volatile organic compounds (VOCs) have been widely used as solvents, coolants, degreasers, petroleum compounds, and as raw materials for other synthetic organic chemicals. The vast majority of Superfund and non-Superfund environmental remediation sites within the public and private sectors suffer from pollution with VOCs and fuel residues, mainly from underground storage tank leaks or improper disposal practices. Many of these sites are potentially amenable to restoration by means of vacuum-induced soil venting, supplanted in many cases by groundwater air stripping methods. However, treatment of off-gases from such remediation systems is problematic. In the many instances where local air quality management districts seek to reduce or eliminate emissions from restoration activities, carbon adsorption is the approach most commonly selected (1). However, the process of VOC adsorption onto carbon merely concentrates the contaminant, creating a solid hazardous waste with its own handling problems. Once breakthrough occurs, the spent carbon must be regenerated on-site or transported to the limited number of disposal or regeneration locations. In addition, this practice can be costly for sites that contain VOCs like methylene chloride or vinyl chloride with low adsorption affinities and that therefore require large amounts of carbon (2).

Ultraviolet light (UV) induced and other radical oxidation processes (ROPs) have been developed for the on-site destruction of organic contaminants in water (3-10).

0097–6156/93/0518–0380$06.00/0

Such processes are considered attractive because, if carried to completion, VOCs are ultimately converted to carbon dioxide, water, and inorganic ions.   Typical UV sources include low- and medium-pressure mercury lamps having peak output at 254 nm, with a smaller (<15%) emission at 185 nm.   The rest of the light energy occurs in the visible and infrared regions, which are not useful for organic photolysis.   Because most VOCs absorb in the sub-250 nm region, mercury UV lamps rely predominantly on hydroxyl radical processes for organic attack.   Thus, these lamps are nearly always used in conjunction with added oxidants, as the 254 line will photolyze ozone or hydrogen peroxide, creating hydroxyl radicals (*10*).   The limitations and trade-offs for aqueous ultraviolet processes include: 1) ozone readily absorbs 254-nm light ($\varepsilon = 3000$ $M^{-1}cm^{-1}$) but is limited by mass transfer to water and requires off-gas control, 2) H_2O_2 is miscible with water but absorbs weakly at 254 nm ($\varepsilon = 20$ $M^{-1}cm^{-1}$), 3) natural organic and inorganic components inhibit the oxidation of VOCs by competing for both photons directly and for hydroxyl radicals once they are produced, and 4) the efficiency of the UV light utilization decreases with decreasing concentration of contaminant. Competition effects are most pronounced for VOCs like chloroform, 1,1,1-trichloroethane, and freons that have relatively low reaction rate constants with hydroxyl radicals.   Thus, not all sites have been amenable to UV processes because performance is site specific for technical as well as for economic reasons.

One approach to minimizing aqueous phase interferences is to strip the VOCs from the water and then carry out the UV photolysis in the resulting air stream (*11,12*). However, full scale applications are in their infancy. Limitations in this method include low rates of direct photolysis with available light sources, difficulties of obtaining gas phase hydrogen peroxide, and permitting requirements when using gaseous ozone. Suppliers of mercury based treatment systems have improved the direct photolysis capability by bolstering the 185 nm line with a lower pressure mercury lamp, albeit at the expense of overall intensity. Other limitations of the 185 nm line are the competitive absorption by oxygen and the need for highly transmissive quartz lamps. We are studying a new UV light source to see if enhancements in VOC destruction rates can be attained. The source is a xenon plasma flashlamp, which generates significant light intensity in the deep UV region (<250 nm) that is better suited for direct photolysis than conventional mercury-based UV lamps.

Background

Pulsed inert-gas lamps have been used for many applications including pump sources for lasers (*13*). Lasers themselves have relatively low overall efficiency and therefore are not likely to find application in large-scale commercial or treatment processes. As developed in the 1930's by Harold E. ("Doc") Edgerton, a flashlamp is an arc lamp that operates in the pulsed mode by alternately storing electrical energy in a capacitor and discharging it through a gas contained in a chamber of UV transmissive quartz (*13*). The discharge quickly heats the gas to a very high temperature ($\geq$ 13,000 K) and pressure, causing ionization and creating a plasma that emits light. The spectral properties of the plasma approach those of an ideal black body radiator which has a peak emission wavelength defined by its characteristic temperature. Increasing energy discharged into the plasma increases its temperature, which lowers the wavelength of its emission maximum. Since, according to the Wien displacement law, the number of photons emitted by a black body is proportional to the temperature to the 4th power, flashlamps produce an exponentially increasing number of photons as the temperature of the plasma increases.

Shortening of lamp life places a practical upper limit to the plasma temperature used. The high temperature plasma can not be contained within any known material for very long; therefore it is only practical to pulse the plasma for microseconds at a time.

The peak power to the lamp determines the ultraviolet content of the spectral emission and the frequency of the pulses determines the average power. The effectiveness of the lamp for photolysis is a function of both the peak power (uv content) and the average power. Commonly used fill gases for flashlamps include xenon, argon, krypton, and other inert gases or mixtures (13). Argon plasmas give shorter wavelengths than xenon at a given power, but xenon generally has greater efficiency for photon production.

The photo-physics of continuous light sources are vastly different. As opposed to non-specific blackbody radiation, most continuous sources emit lines that are characteristic of the electronic transitions of the unionized fill gas. For example, almost all the radiation emitted by a low pressure mercury lamp is resonance radiation of mercury vapor at 254 and 185 nm, arising from the transitions $6^3P_1 \rightarrow 6^1S_0$ and $6^1P_1 \rightarrow 6^1S_0$. Few mercury atoms are excited to levels above 6^1P_1 and therefore the other mercury lines are very weak (14). A major advantage of mercury lamps for low power applications is that they are very efficient at these two wavelengths (up to 60%).

However, a disadvantage of continuous sources is that they can only be operated at relatively low intensities because the number of photons emitted is proportional to the fill gas pressure. Because the photons are in resonance with electronic transitions, they can be reabsorbed by the fill gas. Thus, as the mercury pressure is increased the 254 and 185-nm lines are reabsorbed and re-emitted at longer wavelengths. By contrast, plasma generation allows (and requires) higher power for a given fill gas pressure and emit non-resonant radiation that is not as readily reabsorbed. Flashlamps intensities of several hundred to one thousand Watts per inch are possible compared to a few hundred Watts per inch for continuous mercury lamps (since all lamps are cylindrical, Watts per inch is proportional to overall intensity). This allows smaller reactor dimensions and thus potential savings in capital costs. It also is particularly advantageous for sites that have limited space available.

Experimental

Air mixtures were irradiated in a 208-liter steel cylindrical reactor. A high intensity six-inch xenon flash lamp was inserted in the middle of the reactor through its side. The reactor contained two electric fans inside to facilitate mixing. All photolyses were performed at atmospheric pressure, and the gas temperature ranged from 300 K to approximately 340 K.

Known volumes of reagents were injected into the reactor by syringe, and were given time to evaporate and mix. The chemicals were used without further purification and were obtained from the following sources: from Aldrich trichloroethene 99+%, tetrachloroethene HPLC grade, 1,1,1-trichloroethane 99%, 1,2- dichloroethane HPLC grade, dichloromethane HPLC grade, trichlorofluoromethane 99+% and chloroform HPLC grade were obtained; from Chem Service 1,1,2-trichloro-1,2,2-trifluoroethane, and 1,1-dichloroethene were obtained; and carbon tetrachloride AR grade was purchased from Mallinkrodt.

Analyses of reactant concentrations were performed by gas chromatography after successive exposures of known duration. Samples were drawn from the reactor by gas tight syringe after turning off the lamp. No reaction was observed in laboratory light or in the reactor with the lamp off. Samples were directly injected into an HP5890 model 2 gas chromatograph equipped with a 30-m J&W model 624 fused silica column and using either photoionization or electron capture detection.

Flashlamp emission spectra were measured on an Oriel InstaSpec III 1024 diode-array detector fitted with a 77410 MultiSpec Grating. A Molectron J25 pyroelectric calorimeter was used for absolute intensity determinations.

Results and Discussion

Flashlamp Emission Spectrum. Figure 1 compares the emission spectrum of a 6-inch Purus 3675-W xenon flashlamp with that of a 42-inch Hanovia 4500-W medium pressure mercury lamp taken from the Hanovia literature. The data are presented as the output power integrated over the total area of the lamp and normalized to the same input power. The xenon flashlamp has a maximum at 230 nm and significant output at wavelengths as low as 200 nm, whereas the mercury lamp has most of its output at wavelengths above 250 nm. Recently we determined an overall electrical efficiency of 16.7% for generation of light below 300 nm by the Purus lamp, compared to 11.4% for a commercial Hg lamp used in UV oxidation. For the same power, the flashlamp allows a seven-fold smaller reaction chamber and thus has an advantage in capital costs. Lamp life for both lamps is on the order of 1000 hours.

Figure 2 compares the emission spectrum of the flashlamp with the absorption spectra of several VOCs (*15,16*). Note that the lamp spectrum is given on a linear scale while the absorption spectra are on a logarithmic scale. The halomethanes and TCA are weak absorbers, whereas TCE and other chloroolefins absorb strongly in the deep UV region. The VOCs all absorb strongly below 200 nm. A shift in peak output from 254 to 230 nm is significant because it corresponds to a 1 to 2 order of magnitude increase in absorptivity of many VOCs, thereby greatly enhancing the rates of direct photolysis. A shift to lower wavelength also enhances the rate of hydrogen peroxide photolysis, although to a lesser extent because its absorption band declines less steeply with increasing wavelength.

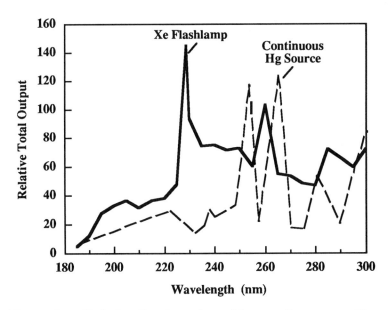

Figure 1.   Emission Spectrum for a Mercury Lamp vs. a Xenon Flashlamp. (All lamps corrected to 3675 W input)

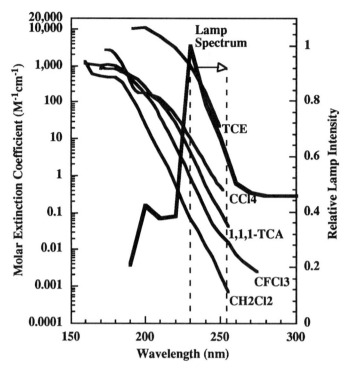

Figure 2.   VOC Absorbance Spectra Compared to Xenon Flashlamp Spectrum

Photolysis Kinetics.   At low total absorbance (< 0.1) the direct photolysis of a compound results in the exponential decrease in concentration of the compound with photolysis time (*17*):

$$- \frac{d[C]}{dt} = 2.3 I_0 \Phi \varepsilon l [C] \tag{1}$$

where I_0 is the incident light intensity, Φ is the apparent disappearance quantum yield, ε is the molar extinction coefficient, l is the light pathlength, and [C] is the contaminant concentration. For a fixed reactor size and light intensity, a simple first-order decrease in concentration of the photoreactant with irradiation time is consistent with a simple
direct photoreaction. A non-exponential decay indicates that the loss of the reactant involves additional reactions whose efficiency is changing during the photolysis period.

The photolysis plots shown in Figures 3 and 4 display both exponential and non-exponential behavior. Compounds whose decay kinetics are first order include carbon tetrachloride, trichlorofluoromethane, 1,1,1-trichloroethane, 1,2-dichloroethane, and 1,1,2-trichloro-1,2,2-trifluoroethane. Non-first order decays were observed for methylene chloride, chloroform, tetrachloroethene, trichloroethene, and 1,1-dichloro-

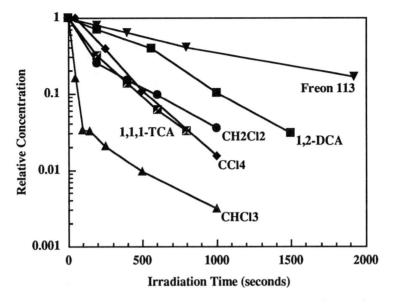

Figure 3.   Photolysis Plots for Individual VOCs

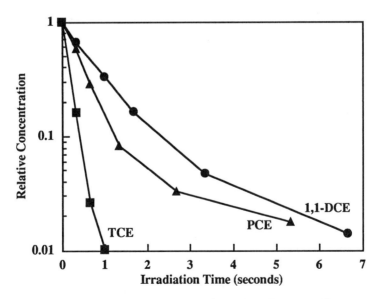

Figure 4.   Photolysis Plots for Chloroolefins

ethene. These differences in the decay behavior provide evidence that the photolysis mechanism for these sets of compounds is different.

Disappearance Quantum Yields. The magnitude of the quantum yield (Φ) for the elimination of the photoreactant also gives some mechanistic information. The ratio of the disappearance quantum yield divided by the primary yield would be unity in the case of a direct photoreaction that contains no subsequent reactions that consumes the reactant. This ratio is greater than unity when later dark reactions consume additional photoreactant, and is substantially greater than unity in the case of a chain reaction.

Table I shows apparent wavelength-averaged quantum yields for the compounds studied determined using carbon tetrachloride as an actinometer. From rate equation 1 for disappearance of the photoreactant in the low absorbance limit, the following expression for the quantum yield can be derived:

$$\frac{\sum I_\lambda \varepsilon_\lambda^{CCl_4}}{\sum I_\lambda \varepsilon_\lambda^{VOC}} \times \frac{k_{VOC}}{k_{CCl_4}} = \Phi_{VOC} \qquad (2)$$

Because a broad band source was used for the excitation light, this expression incorporates the summation over the absorption band of the normalized lamp spectrum multiplied with the extinction coefficient. The disappearance quantum yield for carbon tetrachloride was taken as 1.0, based on the literature data at 214 nm (*18*). For compounds that displayed non-first order decay, the quantum yield was calculated from the initial rate determined from the first time point.

Table I.   First Order Decay Coefficients and Wavelength-Averaged Disappearance Quantum Yields

Compound	k (sec^{-1})*	$\dfrac{\sum I_\lambda \varepsilon_\lambda^{CCl_4}}{\sum I_\lambda \varepsilon_\lambda^{VOC}}$ x	$\dfrac{k_{VOC}}{k_{CCl_4}}$ =	Apparent Φ
CCl4	0.00432	1.0	1.0	1.0
CCl2FCClF2	0.00093	5.09	0.22	1.1
CFCl3	0.0036	1.18	0.84	0.99
CCl3CH3	0.0041	0.79	0.94	0.74
CH2Cl2	0.0070	4.60	1.62	7.5
CHCl3	0.0366	1.79	8.47	15
CH2ClCH2Cl	0.0024	N.D.#	N.D.	N.D.
1,1-DCE	1.24	0.0389	287	11
PCE	1.7	0.0134	394	5.3
TCE	5.5	0.0236	1300	31
TCE + ethylene	0.075	0.0236	17	0.4

* Initial rate constants are taken for non-log-linear curves
# N.D.= not determined

Comparison of the data in Table I and Figures 3 and 4 shows that compounds that display non-first order decays also exhibit apparent quantum yields greater than unity. These compounds include CH_2Cl_2, $CHCl_3$, TCE, 1,1-DCE, and PCE. Furthermore, the other compounds including CCl_4, $CFCl_3$, 1,1,1-trichloroethane, 1,2-dichloroethane, and 1,1,2-trichloro-1,2,2-trifluoroethane, have decays that are first order and apparent quantum yields near unity. We believe that the five compounds in this study that show non-log-linear photooxidation kinetics and apparent quantum yields greater than unity react by a chlorine atom chain mechanism. The other compounds are believed to react purely by direct photolysis.

The ratio of the disappearance quantum yield to that of the primary yield provides a measure of the chain length. Primary quantum yields for C-Cl bond dissociation in chloroalkanes are usually near unity (as verified for $CFCl_3$, 1,1,1-trichloroethane, and 1,1,2-trichloro-1,2,2-trifluoroethane); therefore chain lengths for methylene chloride and chloroform are approximately 8 and 15. The TCE experiment with ethylene added (see below) as a Cl· atom scavenger suggests that the primary quantum yield for TCE is near 0.4, which in turn yields a chain length of about 75.

Chain Photooxidation Mechanism. Heicklen and co-workers have found that chain induced photooxidation occurs for all of the chlorinated ethenes except vinyl chloride (19). Although the reactions in their studies were initiated by Cl_2 sensitization as opposed to direct photodissociation, similar chain propagation steps should occur in both cases. Adopting their mechanism as the most probable, we propose the following mechanisms for VOC photo-oxidation in air:

TCE Photo-induced Chain Reaction

$$HClC=CCl_2 + h\nu \longrightarrow HClC=CCl^\bullet + Cl^\bullet$$
$$Cl^\bullet + HClC=CCl_2 \longrightarrow HCl_2C-CCl_2^\bullet$$
$$HCl_2C-CCl_2^\bullet + O_2 \longrightarrow HCl_2C-CCl_2OO^\bullet$$
$$2HCl_2C-CCl_2OO^\bullet \longrightarrow 2HCl_2C-CCl_2O^\bullet + O_2$$
$$HCl_2C-CCl_2O^\bullet \longrightarrow HCl_2C-CClO + Cl^\bullet \longrightarrow$$

CH_2Cl_2 Photo-induced Chain Reaction

$$CH_2Cl_2 + h\nu \longrightarrow CH_2Cl^\bullet + Cl^\bullet$$
$$Cl^\bullet + CH_2Cl_2 \longrightarrow CHCl_2^\bullet + HCl$$
$$CHCl_2^\bullet + O_2 \longrightarrow CHCl_2OO^\bullet$$
$$2CHCl_2OO^\bullet \longrightarrow 2CHCl_2O^\bullet + O_2$$
$$CHCl_2O^\bullet \longrightarrow HCOCl + Cl^\bullet \longrightarrow$$

TCA Non-chain Photolysis

$$H_3C-CCl_3 + h\nu \longrightarrow H_3C-CCl_2^\bullet + Cl^\bullet$$
$$Cl^\bullet + Cl_3C-CH_3 \longrightarrow Cl_3C-CH_2^\bullet + HCl$$
$$Cl_3C-CH_2^\bullet + O_2 \longrightarrow Cl_3C-CH_2OO^\bullet$$
$$2Cl_3C-CH_2OO^\bullet \longrightarrow 2Cl_3C-CH_2O^\bullet + O_2$$
$$Cl_3C-CH_2O^\bullet + O_2 \longrightarrow Cl_3C-CHO + HO_2^\bullet \longrightarrow$$

These pathways are consistent with the observed decrease in efficiency as the reactant concentration drops, because the fourth steps have a second order dependence on the concentration of the peroxy radical intermediate. As their concentration drops, they become more likely to react in other ways that do not regenerate Cl•.

A common feature of the mechanisms for TCE and methylene chloride is that Cl• atom reacts with them to generate a carbon-centered radical that has chlorine substitution. These carbon-centered radicals are then oxidized in two steps to alkoxy radicals that can then cleave a Cl• atom and propagate the chain. By contrast, the carbon radical formed from TCA has no chlorine substitution and upon oxidation it can only cleave HO_2• instead of Cl•. HO_2• is a much less reactive radical and engages predominantly in termination steps with itself and other radicals. All the compounds listed in Table I fit this general rule, with the exception of 1,2-DCA. 1,2-DCA appears to deviate from the rule because it gave first-order photooxidation kinetics, but the quantum yield is yet undetermined and may still prove to be greater than one. Cl• atoms do not react with C-Cl or C-F bonds and therefore freons and perchloroalkanes will not form a chain.

Photolysis of Chloro-olefin Mixtures. As a further test of the occurrence of a chain reaction with TCE, we photolyzed TCE in the presence of a 500-fold molar excess of ethene as a Cl• atom scavenger. Because ethene contains no chlorine, it cannot propagate a Cl• atom chain. Figure 5 and Table I show that the ethene decreased the TCE photolysis rate by a factor of about 75. Moreover, the rate constant in the absence of ethene decreased with time until at very long times ([TCE] <1 ppmv) it approached the rate constant in the presence of ethene (data not shown). Under both of these conditions chain propagation is inefficient and therefore we believe that the reaction observed in these cases is predominantly direct photolysis.

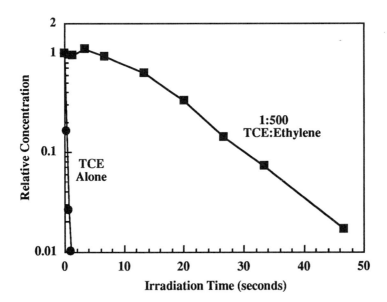

Figure 5.   Chlorine Atom Scavenging With Ethylene

Figure 6 show a photolysis experiment with an equimolar mixture of TCE and 1,1-DCE. Comparing these results with those presented in Figure 4 shows that in the mixture the rate of TCE loss is slightly reduced but the rate of 1,1-DCE loss is considerably enhanced.   We interpret this to be a result of photo-initiation predominantly by TCE because it absorbs more strongly, but with chain propagation predominantly by 1,1-DCE because it reacts with chlorine atoms more rapidly (*20*):

$$Cl^\bullet \ + \ H_2C{=}CCl_2 \ \longrightarrow \ HClC{-}CCl_2^\bullet \quad k_{DCE} \ = \ 14 \times 10^{-11} \ cm^3 \ molec^{-1} \ s^{-1}$$

$$Cl^\bullet \ + \ HClC{=}CCl_2 \ \longrightarrow \ HCl_2C{-}CCl_2^\bullet \quad k_{TCE} \ = \ 8.1 \times 10^{-11} \ cm^3 \ molec^{-1} \ s^{-1}$$

When chains are long compared to the primary photolysis event, the relative rates of loss of 1,1-DCE and TCE should be in the same ratio as their $Cl^\bullet$ atom reaction rate constants of $14/8.1 = 1.7$. This value is in good agreement with the observed ratio of rates of 1.5 and presents further evidence that $Cl^\bullet$ atom is the rate-limiting oxidant in the system.

The experiments with olefin mixtures demonstrate that co-contaminants can cause both sensitization and inhibition of photolysis. Increasing chlorine substitution on the olefin results in stronger light absorption and thus more rapid chain initiation. On the other hand, greater chlorine substitution decreases the rate of the propagation reaction with $Cl^\bullet$ atom thereby reducing chain length. Finally, as chlorine substitution decreases, the statistical chance of regenerating a free $Cl^\bullet$ atom and propagating a chain drops sharply. Thus, TCE and PCE can be expected to sensitize the photo-oxidation of the dichloroethene isomers and vinyl chloride because the former are better light absorbers and the latter react with $Cl^\bullet$ atom more readily. On the other hand, addition of chloroolefins will not sensitize the photoreactions of the chain promoters chloroform

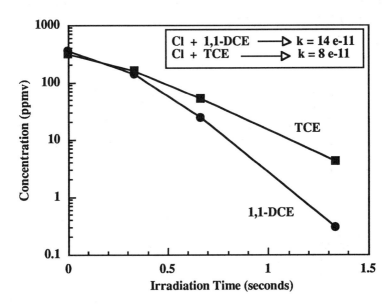

Figure 6.   Photolysis of TCE and 1,1-DCE as a Mixture

and dichloromethane, because the chloroolefins enhance the rate of Cl• atom scavenging as well as production. In principle, any chlorocarbon that photolyzes to form Cl• atom can be expected to sensitize the oxidation of another hydrocarbon, although the effect will be small if a chain reaction is not sustained or the reaction rate constant between Cl• atom and the hydrocarbon is low compared to that between Cl• atom and its precursor.

Photo-oxidation Products. Preliminary product analyses flashlamp photolysis have shown that the disappearance of TCE results in the formation of predominantly dichloroacetyl chloride, as expected from the mechanistic schemes presented above. Phosgene is also a major product and appears to become more important at the expense of dichloroacetyl chloride with increasing photolysis time. We found no TCA, DCE, DCA, vinyl chloride or other VOCs that are measureable using a 624 gas chromatography column and ECD/PID detectors. After long exposures, CO_2 was formed and the chloride ion yield following hydrolysis of the gaseous products approached 80%, indicating that organic halogens are largely destroyed.

The acyl halides hydrolyze to HCl, CO_2 and organic acids on reaction with moisture, and a water scrubber has been demonstrated to effectively remove these compounds. Operated in this mode, phosgene is the most desirable product because it leaves only HCl and CO_2 in the water. Hydrolysis of dichloroacetyl chloride yields dichloroacetic acid, which must be removed from the water. Further work is presently being directed at minimizing the formation of dichloroacetyl chloride in the gas phase and optimizing its removal/destruction in water (as dichloroacetic acid).

Field Demonstration

A field demonstration is being conducted under the auspices of the U. S. Environmental Protection Agency (EPA) Emerging Technology Program to study the effectiveness of the xenon flashlamp in vapor phase treatment of VOCs in off-gas emissions from soil vacuum extraction (SVE) and air stripping systems. The Superfund site selected for the project is Lawrence Livermore National Laboratory, located in Livermore California. A soil venting system has been installed at this site that is capable of generating air flows of better than 300 standard cubic feet per minute (scfm). Off-gas TCE concentrations range up to 500 ppmv.

The overall objectives of the field demonstration are:

1) Measure gas phase VOC destruction rates using pulsed plasma xenon flashlamps.
2) Compare gas phase rate data to water phase treatment rates.
3) Determine optimum operating conditions for the flashlamp technology in terms of power inputs, pulse frequency, and lamp life expectancy.
4) Generate capital, operating, and maintenance costs for the gas phase treatment system.
5) Determine organic by-products, if any.

Conclusions

1) Direct photolysis of VOCs in air with advanced xenon flashlamps has been demonstrated.
2) Chain reaction mechanisms were demonstrated for TCE, PCE, DCE, chloroform, and methylene chloride, i.e. those compounds that yield chlorine substituted carbon-centered radicals upon reaction with chlorine atoms.

3) Of the compounds undergoing chain reactions, the chloro-olefins (TCE, PCE, and DCE) have the highest destruction rates. This is because these compounds have the highest UV absorbances (higher extinction coefficients) which promotes greater numbers of chain initiation events.
4) Mixtures of VOCs (e.g. TCE and DCE) may exhibit synergism if one compound with higher absorbance induces a chain carried on by a compound of lower UV absorbance.
5) A mass balance between starting compound and oxidized products needs to be established. Analyses to date have shown no formation of GC-detectable by-products in the effluent vapors.
6) VOCs other than TCE, PCE, and DCE will require improvements in rate constants before practical treatment processes can be achieved. Trapping and condensing these VOCs from the air stream followed by batch processing is a possible solution.
7) Application of this technology to groundwater treatment via air stripping is possible. Such treatment could be more efficient than treatment in water where interfering compounds often exist.
8) A full scale (500 scfm) soil venting system for TCE removal has been designed and built. Field testing of the system at the Lawrence Livermore National Laboratory Superfund Site has begun.

Acknowledgments

We thank Jim Brucks and Ken Kramasz of Purus, Inc. for engineering support, and Marc van den Berg for measuring the xenon lamp spectrum. This work, in part, has been sponsored by the U.S. EPA's Emerging Technology Grant Program (Project Officer, Norma Lewis). Although the research described in this article has been funded in part by the EPA, it has not been subject to agency review and therefore does not necessarily reflect the views of the agency and no official endorsement should be inferred.

Literature Cited

1. *Air Stripping of Contaminated Water Sources - Air Emissions and Controls* ; U.S. EPA Report EPA-450/3-87-017, Washington, DC, **1987**.
2. Adams, J.Q.; Clark, R.M. *J. Amer. Water Works Assoc.*, **1989**, *81(4)*, 132-140.
3. *Proceedings of the Symposium on Advanced Oxidation Processes*; Toronto, Canada, June 4-5, **1990**; Wastewater Technology Centre: Burlington, Ontario.
4. Glaze, W.H.; Kang, J.W.; Chapin, D.H. *Ozone Sci. Eng.* **1987**, *9*, 335-352.
5. Symons, J.M.; Belhateche, D.; Prengle, W.H. *Hydrogen Peroxide and Visible-Ultraviolet Irradiation for the Oxidation of Organic Environmental Contaminants;* AWWA Annual Conference; Cincinnati, OH, **1990**.
6. *Technology Evaluation Report - Site Program Demonstration of the Ultrox International UV Radiation/Oxidation Technology*; PRC Environmental Management Final Report to EPA under Contract No. 68-03-3484. PRC, Inc: Chicago, IL, **1989**.
7. Hager, D.G.; Loven, C.G.; Giggy, C. *On-Site Chemical Oxidation of Organic Contaminants in Groundwater Using UV Catalyzed Hydrogen Peroxide*; Proceedings, AWWA Annual Conference, Orlando, FL; June 19-23, **1988**, pp. 1149-1165.
8. Ollis, D.F. *Environ. Sci. Technol.* **1985**, *19*, 480-484.
9. Ollis, D.F., Pelizzetti, E., Serpone, N. *Environ. Sci. Technol.* **1991**, *25*, 1522-1529.
10. Baxendale, J.H.; Wilson, J.A. *Trans. Farad. Soc.* **1957**, *53*, 344-356.

11. AWWA Research Foundation **1989**, *Advanced Oxidation Processes for Control of Off-Gas Emissions from VOC Stripping* ; ISBN 0-89867-482-4
12. Cobiella, R. *A Method for Removing Volatile Substances from Water, Using Flash Volatilazation*;   Forum on Innovative Hazardous Waste Treatment Technologies, Atlanta, GA, **1989**.
13. EG&G Publication   **1988**. *Flashlamp Applications Manual* .
14. Phillips, R. *Sources and Applications of Ultraviolet Radiation*; Academic Press, **1963**.
15. Baulch, D.L.; Cox, R.A.; Hampson, R.F.; Kerr, J.A.; Troe, J.; Watson, R.T. *J. Phys. Chem. Ref. Data* **1980**, *9*, 295-471.
16. Hubrich, C.; Stuhl, F. *J. Photochem.* **1980**, *12*, 93-107.
17. Zepp, R.G.; Cline, D.M. *Environ. Sci. Technol.* **1977**, *11*, 359-366.
18. Rebbert, R.E.; Ausloos, P.J. *J. Photochem.* **1976/77**, *6*, 265-276.
19. Sanhueza, E.; Hisatsune, I.C.; Heicklen, J. *Chem. Rev.* **1976**, *76*, 801-826.
20. Atkinson, R.; Aschmann, S.A. *Int. J. Chem. Kinet.* **1987**, *19*, 1097 - 1105.

RECEIVED September 3, 1992

Chapter 19

Chemical Detoxification of Trichloroethylene and 1,1,1-Trichloroethane in a Microwave Discharge Plasma Reactor at Atmospheric Pressure

T. R. Krause and J. E. Helt

Chemical Technology Division, Argonne National Laboratory, 9700 South Cass Avenue, Argonne, IL 60439–4837

Microwave discharge plasma technology is a potential alternative to traditional incineration for the chemical detoxification of low concentrations of chlorinated hydrocarbons in vapor streams. To assess this potential, we have investigated the reactions of trichloroethylene and 1,1,1-trichloroethane with water vapor or molecular oxygen in an argon discharge at atmospheric pressure. Experiments have been conducted in a microwave discharge plasma contained in a tubular flow reactor. Destruction removal efficiencies ranging from 95 to >99% are observed. For the reaction with water vapor, analyses of the effluent stream indicated the primary products are CO_2, CO, H_2, and HCl, with lesser amounts of unreacted parent compound and soot. For the reaction with molecular oxygen, the primary products are CO_2 and HCl, with lesser amounts of unreacted parent compound, CO, H_2O, and unidentified chlorinated hydrocarbons.

Chlorinated hydrocarbons, in particular 1,1,1-trichloroethane (TCA) and trichloroethylene (TCE), are found in more than 10% of all hazardous waste streams at Department of Energy-Defense Production (DOE-DP) facilities (1). These waste streams include mixed waste, contaminated soils and groundwater, off-gases, and bulk liquids. This widespread problem is the result of the extensive use of chlorinated hydrocarbons as cleaning and degreasing solvents. Due to the human and environmental hazards associated with chlorinated hydrocarbons, the safe and efficient remediation of these waste streams has become an important aspect of DOE's environmental restoration program.

Thermal combustion, i.e., incineration, is the primary technology recommended by the Environmental Protection Agency (EPA) for the destruction of chlorinated hydrocarbons. Incinerator systems (e.g., rotary kilns, liquid injection combustors, fixed hearths, and fluidized bed combustors) can meet current EPA regulations (2). However, due to the flame-inhibiting properties of halogenated hydrocarbons, these incinerator systems require the addition of

0097–6156/93/0518–0393$06.00/0

auxiliary fuels, higher operating temperatures, and longer residence times than comparable systems designed for hydrocarbon waste streams (3). In addition, it is difficult to avoid products of incomplete combustion (PIC's) due to a wide variation in the actual residence time and nonuniform temperature profiles and flow distributions in these systems (4). The net result is incinerator systems designed for chlorinated hydrocarbon waste streams have a higher capital equipment and operating costs than comparable systems designed for hydrocarbon waste streams (5).

The nature of the waste stream (e.g., low level contamination in soils or groundwater, bulk liquid storage) is an important criteria for selecting the most efficient and economical remediation technology. While incineration is a viable treatment technology for highly concentrated hazardous chemicals, it is less so for dilute concentrations found in waste streams such as groundwater, soil, and off-gases. Often the hazardous chemical must be separated from the primary waste stream and concentrated prior to incineration which increases both the overall capital and operating costs.

Consequently, there is a considerable research effort seeking alternative technologies for destroying chlorinated hydrocarbon wastes. Examples of alternative technologies include thermal processes, such as catalytically stabilized thermal combustion (4) and supercritical water oxidation (6), chemical and physical processes, such as photocatalyzed oxidation (7,8) and plasma technology, and bioremediation (9).

Although plasma processing has been successfully applied in the metals (10) and microelectronics industries (11), the application of plasma technology to hazardous waste treatment has only recently begun (12). The application of plasma technology to hazardous waste treatment has focussed on two methods of plasma generation, arc discharge and high frequency discharge.

In an arc discharge, the thermonic emission of electrons from metallic electrodes is used to sustain the plasma (13). Much of the initial development work has focussed on arc discharge technology (10,12,14,15). Some examples of commercial applications of arc discharge technology for hazardous waste treatment are a portable system developed by Pyrolysis Systems of Wellington, Ontario, for treating chemical waste removed from Love Canal, New York (10,12) and a pilot plant for treating unsorted municipal waste by SKF Steel Engineering in Hofors, Sweden (14).

In a high frequency discharge, microwave or radiofrequency radiation is used to break down and sustain the plasma discharge. Typically, a frequency greater than 1 MHz is required to sustain the plasma (16). A high frequency discharge differs from an arc discharge in that there are no electrodes, which are a source of possible contamination, present inside the reactor. Microwave or radio-frequency plasma discharge degradation of halogenated hydrocarbons has been investigated by a limited number of research groups (17-20).

Hertzler (17) investigated the oxidation of halogenated hydrocarbons with molecular oxygen in a low-pressure tubular flow microwave-induced discharge plasma reactor. Although conversion of parent hydrocarbon exceeded 99.99%, a complete product analysis was not provided, and, therefore, effluent toxicity could not be determined.

Barat and Bozzelli (18,19) have investigated the microwave-induced plasma degradation of chloroform, trichloroethylene, and chlorobenzene with water vapor or hydrogen at low pressures (~5-10 torr) in a tubular flow reactor. Conversion of the parent chlorinated hydrocarbons ranged from 50 to almost 100%. At conversions of >80%, at least 85% of the chlorine was converted to HCl, with the remaining chlorine in the form of nonparent chlorinated hydrocarbons.

Wakabayshi et al., (20) have investigated the radiofrequency-induced, 4 MHz, plasma degradation of trichloroethylene, trichloro-fluoromethane, trichlorotrifluoroethylene, carbon tetrachloride, and chloroform with water vapor at atmospheric pressures in a tubular flow reactor. Conversions of the parent halogenated hydrocarbon ranged from 89 to 100% with the primary products being the halogenated acid and nonparent halogenated hydrocarbons.

At Argonne National Laboratory, we are investigating the viability of a microwave discharge plasma reactor operating at atmospheric pressure as a remediation technology for chemically detoxifying dilute concentrations of volatile organic compounds, in particular TCA and TCE, and potentially nonchlorinated hydrocarbons found in off-gas waste streams and in air stripping operations. By chemical detoxification, we mean the chemical conversion of chlorinated hydrocarbons to less toxic compounds such as carbon dioxide, water, and hydrogen chloride.

Experimental

The microwave discharge plasma reactor is shown schematically in Figure 1.

Microwave radiation, at a frequency of 2450 MHz and a variable forward power range of 0-7 kW, is generated by a Cober Electronics, Inc. S6F Industrial Microwave Generator. The microwave radiation is directed to the reactor by a waveguide, operating in the TE_{10} mode, manufactured by Associated Science Research Foundation, Inc. (TE_{10} refers to a waveguide in which the electric field is everywhere transverse to the direction of propagation of the microwave radiation but the magnetic field has both transverse and parallel components.) A tuneable "short" is used to couple the microwave radiation with the plasma. Reflected power is measured with a Hewlett Packard Model 435B Power Meter.

The reactor, a 1-in. OD quartz tube, intersects the waveguide at a 90° angle to the direction of microwave radiation propagation. A second quartz tube, approximately 18-mm in OD and 5-in. in length, is inserted in the quartz reactor tube. The position of this quartz liner is adjusted so that the plasma is confined within it. The plasma is approximately 2 inches in length under normal operating conditions. The purpose of this inner quartz liner is to protect the surface of the quartz reactor tube from electron bombardment. Electron bombardment of the quartz surface results in a phase transformation from quartz to christobalite, as determined by x-ray diffraction. The reactor is cooled during operation by force convection of air at ambient temperature which limits the upper

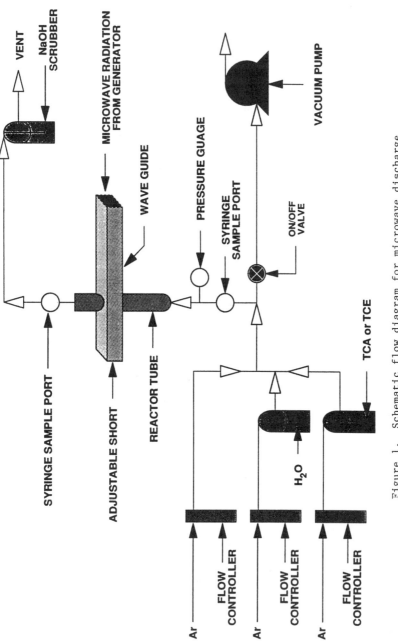

Figure 1. Schematic flow diagram for microwave discharge plasma reactor.

forward power range to approximately 1-kW for continuous operation (~8 hrs). Forward power inputs as high as 3-kW have been applied for shorter operating periods (generally 1 hr or less).

The plasma is initiated by evacuating the system to approximately 50 torr at a forward input power of 600 watts and contacting the reactor wall with a vacuum leak detector (Electro-Technic Products). The reactor is slowly brought to operating pressure, 1.0 to 1.1 atm, with either argon (Matheson or AGA UHP Grade - 99.999%) or an oxygen-argon mixture (20% Oxygen Matheson or AGA Certified Standard). These carrier gases are used without further purification. The inlet gas stream composition and flowrate is controlled using Matheson Model 601, 602, 603, and 610 series flow controllers equipped with metering valves. The forward input power is then adjusted to the desired level (Table I).

No determination or estimation of the gas-phase temperature within the plasma zone was made. The plasma zone had an estimated volume of 7.7 cm^3 based on the diameter of the quartz tube liner and the length of the reactor tube within the waveguide. By visual inspection, the plasma occupied only the region of the reactor tube within the waveguide. Residence times are based on the gas flowrate and the volume of the plasma zone. No attempts were made to determine if the gas was well-mixed or if channeling occurred.

The liquid reactants, either 1,1,1 trichloroethane or trichloro-ethylene (Aldrich, Research Grade, or Mallinckrodt Chemicals, Analytical Grade) and water (Aldrich, HPLC Grade, or Mallinckrodt Chemicals, ChromAR HPLC grade) are used without further purification. The desired reactants are introduced to the reactor as a vapor by bubbling the carrier gas through a liquid reservoir contained in a gas washing bottle.

Chemical analysis of the reactor feed and effluent streams is accomplished primarily by gas chromatography by periodically removing a small volume of the stream (0.5-5.0 ml) with a gas-tight syringe manufactured by Unimetrics Corporation. Chlorinated hydrocarbons are analyzed using a 6' x 1/8" 10% SP-2100 supported on Chromosrob WAW (Supleco) with an helium carrier at a flowrate of 37 ml/min and an inlet pressure of 40 psig, operated isothermally at either 40°C or 70°C, in a Varian Model 3700 Gas Chromatograph equipped with a thermal conductivity detector. A Varian Model CDS111 Integrator is used to quantify the GC peaks. Carbon dioxide, carbon monoxide, and methane are analyzed using a 10' x 1/4" CTR-II column (Alltech) with a helium carrier at a flowrate of 36 ml/min and an inlet pressure of 25 psig, and hydrogen is analyzed using a 6' x 1/8" 5A Molecular Sieve column with an argon carrier at a flowrate of 14 ml/min and an inlet pressure of 40 psig, both columns operated isothermally at 40°C, in a Hewlett-Packard Model 5890 Series II Gas Chromatograph equipped with two thermal conductivity detectors. A Hewlett Packard Model HP3396A Integrator is used to quantify the GC peaks. Nonchlori-nated hydrocarbons are analyzed using a 6' x 1/8" Poropak Q column, operated from 100°C to 200°C at a rate of 10°C/min, in a Varian Model 3700 Gas Chromatograph equipped with a flame ionization detector and a Hewlett Packard Model HP3396A Integrator. Identification of all GC peaks is accomplished by comparison of the retention time of a pure sample with that of an unknown peak in the chromatograph.

TABLE I

MINIMUM FORWARD POWER INPUT REQUIRED
TO SUSTAIN PLASMA IN SOME GASEOUS ATMOSPHERES
AT 1.0-1.1 ATM USING MICROWAVE RADIATION AT 2450 MHz

GAS	FORWARD POWER (watts)
Argon	150
2.45% H_2O/Argon	220
20.13% Oxygen/Argon	450
Nitrogen	450

Due to experimental difficulties, only an estimation of the chlorine and hydrogen chloride yields can be made. This estimation is accomplished by the following procedure. The amount of hydrogen chloride removed from the effluent stream by a sodium hydroxide solution during a given time interval is determined by measuring both the changes in pH, by acid-base titration, and the Cl^- concentration by AgCl precipitation. By comparing the amount of the HCl removed from the effluent stream to the amount of TCA or TCE introduced to the reactor during a given time period, a qualitative estimation of the HCl yield can be obtained. At high flowrates or TCA/TCE feed concentrations; however, the caustic scrubbers are not completely effective at removing HCl from the effluent stream. The Cl_2 yield is estimated from an overall mass balance. A number of samples of the effluent stream were analyzed by mass spectroscopy in an attempt to determine the HCl and Cl_2 concentrations.

Results And Discussion

The chemical detoxification of chlorinated hydrocarbons in an thermal incineration can be represented stoichiometrically by equation 1 (21):

$$C_nH_{m+p}Cl_p + (n + m/4)O_2 \longrightarrow nCO_2 + (m/2)H_2O + pHCl \qquad (1)$$

This equation describes the predominant product formation as long as there is sufficient oxygen to convert all the carbon and hydrogen to CO_2 and H_2O, respectively, and the hydrogen to chlorine ratio is greater than or equal to 1. For chlorinated hydrocarbons such as trichloroethylene, which has a hydrogen to chlorine ratio of 1/3, Cl_2 is produced in addition to HCl (22).

Since HCl is the desired product for chlorine, it is necessary to increase the hydrogen concentration for incineration systems for hydrogen-deficient chlorinated hydrocarbons. One approach for increasing the hydrogen content in an incinerator is to add a hydrocarbon, such as propane, to the feed stream.

An alternative approach to oxidative degradation is to react the chlorinated hydrocarbon with hydrogen in a reducing atmosphere. As Bozzelli and Barat (18) have discussed, the reaction of chlorinated hydrocarbons with hydrogen to yield hydrocarbons and hydrogen chloride is thermodynamically favorable. However, such a process is not viable for many remediation waste streams.

Water, H_2O, is an alternative to hydrogen for chemically detoxifying chlorinated hydrocarbons. Water can serve as both a hydrogen source for the elimination of chlorine as HCl and an oxygen source for the elimination of carbon as CO_x. As in the case of the reaction of chlorinated hydrocarbons with hydrogen, the reaction with water is thermodynamically favorable (18). While earlier investigations have focussed on the reaction of chlorinated hydrocarbons with water in the absence of oxygen (18-20), we are investigating the reaction of chlorinated hydrocarbons in an oxygen-water atmosphere.

Reaction of 1,1,1-Trichloroethane with Oxygen. The reaction of
1,1,1-trichloroethane (TCA) with oxygen (O_2) in a microwave discharge
plasma was investigated under various conditions, including varying
the O_2/TCA molar feed ratio and the forward power input to the
reactor, to investigate the effect of these operating parameters on
the overall TCA conversion and the resulting product distribution.
The results of these experiments are summarized in Table II.

The primary reaction of TCA and O_2 in a microwave discharge
plasma can be described by a global stoichiometry which is identical
to that found in thermal combustion (equation 2):

$$C_2H_3Cl_3 + 2O_2 \longrightarrow 2CO_2 + 3HCl \qquad (2)$$

This equation is applicable in the presence of excess oxygen, O_2/TCA
ratio greater than 2, and a forward power input of 600 watts which is
above the minimum required to sustain the plasma (Table I). When
these operating conditions are met, in excess of 98.3% of the initial
TCA loading is observed to react yielding CO_2 and HCl as the major
products. No H_2 or nonparent chlorinated hydrocarbons are observed
within our detection limits by GC analysis. Recovery of chlorine as
HCl is nearly 100%, although these results are qualitative as
previously discussed. At a O_2/TCA ratio of 1.5, which is less than
that required for complete oxidation of the carbon, both CO and CO_2
are formed and soot formation is observed.

As the forward power input to the reactor is decreased to 450
watts, which is the minimum power input required to maintain a 20%
oxygen/argon plasma (Table I), the primary carbon oxide product is CO
and formation of soot particles is observed indicating incomplete
oxidation of the carbon even in the presence of excess oxygen (O_2/TCA
molar feed ratio of 165). The formation of nonparent chlorinated
hydrocarbons, such as methylene chloride and trichloroethylene, are
also observed. Similar results are observed under oxygen-deficient
conditions as the forward power input approaches the minimum required
to sustain the plasma.

The soot particles are deposited on the inner wall of the quartz
liner in a nonuniform distribution during the course of the reaction.
These soot particles absorb microwave radiation at 2450 MHz, which
was confirmed by observing both a decrease in the reflected power as
soot formation occurred and that the soot particles "glowed red" when
microwave radiation was admitted to the reactor in the absence of a
plasma.

The ability of soot particles to absorb microwave radiation
decreases the amount of energy available to the plasma at a given
forward power input. As soot formation increases, a point is reached
where there is an insufficient amount of energy available to sustain
the plasma and the plasma will be quenched. If the introduction of
TCA to the reactor is halted before the plasma is quenched, the
formation of some CO and CO_2 is observed. This suggests that under
certain conditions, the reaction of TCA and O_2 may involve both
homogeneous and heterogeneous reaction mechanisms.

The carbon in the soot particles is amorphous as determined by x-
ray diffraction. Chlorine is present, as determined by x-ray phosphorescence, although a carbon/chlorine ratio could not be determined.

TABLE II

REACTANT FEED CONCENTRATION AND PRODUCT DISTRIBUTION
FOR THE REACTION OF 1,1,1-TRICHLOROETHANE AND OXYGEN

FORWARD POWER (watts)	FLOWRATE (l min^{-1})	O_2/TCA RATIO	MOLE FRACTION ($\times 10^3$) TCA (1)	TCA CONV (2)	MOLAR RATIO CO_x/TCA (3)	CO_2/CO_x (4)	PERCENT HCl
600	1.59	150	1.31	0.995	2.0	1.00	~100
600	0.23	4.4(5)	43.0	0.983	2.0	0.965	92
600	1.59	1.5	1.35	0.990	--(6,7)	0.534(8)	---
450	1.59	165	1.12	0.994(6)	1.3	0.099	---
400	1.59	1.5	1.45	0.962	--(6,7)	0.589(8)	---
300	1.59	1.5	1.36	0.753(9)	--(6,7)	0.372(8)	---

(1) Mole fraction of TCA in feed stream.
(2) If no TCA is observed in the effluent stream, then the minimum amount of TCA detectable by GC is substituted.
(3) Ratio of the concentration of CO and CO_2 in effluent stream to that of TCA in feed stream.
(4) Defined as CO_2/(CO + CO_2) in effluent stream.
(5) This sample analyzed by mass spectroscopy.
(6) Carbonaceous deposits formed.
(7) Due to fluctuations in CO_x yields, a mass balance was not obtainable.
(8) Average value observed during ~1 hr period.
(9) Effluent stream contained CH_2Cl_2, C_2HCl_3 and other unidentified chlorinated in small yields (total yield <1% initial TCA feed).

There is no evidence for the presence of any molecular compounds in the soot particles.

Reaction of Trichloroethylene with Oxygen. The results of the reaction of trichloroethylene (TCE) and oxygen (O_2) in a microwave discharge plasma are summarized in Table III. The global stoichiometry for the reaction of TCE with O_2 in a microwave discharge plasma (equation 3) is identical to that observed in thermal combustion.

$$C_2HCl_3 + 2O_2 \longrightarrow 2CO_2 + HCl + Cl_2 \qquad (3)$$

Conversion of TCE in excess of 99% is observed for O_2/TCE ratios ranging from 20.4-190 and a forward power input of 800 watts. Qualitatively, approximately 40% of the chlorine is recovered as HCl which is in agreement with the theoretical yield of 33% based on equation 3. The primary carbon product is CO_2, although CO formation is observed at the lower O_2/TCE ratio.

It is known from the investigation of the combustion of TCE that the reaction proceeds in two steps (22). The first step involves the rapid oxidative decomposition of TCE as described in equation 4.

$$C_2HCl_3 + O_2 \longrightarrow 2CO + HCl + Cl_2 \qquad (4)$$

The second step involves the subsequent oxidation of CO to CO_2 as shown in equation 5.

$$2CO + O_2 \longrightarrow 2CO_2 \qquad (5)$$

Although this oxidation may possibly be quite fast, it is inhibited by HCl and Cl_2 and is subsequently very slow in thermal combustion processes involving chlorinated hydrocarbons.

Although we have not performed a mechanistic investigation of the reaction of TCE and O_2 in a plasma, the formation of CO at lower O_2/TCE ratios, i.e., increasing ($HCl + Cl_2$)/O_2 ratios, would suggest that a similar two step process is occurring. This result would not be surprising since the destructive oxidation of TCE with O_2 in a thermal incinerator or a plasma reactor both involve free radical chemistry.

For the reaction of TCA or TCE with O_2, the primary sink for hydrogen is HCl and not H_2O despite the fact that the O-H bond strength (119 kcal/mole in H_2O) is greater than the H-Cl bond strength (103.1 kcal/mole). This can be explained by the Deacon reaction, equation 6, which favors the formation of HCl at elevated temperatures (23).

$$H_2O + Cl_2 \longrightarrow 2HCl + 1/2\ O_2 \qquad (6)$$

Reaction of 1,1,1-Trichloroethane with Water. The reaction of TCA with H_2O in a microwave discharge plasma was investigated under various conditions, including varying H_2O/TCA molar feed ratio and forward power input to the reactor. The objective was to investigate the effect of these operating parameters on the overall TCA conver-

TABLE III

REACTANT FEED CONCENTRATIONS AND PRODUCT DISTRIBUTION
FOR THE REACTION OF TRICHLOROETHYLENE AND OXYGEN

FORWARD POWER (watts)	FLOWRATE ($l\ min^{-1}$)	O_2/TCE RATIO	MOLE FRACTION ($\times 10^3$) TCE(1)	TCE CONV(2)	MOLAR RATIO CO_x/TCE(3)	CO_2/CO_x(4)	PERCENT HCl	HCl/Cl_2 RATIO
800	1.59	190	0.990	0.995	2.0	1.0	~40	0.7
800	1.78	20.4	8.25	0.999(5)	2.0	0.997	~40	0.7
600	0.200	12.9(6)	15.0	0.967	2.0	0.97	27	0.4

(1) Mole fraction of TCE in feed stream.
(2) If no TCA is observed in the effluent stream, then the minimum amount of TCE detectable by GC is substituted.
(3) Ratio of the concentration of CO and CO_2 in effluent stream to that of TCE in feed stream.
(4) Defined as $CO_2/(CO + CO_2)$ in effluent stream.
(5) A small quantity (<<1% based on initial feed concentration of TCE) of an unidentified chlorinated hydrocarbon was observed by GC analysis. The retention time did not correlate with $CH_{3-x}Cl_{1+x}$ (where x = 0 to 3). Possibly phosgene.
(6) This sample analyzed by mass spectroscopy.

sion and resulting product distribution. The results of these experiments are summarized in Table IV.

Ideally, one would prefer for the degradation of TCA to occur according to the stoichiometry described by equation 7a.

$$C_2H_3Cl_3 + 4H_2O \longrightarrow 2CO_2 + 3HCl + 4H_2 \qquad (7a)$$

The stoichiometric amount of H_2O is defined as the amount required to convert all the carbon to CO_2 and the chlorine to HCl. However, under no conditions is the complete conversion of carbon to CO_2 observed; CO is always observed.

For H_2O/TCA molar feed ratios ranging from 10.4 to 62.3 and a forward power input of 600 watts, conversions of TCA in excess of 98.3% were observed with the primary products being CO, CO_2, H_2, and HCl. CO_2 selectivity, defined as $CO_2/(CO + CO_2)$, is observed to decrease, from 0.694 to 0.274, as the H_2O/TCA molar feed ratio decreases from 62.3 to 10.4. A more appropriate stoichiometry equation would be equation 7b,

$$C_2H_3Cl_3 + 3H_2O \longrightarrow CO + CO_2 + 3HCl + 3H_2 \qquad (7b)$$

Note as the CO yield increases, the H_2/CO_x ratio decreases from 2 (no CO) to 1 (no CO_2). The observed H_2/CO_x ratios range from 3.73 to 1.96 under the above conditions, which indicates excess H_2 is being produced.

One possible explanation for the incomplete conversion of carbon to CO_2 is the water-gas shift reaction (equation 8).

$$CO_2 + H_2 \rightleftharpoons CO + H_2O \qquad (8)$$

Note that the sum of equations 7a and 8 is equation 7b. As observed, the CO_2 selectivity decreases with a decrease in the H_2O/TCA molar feed ratio as expected according to Le Chatelier's principle.

As the forward power is decreased at a near constant H_2O/TCA ratio, 19.0-22.1, the conversion of TCA decreases from 99.9% at 600 watts to 84.8% at 300 watts, which is slightly above the minimum power required to sustain the plasma (Table I). The CO_2/CO_x ratio is relatively constant ranging from 0.452 at 600 watts to 0.487 at 300 watts. At 300 watts, considerable soot formation is observed with the formation of simple alkanes, CH_4, C_2H_4, and C_2H_6. As discussed previously, soot formation adversely affects the overall reactant conversion and the product distribution by reducing the effective power available to the plasma at a given forward power input.

Reaction of Trichloroethylene with Water. The reaction of TCE and H_2O in microwave discharge plasma was investigated under various conditions, including varying the H_2O/TCE molar feed ratio, forward power input to the reactor, and residence time, to investigate the effect of these operating parameters on the overall TCE conversion and the resulting product distribution. The results of these experiments are summarized in Table V.

Ideally, one would like for the reaction of TCE by H_2O to occur according to a stoichiometry similar to that discussed for TCA (equation 7a). However, since TCE is hydrogen-deficient for the

TABLE IV

REACTANT FEED CONCENTRATIONS AND PRODUCT DISTRIBUTION
FOR THE REACTION OF 1,1,1-TRICHLOROETHANE + H_2O

FORWARD POWER (watts)	FLOWRATE (l min⁻¹)	H_2O/TCE RATIO	MOLE FRACTION (x 10³) TCA[1]	TCE CONV[2]	CO_x/TCA[3]	MOLAR RATIO CO_2/CO_x[4]	H_2/CO_x	PERCENT HCl
600	2.33	62.3	0.388	0.997	2.0	0.694	3.73	>90
600	3.11	36.4	0.332	0.983	2.0	0.633	2.58	>90
600	2.33	24.2	0.990	0.999	2.0	0.559	2.97	>90
600	3.11	19.9	0.604	0.999	1.9[5]	0.452	2.03	>90
600	3.11	17.0	0.708	0.992	1.9[5]	0.432	1.96	>90
600	2.38	10.4	2.27	0.997	1.9[5]	0.274	1.96	---
400	3.11	19.0	0.633	0.999	1.9[5]	0.391	2.06	---
300	3.11	22.1	0.542	0.848	1.4[6]	0.487	2.74	---

(1) Mole fraction of TCA in feed stream.
(2) If no TCA is observed in effluent stream, then the minimum amount of TCA detectable by GC is substituted.
(3) Ratio of the concentration of CO and CO_2 in effluent stream to that of TCA in feed stream.
(4) Defined as CO_2/(CO + CO_2) in effluent stream.
(5) Carbonaceous deposits formed.
(6) Considerable carbon deposits, with the observance of CH_4, C_2H_4, and C_2H_6 in the effluent stream. Plasma extinguished after approximately 40 minutes at these conditions.

TABLE V

REACTANT FEED CONCENTRATIONS AND PRODUCT DISTRIBUTION
FOR THE REACTION OF TRICHLOROETHYLENE + H_2O

FORWARD POWER (watts)	FLOWRATE (l min^{-1})	H_2O/TCE RATIO	MOLE FRACTION ($\times 10^3$) TCE[1]	TCE CONV[2]	CO_x/TCE[3]	MOLAR RATIO CO_2/CO_x[4]	H_2/CO_x	PERCENT HCl
600	1.59	173.	0.14	0.99	2.0	0.905	3.33	>90
600	1.59	42.8	0.56	1.00	2.0	0.836	1.98	>90
600	1.59	27.8	0.846	0.998	2.0	0.570	1.11	>90
600	1.59	22.9	1.02	0.998	2.0	0.630	1.52	>90
600	1.71	10.3	2.28	0.992	~2[5]	0.528	1.27	>90
600	0.185	16.3	1.43	0.997	~2[5]	0.698	1.58	---
500	0.185	14.8	1.58	0.997	~2[5]	0.695	1.49	---
400	0.185	10.8	2.17	0.998	~2[5]	0.652	1.54	---
800	1.59	25.2	0.938	0.995	2.0	0.783	1.31	---
600	1.59	33.7	0.701	0.994	2.0	0.769	1.54	---
500	1.59	32.6	0.723	0.994	2.0	0.805	1.50	---
400	1.59	31.8	0.743	0.994	2.0	0.674	1.12	---

(1) Mole fraction of TCE in feed stream.
(2) If no TCE is observed in the effluent stream, then the minimum amount of TCE detectable by GC is substituted.
(3) Ratio of the concentration of CO and CO_2 in effluent stream to that of TCE in feed stream.
(4) Defined as $CO_2/(CO + CO_2)$ in effluent stream.
(5) Carbonaceous deposits formed.

complete conversion of the chlorine to HCl, H_2O would provide both the oxygen for CO_2 and hydrogen for HCl formation (equation 9a).

$$C_2HCl_3 + 4H_2O \longrightarrow 2CO_2 + 3HCl + 3H_2 \qquad (9a)$$

Consequently, one less mole of H_2 would be produced for each mole of TCE consumed compared to TCA if the reaction of H_2O with TCA or TCE was to proceed as described by equation 7a or equation 9a, respectively.

As discussed for the reaction of TCA with H_2O, the complete conversion of carbon to CO_2 is not observed. For H_2O/TCE molar feed ratios ranging from 173 to 10.3 and a forward input power of 600 watts, the CO_2/CO_x ratio decreases from 0.905 to 0.528. As with TCA, a more appropriate equation is equation 9b.

$$C_2H_cl_3 + 3H_2O \longrightarrow CO + CO_2 + 3HCl + 2H_2 \qquad (9b)$$

The H_2/CO_x ratio should range from 1.5 (no CO) to 0.5 (no CO_2). The observed H_2/CO_x ratios range from 3.33 to 1.11 for the above described conditions. As expected, the H_2/CO_x ratio for the reaction of TCE with H_2O is less than that for the reaction of TCA with H_2O under similar conditions.

Increasing the residence time by a factor of 9 did not result in any significant improvement in the CO_2 selectivity. For example, at a residence time of 290 ms, the CO_2/CO_x ratio observed is 0.528 at a H_2O/TCE molar feed ratio of 10.3 and a forward power input of 600 watts. At a residence time of 2500 ms, the CO_2/CO_x ratio is 0.698 at a H_2O/TCE ratio of 16.3 and 600 watts forward power input. A small amount of soot formation was observed in both cases. No increase in the overall TCE conversion was observed due to the limitations of the GC analysis.

Reaction of Trichloroethylene with a Mixture of Oxygen and Water. As has been demonstrated, chlorinated hydrocarbons react with O_2 in a microwave discharge plasma yielding primarily CO_2 and HCl if the H/Cl ratio of the reactant molecule is greater than or equal to 1 or CO_2, HCl, and Cl_2 if the H/Cl ratio is less than 1. Water will react with chlorinated hydrocarbons, serving as both an oxygen source for the formation of CO_x and hydrogen source for the formation of HCl; however, CO is formed in preference to CO_2.

Alternatively, if the chlorinated hydrocarbon is reacted with a mixture of H_2O and O_2, it might be possible to eliminate all the chlorine as HCl and the carbon as CO_2 as long as there is a stoichiometric excess of H_2O and O_2.

The results of the reaction of TCE with a mixture of O_2 and H_2O in a microwave discharge plasma are summarized in Table VI. Conversions of TCE of >99% are observed with the primary products being CO_2 and HCl. Qualitatively, all the chlorine is recovered as HCl. A maximum Cl_2 concentration of 4 ppm is observed by mass spectroscopy in the effluent stream for a H_2O/TCE ratio of 2.40 and a O_2/TCE ratio of 18.8. No CO is detected by gas chromatography.

TABLE VI

REACTANT FEED CONCENTRATIONS AND PRODUCT DISTRIBUTION
FOR THE REACTION OF TRICHLOROETHYLENE + H_2O + O_2

FORWARD POWER (watts)	FLOWRATE (l min^{-1})	H_2O/TCE RATIO	O_2/TCE RATIO	MOLE FRACTION (x 10^3) TCE (1)	TCE CONV (2)	MOLAR RATIO CO_x/TCE (2)	CO_2/CO_x (4)	Cl_2 (5)	PERCENT HCl
800	1.59	25.7	202	0.890	0.990	2.0	1.0	$<2\times10^{-6}$	~100
800	1.71	6.37	49.9	3.44	0.997	2.0	1.0	---	~100
800	1.78	2.40	18.8	8.56	0.999	2.0	1.0	$<4\times10^{-6}$	~100

(1) Mole fraction of TCE in feed stream.
(2) If no TCE is observed in the effluent stream, then the minimum amount of TCE detectable by GC is substituted.
(3) Ratio of the concentration of CO and CO_2 in effluent stream to that of TCE in feed stream.
(4) Defined as $CO_2/(CO + CO_2)$ in effluent stream.
(5) Determined by mass spectroscopy.

Conclusions

We have demonstrated a microwave discharge plasma reactor can effectively detoxify dilute vapor concentrations of TCA or TCE using either O_2, H_2O vapor, or a O_2-H_2O vapor mixture as the coreactant in an argon gas stream at atmospheric pressures. Both the conversion of either TCA or TCE and the resulting product distribution are a function of the forward power input to the reactor. The primary products are CO, CO_2, HCl, Cl_2, and H_2, with the actual selectivity and yield depending on the reactant feed mixture and forward power input to the plasma reactor. As the forward power input approaches the minimum required to sustain the plasma, the overall TCA or TCE conversion decreases, incomplete oxidation of carbon to CO_2 occurs, soot formation occurs, and the products include nonparent chlorinated hydrocarbons and/or simple hydrocarbons.

Acknowledgments

We would like to acknowledge B. Tani, who provided both the x-ray diffraction and x-ray phosphorescence results, and A. Engelkemeir, who provided the mass spectroscopy results, both of the Analytical Chemistry Laboratory. This work is supported by the U.S. Department of Energy, Office of Technology Development, Environmental Restoration and Waste Management Division, under Contract No. W-31-109-ENG-38.

Literature Cited

(1) Fore, C. S.; Post, R. G. *Waste Management 86. Volume 1: General Interest*; University of Arizona, Tucson, AZ, 1986; pp 495-498.
(2) Keitz, E.; Vogel, G.; Holberger, R.; Boberschmidt, L. A Profile of Existing Hazardous Waste Incineration Facilities and Manufacturers in the United States. Project Summary, U. S. Environmental Protection Agency, April 1984; EPA-600/2-84-052.
(3) Oppelt, E. T. *JAPCA* **1987**, *37*, 558-586.
(4) Hung, S. L.; Pfefferle, L. D. *Environ. Sci. Technol.* **1989**, *23*, 1085-1091.
(5) Santoleri, J. J. In *Standard Handbook of Hazardous Waste Treatment and Disposal*; Freeman, H. M., Ed.; McGraw-Hill: New York, NY, 1988; pp 8.3-8.18
(6) Rofer, C. K.; Streit, G. E. Phase II Final Report: Oxidation of Hydrocarbons and Oxygenates in Supercritical Water. Los Alamos Report LA-11700-MS (DOE/HWP-90); September 1989
(7) Lawrence Livermore National Laboratory. Environmental Restoration: Dynamic Planning and Innovation. Lawrence Livermore National Laboratory Report UCRL-52000-90-4; 1990.
(8) Pacheco, J. E.; Carwile, C.; Magrini, K. A.; Mehos, M. Developments in Solar Photocatalysis for Destruction of Organics in Water. Sandia National Laboratory Report SAND89-2236C; 1989.
(9) *Chem. Eng. News* **1989**, *67*(40), 18.
(10) Zanetti, R. J. *Chemical Engineering* **1983**, *90*(26), 14-16.
(11) Fraser, D. B.; Westwood, W. D., In *Handbook of Plasma Processing Technology*; Rossnagel, S. M.; Cuomo, J. J.; Westwood, W. D., Eds.; Noyes: Park Ridge, NJ, 1990, p 2.

(12) Lee, C. C. In *Standard Handbook of Hazardous Waste Treatment and Disposal*; Freeman, H. M., Ed.; McGraw-Hill: New York, NY, 1989, pp 8.169-8.178.
(13) Bell, A. T. In *The Application of Plasmas to Chemical Processing*; Baddour, R. F.; Timmins, R. S., Eds.; The M.I.T. Press: Cambridge, MA, 1967; p 6.
(14) Hertzler, B. C. Development of Microwave Plasma Detoxification Process for Hazardous Wastes (Phase III). U.S. EPA Contract 68-03-2190, 1979.
(15) Copsey, M. *Physics World* **1991**, *4*(5), 23-24.
(16) Shohet, J. L. In *Encylcopedia of Physical Sciences*; Academic Press: New York, NY, 1987, Vol. 10; p 723.
(17) Herlitz, H. G. *Environ. Sci. Technol.* **1986**, *20*, 1102-1103.
(18) Bozzelli, J. W.; Barat, R. B. *Plasma Chem. Plasma Process* **1988**, *8*, 293-341.
(19) Barat, R. B.; Bozzelli, J. W. *Environ. Sci. Technol.* **1989**, *23*, 666-671.
(20) Wakabayashi, T.; Mizuno, K.; Imagowa, T. I.; Amano, T.; Hirakawa, S.; Komaki, H.; Kobayashi, S.; Kushiyama, S.; Aizawa, R.; Koinuma, Y.; Ohuchi, H. Decomposition of Halogenated Organic Compounds by r.f. Plasma at Atmospheric Pressure. paper presented at 9th International Symposium on Plasma Chemistry, Pugnochiuso, Italy September 4-8, 1989.
(21) Tsang, W. *Waste Management* **1990**, *10*, 217-225.
(22) Bose, D; Senkan, S. M. *Combust. Sci. Tech.* **1983**, *35*, 187-202.
(23) Chang, W.-D., Karra, S. B., Senkan, S. M. *Environ. Sci. Technol.* **1986**, *20*, 1243-1248.

RECEIVED October 6, 1992

Chapter 20

Bioprocessing of Environmentally Significant Gases and Vapors with Gas-Phase Bioreactors

Methane, Trichloroethylene, and Xylene

William A. Apel, Patrick R. Dugan, Michelle R. Wiebe, Earl G. Johnson, James H. Wolfram, and Robert D. Rogers

Center for Biological Processing Technology, Idaho National Engineering Laboratory, EG&G Idaho Inc., Idaho Falls, ID 83415–2203

Fixed thin film, gas/vapor phase bioreactors were assessed relative to their potential for the bioprocessing of methane, trichloroethylene (TCE), and *p*-xylene. Methanotrophic bacteria were used to process the methane and TCE while a xylene resistant strain of *Pseudomonas putida* was used to process the *p*-xylene. Comparisons between the gas phase bioreactors and conventional shaken cultures and sparged liquid bioreactors showed that the gas phase bioreactors offer advantages over the other two systems for the degradation of methane in air. Rates of methane removal with the gas phase bioreactors were 2.1 and 1.6 fold greater than those exhibited by the shaken cultures and sparged liquid bioreactors, respectively. The gas phase bioreactors were shown to have application for the removal of TCE vapors from air with a removal rate of approximately 9 μg TCE d^{-1} bioreactor^{-1}. Xylene vapors were also scrubbed from air using gas phase bioreactors. At a feed rate of 140 μg of xylene min^{-1}, approximately 46% of the xylene was mineralized to carbon dioxide in a single pass through a bench scale gas phase bioreactor.

The biological processing of environmentally significant gases and vapors such as volatile halocarbons and hydrocarbons is becoming a topic of ever-increasing importance. Unacceptably slow degradation rates are a significant problem in the bioprocessing of these types of compounds due, at least in part, to the gas/vapor-in-water solubility limitations inherent in conventional aerated liquid phase bioreactors. Fixed thin film, gas/vapor phase bioreactors designed specifically for the bioprocessing of gases and vapors offer a potential means to partially combat this gas-in-liquid solubility problem. This chapter discusses results obtained with

0097–6156/93/0518–0411$06.00/0

prototype laboratory scale gas/vapor phase bioreactors used for bioprocessing three different gases/vapors: methane, trichloroethylene (TCE) and xylene.   The methane and TCE were bioprocessed with methanotrophic bacteria, and the xylene with a xylene metabolizing strain of *Pseudomonas putida*.

Background.   Methanotrophic bacteria have been known and studied for the past 85 years (*1*).   During this period, the basic physiological capabilities of these organisms were elucidated with their ability to sequentially oxidize methane in the presence of air (O_2) to carbon dioxide and water being particularly well defined (*2*).

     In recent years increased emphasis has been placed on exploiting the physiological potential of the methanotrophs.   Areas of interest include bioconversion of methane to alternate and potentially valuable products such as single cell protein, methyl ketones, etc. (*3*), control of methane emissions resulting from coal mining (*4*) (Apel, W. A.; Dugan, P. R.; Wiebe, M. R; "Influence of Kaolin on Methane Oxidation by *Methylomonas methanica* in Gas Phase Bioreactors" Fuel, in press), and degradation of environmentally significant low molecular weight halocarbons like trichloroethylene (*5-9*).   As a result of these interests, development of bioreactor systems allowing more efficient methanotrophic conversion of gases and vapors are of considerable relevance.

     Traditionally, production of the large amounts of methanotrophic bacteria required for the above applications has been accomplished by growing the organisms on methane/air mixtures added to liquid cultures (*10,11*).   An inherent limitation of this method is the relatively limited solubility of methane and air in the liquid phase so that these gases are not readily available to the bacteria in the quantities necessary to sustain maximum reaction rates.   Various techniques have been employed to combat this problem including mechanical agitation and sparging of methane/oxygen or methane/air mixtures through the cultures in an attempt to continuously saturate the liquid culture medium with the necessary gases.

     One approach to increasing gas delivery to the methanotrophic bacteria is culturing the organisms as a thin biofilm on inert supports suspended in a gas or vapor phase.   In such a system, gas and/or vapor availability to the cells can be increased.   As such, it should be theoretically possible to increase methanotrophic gas and vapor removal rates by making the necessary gases or vapors more readily available to the organisms.

     While this approach is particularly well suited to methanotrophs due to their use of a gas (i.e. methane) as a carbon and energy source, the same approach should be applicable to other bacteria which are capable of bioconverting other gases and vapors.   The biodegradation of volatile, higher molecular weight organic compounds such as those in the BTEX group (benzene, toluene, ethylbenzene and xylene) are of particular interest due to their association with hydrocarbon fuel spills which are often, at least in part, remediated biologically.   Certain Gram (-) heterotrophic bacteria are useful in the conversion of BTEX contaminants since they have been shown to be resistant to high concentrations of compounds like xylene and capable of metabolizing these compounds as their sole carbon and energy source (*12*).   The catabolic processes by which these types of organisms process aromatic hydrocarbons vary but usually involve the oxidation of the aromatic substrate to a diol such as catechol followed by cleavage to a diacid such as *cis, cis* muconic acid (*13*). The diacid is then further metabolized, ultimately to carbon dioxide and water.

This paper reports the results from a series of experiments utilizing fixed, thin film, gas phase bioreactors for (a) culturing methanotrophic bacteria and comparing rates of methanotrophic methane removal in gas phase bioreactors vs. conventional shaken and sparged liquid cultures, (b) evaluating the feasibility of using the methanotrophic gas phase bioreactors for TCE degradation, and (c) the degradation of xylene by *Pseudomonas putida*.

Materials and Methods

Culture Techniques. The *Methylomonas methanica* 739 used in these experiments was provided by the Ohio State University Department of Microbiology (Columbus, OH, U.S.A.) while the *Methylosinus trichosporium* OB3b isolate used was from the Oak Ridge National Laboratory culture collection (Oak Ridge, TN, U.S.A). Both cultures were maintained in CM salts medium as previously described (*4*).

The strain of *Pseudomonas putida* employed for the xylene degradation studies was isolated by two of the authors (J.H.W. and R.D.R.) as previously reported (*12*). Cultures were grown on a mineral salts medium consisting of the following components per liter: 0.7g KH_2PO_4, 0.5g $(NH_4)_2SO_4$, 0.3g $MgSO_4$ with traces of $FeSO_4 \cdot 7H_2O$, $MnCl_2 \cdot H_2O$ and $NaMoO_4 \cdot 7H_2O$. This mineral salts medium was aseptically dispensed into 50 ml aliquots into sterile 125 ml serum bottles which were sealed with teflon coated rubber septa. The bottles were gassed with air which had been saturated with xylene. Xylene was the organism's sole carbon and energy source. Incubations were performed at $22\pm2°C$ on a gyratory shaker at 200 rpm. Cultures were transferred at least biweekly to maintain viability.

Liquid Culture Studies. Liquid culture methane depletion studies were performed using *M. methanica* at stationary phase in 50 ml of CM salts contained in 125 ml serum bottles sealed with teflon coated septa. The bottles were gassed with various concentrations of methane-in-air ranging from 5 to 40% (v/v) and incubated in inverted position at $22\pm2°C$ on a gyratory shaker set at 120 rpm. Head space gas samples were periodically removed and analyzed for methane, oxygen and carbon dioxide as described below.

In a similar manner, methane depletion studies were also performed in liquid cultures contained in columns exactly like those used for the methanotrophic gas phase bioreactor studies (see description below) except that the Pall Rings were eliminated from the column and the column was filled with CM salts medium to a volume equal to that displaced by the Pall Rings in the gas phase bioreactor studies. Methane depletion rates were determined with a variety of methane/air mixtures ranging from 2 to 46% (v/v). In these studies, the methane-in-air mixtures were recirculated in an up flow direction through the column at 200 ml min^{-1}. The bioreactor was incubated at $22\pm2°C$. Head space gas samples from the liquid bioreactor were periodically analyzed for methane, oxygen and carbon dioxide as described below. These experiments were performed to give a direct comparison of methane removal rates by methanotrophic bacteria in similarly configured sparged liquid and gas phase bioreactors.

Gas Phase Bioreactors. The gas phase bioreactors used for experiments with methanotrophic bacteria were constructed by filling a 7.62 X 76.2 cm glass column with 1.6 cm (5/8 in) Pall Rings (Norton Chemical Process Products Division, Akron, Ohio) and sealing the open end with a rubber stopper (Fig. 1). The Pall Rings served as inert supports for the growth of thin layers of biofilm composed of the methanotrophic bacteria. The seal was further secured by over wrapping the boundary area between the stopper and the column with parafilm. A closed loop for gas recirculation through the bioreactor was constructed using flexible 0.4 cm o.d. teflon tubing connected to the upper and lower ends of the bioreactor. Included in the loop was a 1 l Erlenmeyer flask to increase the gas volume of the system and a peristaltic pump to recirculate the gas. The gas was circulated in an up flow direction through the bioreactor at a rate of 200 ml min^{-1}. Approximately 100 ml of CM salts medium was maintained in the base of the bioreactor to humidify the recirculating gas mixture. Total system volume was measured by water displacement and found to be 4.5 l. It should be noted that unlike the xylene studies reported below, all methanotrophic experiments were conducted as closed loop, batch bioreactor studies.

The gas phase bioreactors to be used in methane depletion experiments were inoculated with *M. methanica* and incubated as previously described (4) until the bioreactors reached steady state for methane-oxygen uptake and carbon dioxide evolution. This occurred approximately 6 weeks after inoculation at a methane uptake rate of 40 mg methane hr^{-1} when feeding the bioreactors a 30% methane-in-air gas mixture. This rate was maintained for several months by draining the humidification heel of CM salts medium from the base of the bioreactors and pouring 100 ml of fresh, sterile, CM salts medium over the supports at approximately 1 month intervals. The fresh CM salts solution was allowed to collect as a new humidification heel in the bottom of the reactor. Methanotrophic bacteria grew in this heel, but studies demonstrated no detectable methane uptake could be attributed to these organisms vs. those growing on the supports in the gas phase. With the exception the volume occupied by this heel, the void volume of the reactor was maintained in the gas phase. All methane uptake rate studies were performed in the steady-state gas phase bioreactors at 22 $\pm$ 2°C. For methane depletion studies both the liquid cultures and the gas phase bioreactors were charged with various concentrations of methane in air. Methane, oxygen, and carbon dioxide levels were monitored in the gas phase bioreactors using the methodology described below. Rates of methane uptake per unit of biomass per unit time were calculated from these data.

Gas phase bioreactors used for TCE degradation studies were prepared as indicated above except they were inoculated with *M. trichosporium* instead of *M. methanica*. This change in microorganism was necessary since previous screening studies had shown essentially no TCE degradation by *M. methanica* 739

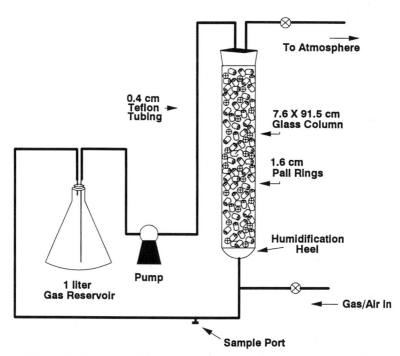

Figure 1. Schematic diagram of gas phase bioreactor configured for bioprocessing with methanotrophic bacteria.

and excellent rates of TCE degradation by *M. trichosporium* OB3b. Copper was deleted from the CM salts medium used. Researchers have shown that the presence of copper inhibits the expression of soluble methane monooxygenase, the enzyme believed to be responsible for TCE degradation by *M. trichosporium* (*14*). For each experiment, the bioreactor was flooded with 5% methane in air (v/v) with TCE added as a vapor to the recirculating gas stream. A second uninoculated bioreactor was run in parallel to the inoculated bioreactor to serve as an abiotic control to account for TCE loss through adsorption, etc. Abiotic control bioreactors used in all experiments described in this paper consisted of bioreactors structurely identical to the inoculated bioreactors. The abiotic bioreactors were steam steriled by autoclaving for 30 min. at 240 p.s.i. and all gas and liquids introduced into the abiotic control were filter steriled directly prior to introduction. Gas samples were removed from the bioreactors, and TCE, methane, oxygen, and carbon dioxide concentrations were analyzed as described below.

The gas phase bioreactors for xylene degradation by *Pseudomonas putida* were constructed in a manner similar to that described above except that the bioreactors were configured for a single pass of the gas/vapor phase (i.e. as a plug flow reactor) and a counter current basal salts medium feed was established (Figure 2). Specifically, the bioreactor consisted of a 7.6 X 91.5 cm glass process pipe column with 5.1 X 7.6 cm glass reducers fitted to each end via stainless steel clamps with teflon seals. Custom machined, teflon plugs were attached to the small ends of each reducer using standard process pipe stainless steel clamps with teflon seals. These plugs were drilled and tapped to accept tubing adapters. Both gases and liquids were transported through 0.4 cm o.d. teflon tubing. The column was filled with 1.6 cm (5/8 in) polypropylene Pall Rings. The column was sterilized and inoculated with *P. putida* by slowly dripping a pregrown stationary phase culture of *P. putida* over the supports for a period of approximately one week. A xylene/air mixture (approximately 2000 μg *p*-xylene min^{-1} at a gas/vapor flow of 55 ml min^{-1}) was then fed in an up-flow direction through the bioreactor for an additional two weeks. The entire void volume of the bioreactor was maintained in a gas phase. During experimental operation, the column was fed with a gas stream consisting of air that had been sparged through a *p*-xylene reservoir. The *p*-xylene carried by the air stream was the sole carbon source entering the bioreactor. Gas samples were removed from sampling ports as noted in Figure 2. These samples were analyzed for xylene, carbon dioxide and oxygen concentrations as described below. In addition, samples were removed from the liquid effluent stream exiting the bioreactor and analyzed for xylene, total inorganic carbon (TIC), and total organic carbon (TOC) as described below. The flow rates of the gas stream entering and exiting the bioreactor as well as the flow rate of the liquid effluent stream were determined at the time samples were withdrawn for chemical analyses. The flow rates together with the results from the chemical analyses were used to calculate carbon mass balances through the bioreactor.

Biomass development in the gas phase bioreactors was periodically monitored by weighing the gas phase bioreactors and comparing changes in weight to an original tare weight established for each of the bioreactors immediately after assembly. The weight of each bioreactor was then subtracted from the original

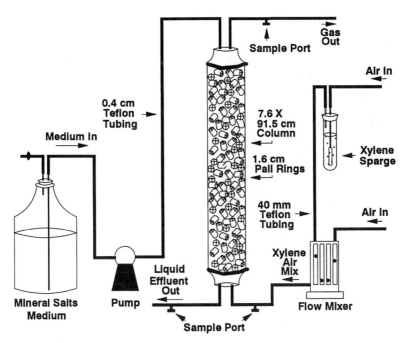

Figure 2. Schematic diagram of gas phase bioreactor configured for single pass bioprocessing of xylene vapors.

tare weight to determine a wet weight value for biomass in the bioreactor. Several differnt samplings of wet biomass were withdrawn from the bioreactors and dried to establish a conversion factor between wet weight biomass and dry weight biomass. This conversion factor was then used to calculate dry biomass weights from the wet weight measurements.

Analytical Methods. Gas levels (methane, oxygen, and carbon dioxide) in the serum bottle cultures and the bioreactors were analyzed using a Gow-Mac Series 550P gas chromatograph equipped with a thermal conductivity detector and an Alltech CTRI column. The gas chromatograph was connected to a Hewlett Packard model 3390A integrator. Samples consisted of 600 μl gas volumes manually injected with a gas tight syringe into the gas chromatograph. Helium was the carrier gas at a flow rate of 60 ml per minute. The gas chromatograph was operated under isocratic conditions at 30° C.

TCE was measured by removing 100 μl samples from the bioreactor which were analyzed using a Hewlett Packard 5890A gas chromatograph (Avondale, PA) equipped with an electron capture detector and an Alltech 624 non-packed column. The detector temperature was 260°C and the injection temperature was 225°C. The carrier gas was helium at 5 ml min^{-1} and the auxiliary gas was nitrogen at 65 ml min^{-1}.

Xylene vapors were analyzed by removing samples from the gas inlet sampling port and the gas outlet sampling port of the bioreactor using a 100 μl gas tight syringe. The xylene was quantified by injecting the 100 μl sample into a Sigma 4B Perkin-Elmer gas chromatograph (Norwalk, CT) equipped with a flame ionization detector and a Supelco SP-1500 Carbopak B column. The carrier gas was helium at a flow rate of 33 ml min^{-1}. The oven temperature was 220°C. The air flow rates through the bioreactor were determined at the sampling points using a bubble flow meter. The flow rates together with the results from the chromatographic analyses were used to calculate flow per unit time of xylene contained in the gas streams entering and exiting the bioreactor.

Xylene was analyzed in the liquid effluent by taking samples from the effluent stream sampling port at the base of the bioreactor. The samples were placed in 2 ml serum vials containing 0.05 ml of 2% $HgCl_2$. The vials were then sealed with a teflon lined septum. These samples were analyzed using the same method as that described for the xylene samples in the vapor phase except a 1 ul sample was injected into the gas chromatograph. The liquid flow rate of the effluent from the column was measured at the time of sampling and used in conjunction with the gas chromatography analyses to calculate the flow per unit time of xylene in the liquid effluent from the bioreactor.

The liquid effluent samples were also analyzed for carbon dioxide and other forms of carbon on a Model 700 TIC/OC Analyzer (O-I Corporation, College Town, TX). A 500 μl sample was introduced by syringe injection. A 200 μl volume of a 5% (v/v) solution of phosphoric acid and 500 μl of a 100 g l^{-1} solution of sodium persulfate (Fisher Scientific) were added to each sample. Potassium biphthalate (Fisher Scientific) was used as the organic standard.

Results and Discussion

Methane removal data over a range of methane concentrations by *M. methanica* in the shaken serum bottles, the liquid bioreactor and the gas phase bioreactor are represented in Figures 3-5, respectively. All methane removal experiments were conducted with the methanotrophic cultures in stationary growth phase, and the data are reported as methane contained in the gas phase of the bottles or the bioreactors at the time in question. The data in Figures 3-5 are on a per bottle or bioreactor basis and represent gross methane removal unadjusted for biomass densities.

The data contained in Figures 3-5 are summarized in Figure 6 in terms of methane removal rates per unit time per g dry wt of biomass over a number of initial methane concentrations ranging from 1% to 53% methane-in-air for each of the three test systems. The shaken serum bottles exhibited relatively constant methane removal rates of approximately 3.8 mg methane hr^{-1} (g dry wt of biomass)$^{-1}$ at initial concentrations of methane-in-air up to approximately 40%. Above 40%, methane removal rates appeared to decrease as initial methane concentration increased.

In general, methane removal rates by *M. methanica* in the sparged liquid bioreactor increased as a function of increasing methane concentration in air to a maximum removal rate of 5 mg methane hr^{-1} (g dry wt biomass)$^{-1}$. This maximum occurred at a 22.5% initial methane-in-air concentration. At higher methane-in-air concentrations the removal rates appeared to decrease as initial methane concentrations increased.

The results obtained with the gas phase bioreactor tended to parallel those noted with the sparged liquid bioreactor. Methane removal rates increased as a function of increasing initial methane concentrations up to a 28% methane-in-air concentration. At this point the methane removal rate was approximately 8 mg methane hr^{-1} (g dry wt biomass)$^{-1}$.

The observed increase in methane removal rates as a function of increasing initial methane concentration was anticipated, and was almost certainly the result of two interrelated factors: (1) the kinetics of the methanotrophic enzyme system, and (2) transport of the substrate across the phase boundary and diffusion through the biofilm to the methanotrophic cells. Until the methanotrophic enzyme system is saturated with substrate, in this case methane, the system would be expected to follow first-order reaction kinetics with rates of reaction increasing as substrate concentration increases. At levels equal to or greater than enzyme saturation, the degradation kinetics would be expected to become zero-order with the degradation rate becoming independent of substrate concentration. Likewise, methane concentration in the gas phase influences the physical availability of methane to the cells in the biofilm. At gas phase levels below those sufficient to saturate the biofilm, transport of the methane across the boundary layer and further diffusion of the methane through the biofilm to the methanotrophic cells could be the rate limiting step. At gas phase methane concentrations equal to, or greater than the biofilm saturation point (i.e. the critical concentration), the biofilm would be fully

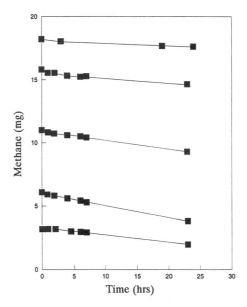

Figure 3. Removal of methane at various initial concentrations by *Methylomonas methanica* grown in serum bottles at $22\pm2°$ C. (125 ml serum bottles containing 50 ml medium, methane measured as gas in head space)

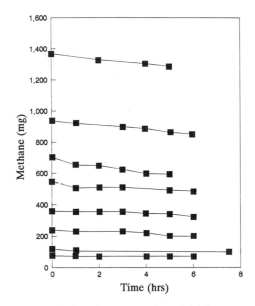

Figure 4. Removal of methane at various initial concentrations by *Methylomonas methanica* grown in liquid phase bioreactors at $22\pm2°$ C.

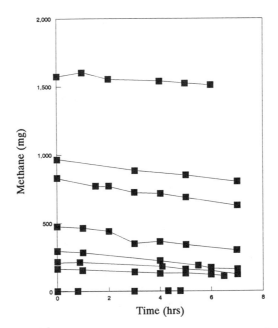

Figure 5. Removal of methane at various initial concentrations by *Methylomonas methanica* grown in gas phase bioreactors at 22±2° C.

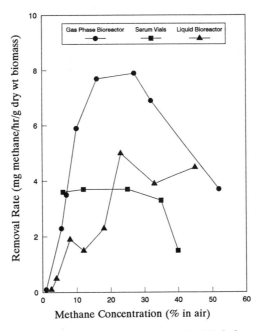

Figure 6. Comparison of methane removal rates by *Methylomonas methanica* grown in recirculating gas phase bioreactors, liquid bioreactors, and shaken serum vials at 22±2° C.

saturated and the metabolic activity of the cells in the biofilm would be the rate limiting step.

As initial methane concentrations further increased above 30%, methane removal rates fell precipitously to a rate of 3.9 mg methane hr^{-1} (g dry wt biomass)$^{-1}$ at 52% initial concentration of methane in air. This decrease in methane removal rates at higher initial methane concentrations may be a result of oxygen limitation. Since all experiments reported above were conducted with various concentrations of methane in air, oxygen levels contained in the bioreactors necessarily decreased as methane levels increased. For example, from Figure 7 it can be determined that at a 52% methane-in-air, the highest concentration tested, the oxygen concentration in the gas phase is only 10%. Since oxygen is an integral substrate in the methanotrophic oxidation of methane, oxygen may become the limiting substrate at higher methane-in-air concentrations leading to the decrease in methane oxidation rates noted in Figure 6. Thus, it is possible if the methane feed to these bioreactors where supplemented with additional oxygen, the maximum methane oxidation rates observed might be significantly increased both by driving the enzyme kinetics of the system towards zero-order and by fully saturating the biofilm. Supplemental oxygen feeds were not used in the experiments reported since there was a specific interest in exploring industrial processing of methane based exclusively on methane in air feed streams. It is apparent, however, that to fully assess the maximum rates of methane removal by methanotrophic bioreactors at higher methane feed concentrations, additional experimentation is needed with oxygen supplemented feed streams where bisubstrate kinetics and phase transport parameters with both methane and oxygen are considered as an integral part of the experimental design.

In net, within the context of the experimental procedures reported here, the gas phase bioreactors exhibited a notable advantage in maximum methane removal rates versus the more conventional shaken liquid cultures and the sparged liquid bioreactor. This advantage was 2.1 and 1.6 fold, respectively. These results indicate that the bioprocessing rates of gases and vapors with relatively low aqueous solubility may be increased by the use of gas phase bioreactors.

Preliminary experiments have also shown the processing of TCE vapors to be feasible using gas phase bioreactors. Figure 8 shows typical TCE removal rates using a gas phase bioreactor identical to those used for methane removal studies with the exception that a biofilm of *M. trichosporium* OB3b was employed in place of *M. methanica*. All data shown in Figure 8 were corrected for abiotic adsorption of TCE, and as such, represent bacterial degradation. Best removal rates were obtained between days 2 and 5 with TCE removal averaging approximately 9 μg d^{-1} bioreactor^{-1} during this period. It was suspected that the relatively long lag phase was a product of excessive initial methane feed concentrations since it is known that while methane serves as the source of reducing power for the generation of the NADH necessary for methanotrophic methane monooxygenase (MMO) to catalyze the degradation of TCE, it at the same time competitively inhibits TCE oxidation by the enzyme. A more successful strategy for TCE degradation using methanotrophic gas phase bioreactors may be to use a continuous, plug flow bioreactor design with very low levels of methane being fed in conjunction with the

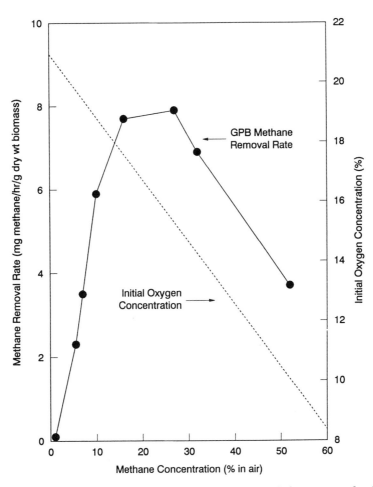

Figure 7. Comparison of methane removal rates by *Methylomonas methanica* grown in recirculating gas phase bioreactors versus initial oxygen concentrations.

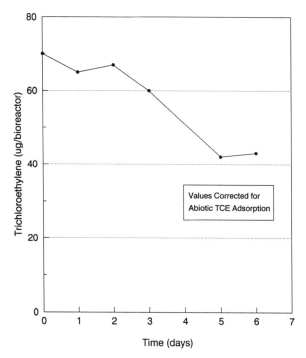

Figure 8. Trichloroethylene removal by *Methylosinus trichosporium* OB3b cultured in a recirculating gas phase bioreactor.

TCE vapors. If competitive inhibition of the MMO by methane is a factor in lowering TCE degradation rates, this may be a viable approach to lessen the inhibition and increase TCE degradation rates.

Experiments with *p*-xylene indicated that the gas phase bioreactor is capable of removing virtually all the xylene from a gas feed stream as noted in Figure 9. This figure shows a mass balance based on carbon for a gas phase bioreactor containing a stationary culture of *Pseudomonas putida*. The gas feed stream rate was 16 ml min^{-1} with a xylene feed in the gas stream of 139.9 μg min^{-1} based on carbon. The liquid medium feed rate was 1.7 ml min^{-1} and the reactor temperature was 22$\pm$2°C. Only 1.2 μg min^{-1} xylene (based on carbon) was detected in the off gas stream exiting the bioreactor, however significant partitioning of the xylene to the liquid phase in the bioreactor did occur with 74.8 μg min^{-1} xylene (based on carbon) being detected in the liquid effluent stream exiting the reactor. Only 1.2 μg min^{-1} of xylene carbon was not accounted for in the mass balance shown in Figure 9 with the remainder of the carbon being either mineralized to carbon dioxide or unidentified non-xylene organic carbon. In net, the degradation rate of the bioreactor operated under the conditions specified above was 46%. It may be possible to recirculate the liquid effluent back through the bioreactor, aerate the liquid effluent, decrease the flow rate through the column, or increase the column length to approach 100% degradation.

Based on the information conveyed in Figure 9 it is possible to calculate a volumetric productivity value for the bioreactor which can be a useful overall gauge of bioreactor conversion efficacy as well as serving as a basis for sizing scaled up reactors. The volumetric productivity value in this instance could be defined as the mass of xylene fed per unit time per unit bioreactor volume times the degradation factor (i.e. 0.46). This calculation reveals that the volumetric productivity value for the gas phase bioreactor operated under the above described conditions is 20.6 μg min^{-1} l^{-1} for *p*-xylene. This productivity value may be increased significantly by developing a secondary treatment for xylene contained in the liquid effluent as discussed above.

In conclusion, the gas phase bioreactors of the type employed in these studies appear to offer significant potential for the enhanced bioprocessing of gases and vapors. This view is supported by the data presented in this paper as well as by considerable practical experience being gained in biofiltration with soil and compost-based bioreactors (*15*), which in reality, are little more than thin film, plug flow, gas phase bioreactors. Clearly, however, more research needs to be performed in this area. A question of particular relevance that remains to be answered concerns applicability of gas phase bioreactors vs. substrate solubility. At some point, a trade-off must exist in which a gas or vapor is sufficiently water soluble that it is better processed using a conventional liquid phase bioreactor. This critical solubility point is yet to be defined.

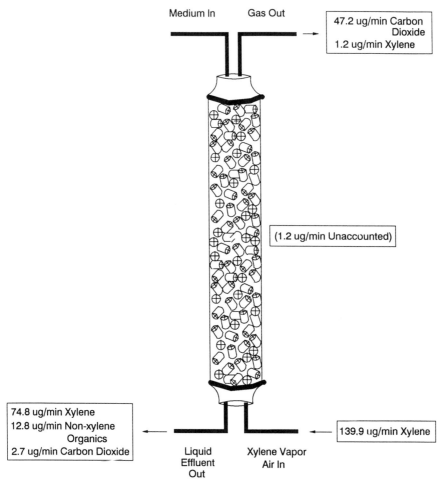

Figure 9. Mass balance of gas phase bioreactor processing xylene. All values shown as mass of carbon per unit time. Conditions were as follows: gas flow rate - 16 ml min⁻¹, liquid effluent flow rate - 1.7 ml min⁻¹, incubation temperature - 22±2° C, organism - *Pseudomonas putida*.

Acknowledgments

This work was supported under contract number DE-AC07-76IDO1570 for the U.S. Department of Energy to the Idaho National Engineering Laboratory/EG&G Idaho, Inc.

Literature Cited

1. Söhngen, N.L.; "Ueber Bakterien, Welch Methan als Kahlenstoffnahrung und Energiequelle Gebrauchen"; *Zentr. Bakt. Parasitenk.* **1906**, 15, pp. 513-517.
2. Haber, C.L.; Allen. L.N.; Zhao, S.; Hanson, R.S.; "Methylotrophic Bacteria:Biochemical Diversity and Genetics"; *Sci* **1983**, 221, pp. 1147-1153.
3. Hou, C.T.; Patel, R.; Laskin, A.I.; Barnabe, N.; Marczak, I.; "Microbial Oxidation of Gaseous Hydrocarbons: Production of Methyl Ketones from Their Corresponding Secondary Alcohols by Methane- and Methanol-Grown Cells"; *Appl. Environ. Microbiol.* **1979**, 38, pp. 135-142.
4. Apel, W.A.; Dugan, P.R.; Wiebe, M.R.; "Use of Methanotrophic Bacteria in Gas Phase Bioreactors to Abate Methane in Coal Mine Atmospheres"; *Fuel* **1991**, 70, pp. 1001-1003.
5. Dugan, P.R.; Apel, W.A.; Wiebe, M.R.; "Degradation of Trichloroethylene by Methane Oxidizing Bacteria Grown in Liquid or Gas/Vapor Phase"; In *Bioprocess Engineering Symposium 1990*; Hochmuth, R.M.; The American Society of Mechanical Engineers, New York, NY, **1990**, Vol. 16, pp. 55-59.
6. Phelps, T.J.; Niedzielski, J.J.; Schram, R.M.; Herbes, S.E.; White, D.C.; "Biodegradation of Trichloroethylene in Continuous-Recycle Expanded-Bed Bioreactors"; *Appl. Environ. Microbiol.* **1990**, 56, pp. 1702-1709.
7. Tsien, H.C.; Brusseau, G.A.; Hanson, R.S.; Wackett, L.P.; (1989) Biodegradation of Trichloroethylene by *Methylosinus trichosporium* OB3b. *Appl. Environ. Microbiol.* **1989**, 55, pp. 3155-3161.
8. Uchiyama, H.; Nakajima, T.; Yagi, O.; Tabuchi, T.; "Aerobic Degradation of Trichloroethylene by a New Type II Methane-Utilizing Bacterium, Strain M"; *Agric. Biol. Chem.* **1989**, 53, pp. 2903-2907.
9. Wilson, J.T.; Wilson, B.H.; "Biotransformation of Trichloroethylene in Soil"; *Appl. Environ. Microbiol.* **1985**, 49, pp. 242-243.
10. Whittenbury, R.; Dalton, H.; "The Methylotrophic Bacteria"; In *The Procaryotes. A Handbook on the Habitats, Isolation, and Identification of Bacteria.* Starr, M.P.; Stolp H.; Trüper, H.G.; Balows, A.; Schlegel, H.G.; Springer-Verlag, Berlin, Germany, **1981**. Chapt. 71, pp. 892-902
11. Whittenbury, R.; Phillips, K.C.; Wilkinson, J.F.; "Enrichment, Isolation and Some Properties of Methane-Utilizing Bacteria"; *J. Gen. Microbiol.* **1970**, 61, pp. 205-218.

12. Wolfram, J.H.; Rogers R.D.; Higdem, D.M.; Nowers, D.A.; "Continuous Biodegradation of Waste Xylene"; In *Bioprocess Engineering Symposium 1990*; Hochmuth, R.M.; The American Society of Mechanical Engineers, New York, NY, **1990**, Vol 16, pp 61-67.
13. Atlas, R.M.; "Microbial Degradation of Petroleum Hydrocarbons: an Environmental Perspective"; *Microbiol. Rev.* **1981**, 45, pp. 180-209.
14. Oldenhuis, R.; Vink, R.L.J.M.; Janssen, D.B.; Witholt, B.; "Degradation of Chlorinated Aliphatic Hydrocarbons by *Methylosinus trichosporium* OB3b Expressing Soluble Methane Monooxygenase"; *Appl. Environ. Microbiol.* **1989**, 55 pp. 2819-2826.
15. Leson, G.; Winer, A.M.; "Biofiltration: an Innovative Air Pollution Control Technology for VOC Emissions"; *J. Air. Manage. Assoc.* **1991**, 41, pp. 1045-1054.

RECEIVED October 1, 1992

SELECTED MIXED-WASTE
TREATMENT APPLICATIONS

Chapter 21

Mediated Electrochemical Process for Hazardous Waste Destruction

R. G. Hickman, J. C. Farmer, and F. T. Wang

Lawrence Livermore National Laboratory, Livermore, CA 94550

There are few permitted processes for mixed waste (radioactive plus chemically hazardous) treatment. We are developing an electrochemical process, based upon mediated electrochemical oxidation (MEO), that converts toxic organic components of mixed waste to water, carbon dioxide, and chloride or chloride precipitates. Aggressive oxidizer ions such as Ag^{2+}, Co^{3+}, or Fe^{3+} are produced at an anode. These can attack organic molecules directly, and may also produce hydroxyl free radicals that promote destruction. Solid and liquid radioactive waste streams containing only inorganic radionuclide forms may be treated with existing technology and prepared for final disposal. The coulombic efficiency of the process has been determined, as well as the destruction efficiency for ethylene glycol, a surrogate waste.

In addition, hazardous organic materials are becoming very expensive to dispose of and when they are combined with transuranic radioactive elements no processes are presently permitted.

Mediated electrochemical oxidation is an ambient-temperature aqueous-phase process that can be used to oxidize organic components of mixed wastes. Problems associated with incineration, such as high-temperature volatilization of radionuclides, are avoided. Historically, Ag(II) has been used as a mediator in this process. Fe(III) and Co(III) are attractive alternatives to Ag(II) since they form soluble chlorides during the destruction of chlorinated solvents. Furthermore, silver itself is toxic heavy metal. Quantitative data have been obtained for the complete oxidation of ethylene glycol by Fe(III) and Co(III). Though ethylene glycol is a nonhalogenated organic, these data have enabled us to make direct comparisons of activities of Fe(III) and Co(III) with Ag(II). Very good quantitative data for the oxidation of ethylene glycol by Ag(II) had already been collected.

0097–6156/93/0518–0430$06.00/0

An ambient-temperature aqueous process, known as mediated electrochemical oxidation (MEO), for the treatment of hazardous and mixed wastes is being developed. Mixed wastes may contain both toxic organic and radioactive components. MEO can be used to oxidize organic components of mixed wastes, thereby reducing the total volume that has to be transported and stored. Treatment of mixed wastes in a condensed phase at ambient temperature prevents high-temperature volatilization of radionuclides particularly as oxyhydroxides, a problem that may be encountered during incineration operating upsets.

In previous studies, electrochemically-generated Ag(II) has been used to completely oxidize dissolved organics such as ethylene glycol and benzene. (*1–4*) Ag(II) is a very effective oxidizing agent for the destruction of nonhalogenated organic compounds. Unfortunately, halide ions liberated during the destruction of halogenated organics react with Ag(II) to form insoluble precipitates. Therefore, it has been necessary to investigate other mediators (powerful oxidizing agents) that are tolerant of halide anions, i.e., their chlorides are soluble. Fe(III) and Co(III) satisfy this criterion.

The mediators are regenerated in a compartmented electrochemical cell in which the anolyte and catholyte are prevented from mixing by an ion permeable membrane. At the cathode, nitrate ion is reduced to nitrous ion and nitrogen oxides. In a large system these reaction products would be run through an air contactor to regenerate the nitrate ion and avoid unwanted emissions to the air. At the anode, the mediator is raised to its higher valence state. The anode potential required for this oxidation should be as high as possible without oxidizing the water. Anodic water oxidation would result in oxygen gas evolution with an increase in H^+, and a corresponding loss in current efficiency. Ag(II) is very close to this threshold potential and it will homogeneously attack water, particularly at temperatures above 50°C. Attack on water by the mediating ion is essential for the destruction of organics, however. For example, the overall homogeneous reaction in the anolyte for destruction of ethylene glycol by Ag(II) may be written

$$10Ag^{++} + (CH_2OH_2)_2 + 2H_2O \longrightarrow 10Ag^+ + 2CO_2 + 10H^+ \ .$$

It is immediately obvious that the source of some of the oxygen required to produce CO_2 is the water. More specifically, hydroxy free radicals are probably generated by reaction of the mediator with water. It is believed that ethylene glycol is converted to carbon dioxide via two intermediates, formaldehyde and formic acid. (*1–2*) The Ag(II) ion is actually present as a well-known nitrate complex. This mechanism is shown schematically in Figure 1. A comparison of the time dependent destruction predicted by this mechanism with that actually observed is shown in Figure 2.

As a step in the on-going investigation, we have studied the MEO of ethylene glycol by Fe(III) and Co(III). Quantitative data already exist for the oxidation of ethylene glycol by Ag(II) and can be used as a basis of comparison. The reactor had a rotating-cylinder anode that was operated well below the limiting current for mediator generation (the current at which O_2 would be evolved). Rates of carbon dioxide evolution were measured. Gas chromatography with mass spectrometry (GC/MS) was used to determine reaction intermediates.

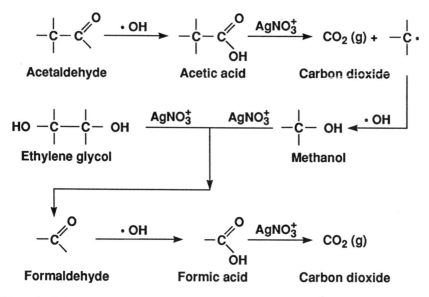

Figure 1. Schematic mechanism of aliphatic destruction by Ag^{+2}.

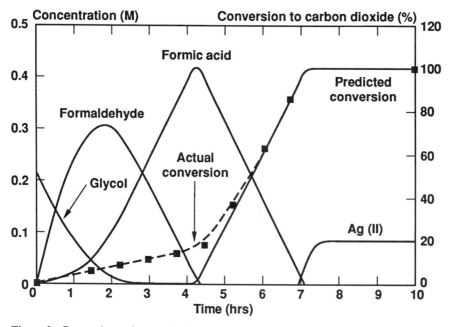

Figure 2. Comparison of numerical model predictions with experimental CO_2 evolution data for Ag(II).

Measurements with Co and Fe Mediators

Ethylene glycol was oxidized to carbon dioxide at near-ambient temperature in an electrochemical reactor charged with 40 ml of anolyte. Three different anolyte compositions were explored: (a) 0.5 M Co(NO$_3$)$_2$, 3.25 M HNO$_3$, and 0.22 M ethylene glycol; (b) 0.5 M Fe(NO$_3$)$_3$, 3.25 M HNO$_3$, and 0.22 M ethylene glycol; and (c) 0.5 M Fe(NO$_3$)$_2$, 8 M HNO$_3$, and 0.22 M ethylene glycol. The anode was a platinum rotating-cylinder having a diameter of 1.2 cm and a length of 1.78 cm. The rotation speed of the anode was maintained at 1500 rpm. The cell current was maintained well below the limiting current for mediator generation, approximately 673 mA. Elevations of the anolyte temperature above ambient (38–64°C) were due to ohmic heating in the cell. A stationary platinum cathode was separated from the anode by a perfluorinated sulfonic acid action-exchange membrane (Nafion 117). Reactions were conducted in a closed vessel so that all carbon dioxide could be captured. The volume of this container was 38.2 liters, which corresponds to approximately 1.5 moles of gas at ambient temperature and pressure. Carbon dioxide concentrations in the gas phase were periodically determined by mass spectrometry and used to calculate percentage conversion of the organic substrate.

Cobalt System. Measurements of carbon dioxide evolution during mediated electrochemical oxidation of ethylene glycol by Co(III) were made during two experiments and are shown in Figures 3a and 3b. Calculated conversions ranged from 90 to 120%. The uncertainty may be due to errors in calibration of the mass spectrometer that was used to measure gas-phase concentrations of carbon dioxide or sampling errors. We conclude that all of the ethylene glycol charged to the reactor was converted to carbon dioxide. If the process were 100% efficient, the ultimate conversion should have occurred after 3.8 hours. In the second experiment, almost 10 hours were required to reach the ultimate conversion (calculated as 120%). This is indicative of relatively poor process efficiency (approximately 40%). In similar experiments with Ag(II), complete conversion was achieved after 3.8 hours, which indicated 100% process efficiency (utilization of cell current).

Anolytes containing Co(NO$_3$)$_2$ are pink or rose in color. In the absence of organics, the anolyte color changes to turquoise or dark blue color after conversion of Co(II) to Co(III) by anodic oxidation. In the presence of organics, the anolyte color remains pink or rose. This observation leads us to believe that Co(III) must be completely consumed by reaction with ethylene glycol in a thin layer of anolyte near the anode.

Some Co(II) diffuses through the Nafion membrane that separates the anolyte and catholyte; the catholyte turns pink during experiments. The concentration of Co(II) at the beginning of the run was determined to be 0.44 M by inductively coupled plasma with mass spectrometry (ICP-MS). Recall that the anolyte was believed to have been 0.5 M Co(II), based upon the weight of Co(NO$_3$)$_2$ added to the electrolyte. The concentration of Co(II) in the anolyte dropped to 0.3 M by the end of the experiment. Apparently, 32% of the Co(II) diffused through the Nafion membrane during 20–25 hours of operation. The concentration of Co(II) in the catholyte was determined to be 0.17 M at the end of the first experiment. At the end of the second experiment, the concentration of Co(II) in the catholyte was determined to be 0.10 M. A very small volume of unidentified immiscible liquid was observed.

In experiments with Co(II)/Co(III), very little NO$_x$ was generated in the cathode compartment. Since NO$_x$ is dark brown, it is easily visible. In contrast, large amounts of NO$_x$ were generated during experiments with both Ag(I)/Ag(II) and Fe(II)/Fe(III).

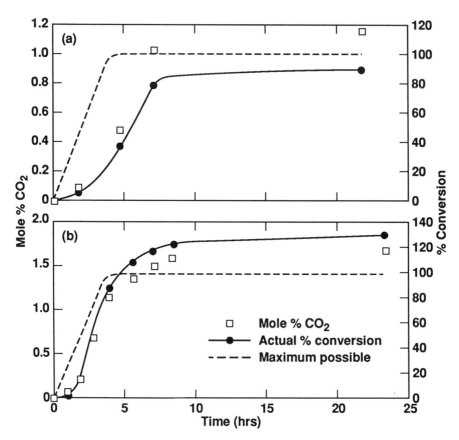

Figure 3. (a) MEO of ethylene glycol by Co(III) at 38°C. (b) This experiment was identical to 3a, except temperature was 42°C.

Iron System. During the first experiment with Fe(III), the anolyte concentration was 0.5 M Fe(NO₃)₂, 3.25 M HNO₃, and 0.22 M ethylene glycol. Measurements of carbon dioxide evolution are shown in Figure 4a. An ultimate conversion of 67–71% was reached after approximately 8 hours of operation. Approximately 3.8 hours would be required at 100% efficiency. Process efficiency appears to be lower with Fe(III) than with Ag(II). Substantial quantities of Fe(II) migrated through the Nafion membrane to the catholyte. During approximately 30 hours of permeation, the concentration of Fe(II) in the anolyte dropped from 0.4 M to 0.17 M (both determined with ICP-MS). The pH of the catholyte approached neutrality during the experiment. Consequently, most of the Fe(II) in the catholyte precipitated as oxide, believed to be Fe₂O₃. Some Fe₃O₄ was found deposited on the cathode. Low efficiency and low ultimate conversion observed during the first experiment with Fe(III) may be due to the loss of Fe(II) from the anolyte.

During the second experiment with Fe(III), the anolyte concentration was 0.5 M Fe(NO₃)₂, 8 M HNO₃, and 0.22 M ethylene glycol. By increasing the acid concentration, migration of Fe(II) was reduced and problems associated with precipitation were

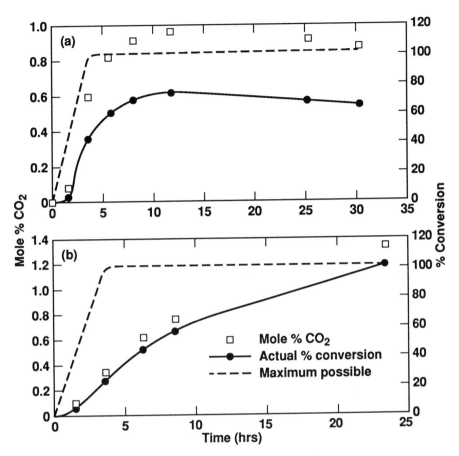

Figure 4. (a)MEO of ethylene glycol by Fe(III) at 673 mA and 38–64°C. (b) This experiment was identical to 4a, except for 8M HNO₃.

eliminated. Measurements of carbon dioxide evolution are shown in Figure 4b. An ultimate conversion of 101% (calculated) was reached after approximately 24 hours of operation. The time required to reach the ultimate conversion is longer at 8 M HNO₃ than at 3.25 M HNO₃. The oxidation of ethylene glycol probably relies on hydroxyl free radical (·OH). By increasing the concentration of acid, the rate of reduction of Fe(III) by water may be reduced. The rate of oxidation of water may also be slowed since the reaction of Fe(III) with water is probably responsible for the formation of ·OH.

No Mediator. Two experiments were conducted without mediator [Fe(III) or Co(III)]. See Figures 5a and 5b. Initial anolyte concentrations were 3.25 M HNO₃ and 0.22 M ethylene glycol. During the first experiment, a gold rotating-cylinder anode was used. After operating 4 hours at 673 mA, a final conversion of 36% was achieved. A steady-state anolyte temperature of 36°C was maintained. In a companion experiment, conditions were

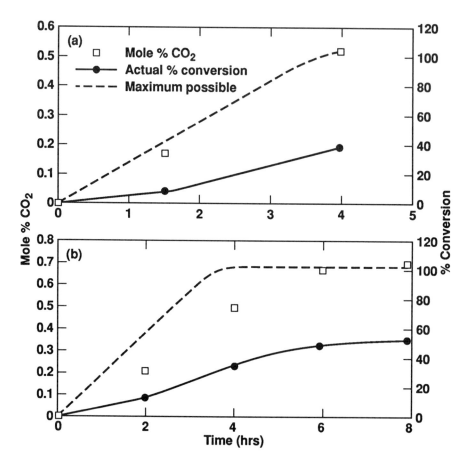

Figure 5. (a) Direct anodic oxidation of ethylene glycol without mediators. (b) This experiment was identical to 5a, except a platinum anode was used.

identical except that a platinum anode was used. Significant corrosion of the platinum anode was not observed. Organics were destroyed by directed anodic oxidation, confirming the work of others.

GC/MS Analyses of Intermediates. Gas Chromatography with Mass Spectrometry (GC/MS) was used in an attempt to identify reaction intermediates in the anolyte. The conversion of ethylene glycol to carbon dioxide is believed to involve both formaldehyde and formic acid as intermediates. GC/MS analyses on sample extracts were performed on a Hewlett-Packard (H-P) 5890/5970 benchtop GC/MS system running on RTE/6, a software package used for data acquisition and spectra library searching. The column was a Stabilwax-DA with an inner diameter of 0.25 mm, a length of 30 m, and a film thickness of 0.5 microns. Ethylene glycol was partially oxidized at 673 mA and 38°C. An anolyte sample without preparation (neat) was injected directly into the instrument. The temperature of the chromatography was ramped from 40°C to 240°C at a rate of 2°C/min. A CHO fragment, due to formaldehyde, was detected. Unfortunately it was obscured by CO_2 and N_2O which elute at 1.8–3.0 minutes as shown in Figure 6a. Formic and acetic acids were detected at 43.52 and 39.79 minutes, respectively as shown in Figure 6b. Clearly, formic acid is a major constituent.

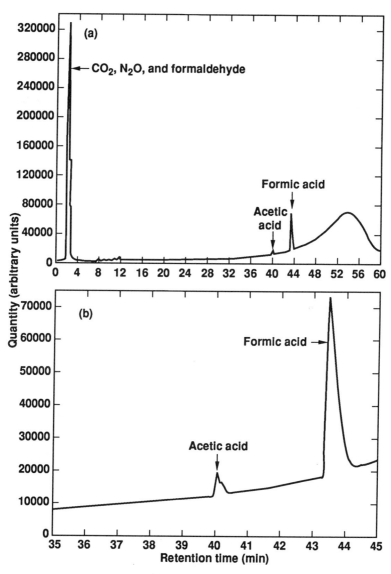

Figure 6. (a) Total ion chromatogram (TIC) of neat anolyte showing species eluting from 0 to 60 minutes. (b) Expanded TIC showing region from 35 to 45 minutes where formic and acetic acids elute.

Summary

Several important preliminary conclusions have been drawn from these experiments. These are enumerated below:

1. The complete conversion of ethylene glycol to carbon dioxide by Fe(III) and Co(III) has been demonstrated. Unfortunately, the efficiency of MEO with Fe(III) and Co(III) appears to be lower than with Ag(II). Better CO_2 evolution data are needed to quantify differences. In a large scale application this would increase costs, but this increase may be more than offset by avoiding the use of silver.
2. Both formic and acetic acids were detected in the anolyte prior to complete conversion of the ethylene glycol to carbon dioxide. Acetic acid was present at a relatively low concentration. Its formation mechanism was not identified, but it was subsequently oxidized. Formic acid is an expected intermediate formed during the MEO of ethylene glycol and was present at a high concentration.
3. The conversion of ethylene glycol to carbon dioxide by anodic oxidation without any mediator was also demonstrated.
4. Gold anodes undergo significant corrosion in the presence of ethylene glycol and nitric acid.
5. The inadvertent presence of Co(II) in the catholyte was found to suppress NO_x generation.
6. Fe(II) migration though Nafion cation exchange membranes was reduced by increasing the concentrations of nitric acid in the anolyte and catholyte.

---

*This work was performed under the auspices of the U.S. Department of Energy by the Lawrence Livermore National Laboratory under contract number W-7405-ENG-48.

References

1. Farmer, J. C., Hickman, R. G., Wang, F. T., Lewis, P. R., and Summers, L. J., "Electrochemical Treatment of Mixed and Hazardous Wastes: Oxidation of Ethylene Glycol by Ag(II)," U. Calif., Lawrence Livermore National Laboratory, UCRL-JC-106947; Rev. 2 (1991).
2. Farmer, J. C., Wang, F. T., Hawley-Fedder, R. A., Lewis, P. R., Summers, L. J., and Foils, L., "Electrochemical Treatment of Mixed and Hazardous Wastes: Oxidation of Ethylene Glycol and Benzene by Ag(II)," U. Calif., Lawrence Livermore National Laboratory, UCRL-JC-107043; Rev. 3 (1991); *J. Electrochem. Soc.* **139**, No. 3, pp. 654–662 (1992).
3. Farmer, J. C., Hickman, R. G., Wang, F. T., and Lewis, P. R., "Mediated Electrochemical Oxidation of Ethylene Glycol: U. Calif., Lawrence Livermore National Laboratory, UCRL-JC-105357 (1990); 179th Electrochem. Soc. Meeting, Washington, DC, May 5–10, 1991, Ext. Abs., Vol. 91-1, pp. 799–800, *Electrochem. Soc.*, Pennington, NJ (1991).
4. Farmer, J. C., Hickman, R. G., Wang, F. T., Lewis, P. R., and Summers, L. J., "Initial Study of the Complete Mediated Electrochemical Oxidation of Ethylene Glycol," U. Calif., Lawrence Livermore National Laboratory, UCRL-LR-106479 (1991).

RECEIVED September 3, 1992

Chapter 22

Long-Term Durability of Polyethylene for Encapsulation of Low-Level Radioactive, Hazardous, and Mixed Wastes

P. D. Kalb, J. H. Heiser, and P. Colombo

Brookhaven National Laboratory, Upton, NY 11973

The durability of polyethylene waste forms for treatment of low-level radioactive, hazardous, and mixed wastes is examined. Specific potential failure mechanisms investigated include biodegradation, radiation, chemical attack, flammability, environmental stress cracking, and photodegradation. These data are supported by results from waste form performance testing including compressive yield strength, water immersion, thermal cycling, leachability of radioactive and hazardous species, irradiation, biodegradation, and flammability. Polyethylene was found to be extremely resistant to each of these potential failure modes under anticipated storage and disposal conditions.

A polyethylene encapsulation process for treatment of low-level radioactive (LLW), hazardous, and mixed wastes has been developed at Brookhaven National Laboratory (BNL). Mixed wastes are LLW that also contain hazardous constituents above allowable limits, as defined by the Environmental Protection Agency (EPA) under the Resource Conservation and Recovery Act. The system has successfully undergone bench-scale development/testing and a production-scale technology demonstration of this process is planned. Polyethylene encapsulation has been applied to a wide range of waste types including evaporator concentrate salts, sludges, incinerator ash, and ion exchange resins. (1) This study contains test data from the encapsulation of nitrate salt wastes generated in large volume

0097–6156/93/0518–0439$06.00/0

at several DOE facilities. These salts contain toxic heavy metals, and as such, are defined as mixed wastes.

Polyethylene is a thermoplastic polymer that is heated above its melting point, combined with waste to form a homogeneous mixture, and allowed to cool resulting in a monolithic solid waste form. In contrast to conventional solidification agents, such as hydraulic cement, no chemical reaction is required for solidification. This provides a number of advantages. A wide range of waste streams are compatible with polyethylene since constituents present in the waste will neither inhibit nor accelerate solidification. Processing is simplified since variations in waste composition do not require adjustment of the solidification chemistry. Polyethylene encapsulation results in higher loading efficiencies (i.e., more waste encapsulated per drum) and better waste form performance, compared with hydraulic cements. For example, up to 70 wt% nitrate salt waste can be encapsulated in polyethylene, compared with a maximum of 13 - 20 wt% using portland cement.

The polyethylene encapsulation process uses a single-screw extruder to mix, heat, and extrude the waste-binder mixture. Extrusion is a proven technology that has been used in the plastics industry for over 50 years. Modifications necessary for encapsulation of waste (e.g., use of calibrated dynamic feeders for waste and binder) are described in detail in Reference 1.

Waste Form Durability

Polyethylene was first developed in 1933, but was not widely used until World War II. Little is known about the long-term durability of polyethylene for waste encapsulation since it is a relatively new engineering material and has not been used for this purpose on a production-scale. Waste form durability can directly affect mobility of contaminants in the accessible environment and, ultimately, the health and safety of an exposed population. This paper highlights the results of a literature review and testing program on the properties of polyethylene relevant to the long-term durability of waste forms in storage and disposal. (2) Waste form performance testing included tests required by the Nuclear Regulatory Commission (NRC) in their Branch Technical Position for licensing commercial LLW waste forms in support of 10 CFR 61 (3) and by EPA in 40 CFR 261 for hazardous wastes. (4) Polyethylene waste form test specimens contained 30 - 70 wt% sodium nitrate. Test specimens were fabricated using either a bench-scale screw extruder (32mm diameter screw) or a production-scale extruder (114 mm diameter screw). Additional details on sample preparation and test procedures are contained in Reference 2.

Properties That Impact Waste Form Durability

Compressive Strength. Compressive strength is a measure of a waste form's mechanical integrity and its ability to withstand pressures resulting from stacking or overburden of soil and barrier materials. Polyethylene is a nonrigid plastic with no discrete brittle fracture yield point under compressive load. However, polyethylene waste forms containing dry solid wastes (i.e., aggregates), do exhibit a discernible yield strength when tested for compressive strength. Compressive strength testing of polyethylene waste forms containing sodium nitrate at several waste loadings was conducted in accordance with ASTM D-695, "Standard Method of Test for Compressive Properties of Rigid Plastics." (5) Waste form strength (Figure 1) ranged from a minimum of 7.03 MPa (1,020 psi) for those containing 70 wt% sodium nitrate, to a maximum of 16.3 MPa (2,360 psi) for those containing 30 wt% sodium nitrate. All samples tested had compressive strengths well above NRC minimum criteria of 60 psi (500 psi for cement waste forms) for licensing of commercial low-level radioactive waste forms. (3)

Resistance to Saturated Conditions. Water immersion testing was conducted by submerging waste forms in distilled water for 90 days to determine waste form stability under saturated disposal conditions. Compressive strength measurements taken following immersion revealed no statistically significant changes in waste form strength, as shown in Figure 1.

Resistance to Thermal Cycling. Thermal cycling was performed in accordance with ASTM B-553, "Thermal Cycling of Electroplated Plastics," (6) to determine the effects of extreme temperature variations that waste forms may encounter during storage, transportation, or disposal. Waste form specimens containing 60 wt% sodium nitrate were cycled between temperatures of -40°C and +60°C for a total of thirty, 5-hr periods. Samples were tested for compressive strength on completion of the thermal conditioning. Results given in Table 1 show that no statistically significant changes in yield strength occurred as a result of thermal cycling.

Resistance to Microbial Degradation. Biodegradation is the process by which organic substances are broken down into carbon dioxide, water, and other simple molecules by microorganisms such as molds, fungi, and bacteria when exposed under conditions conducive to their growth. The overwhelming consensus of studies on biodegradation of plastics indicates that the majority of synthetic polymers, including polyethylene, are highly resistant to microbial degradation. (7) In fact, studies conducted to enhance biodegradation of plastics and, thereby, relieve overburdened municipal waste landfills have largely proven unsuccessful. (8,9) Potts, et al., found that only polymers with very low molecular weight linear chains (molecular weights below about 450) were susceptible to microbial degradation. (5) The average molecular

Table 1. Compressive Yield Strength of Cored Pilot-Scale
Polyethylene Waste Forms Containing 60 wt% Sodium Nitrate

Test Description	Compressive Yield Strength, MPa (psi)[a,b]
Initial	14.2 ± 0.3 (2,060 ± 45)
Post Thermal Cycling	13.3 ± 0.5 (1,930 ± 70)
Post Irradiation	16.7 ± 0.7 (2,420 ± 100)
Post Biodegradation[c]	10.1 ± 1.8 (1,460 ± 255)

[a] Based on 6 replicate specimens.
[b] Error expressed as ± one standard deviation.
[c] Based on 12 replicate specimens.

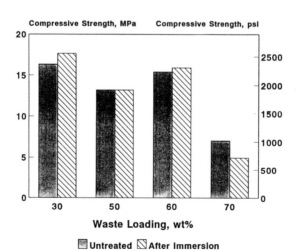

Figure 1.   Compressive yield strength of polyethylene waste forms
containing sodium nitrate salt, untreated and after 90 days
in water immersion.

weight of low density polyethylene is between 10^4 and 10^5, well above the range where growth is supported. Barua, et al., found that even low molecular weight hydrocarbons are not biodegradable when they are branched or cross-linked. (10) Compared to other polymers, low density polyethylene contains many branched chains, further reducing susceptibility to microbes.

Polyethylene waste forms were tested for resistance to microbial degradation to substantiate durability under long-term disposal conditions. Samples containing 60 wt% sodium nitrate and blanks containing no waste were tested in accordance with ASTM G-21, "Standard Practice for Determining Resistance of Synthetic Polymeric Materials to Fungi," (11) and ASTM G-22, "Standard Practice for Determining Resistance of Plastics to Bacteria." (12) These tests expose specimens to microbes under ideal growth conditions for a period of three weeks. No fungal or bacterial growth was observed on any of the test specimens. Mean compressive strength following bio-testing (Table 1) was 10.1 Mpa (1,460 psi), which is well above the minimum NRC standards. Loss in strength for bio-test specimens was an artifact of sample preparation (i.e., coring from a large sample and dissolution of exposed salt in sterilization media) and is not related to microbial degradation.

Resistance to Ionizing Radiation.   Numerous studies indicate that the predominant effect of ionizing radiation on polyethylene is an increase in bonding between and within polymer chains, known as crosslinking. About 10 lateral crosslinks are formed in the main chain for every bond scission that occurs in polyethylene irradiated in the absence of air. (13) Crosslinking affects primary bonds that require large amounts of energy to disturb (200 to 600 kJ/mol). Thus, the impacts of crosslinking are irreversible and can have a positive effect on many mechanical properties of interest including compressive strength, thermal stability, solvent resistance, environmental stress cracking, creep, and permeability. (14,15)

The effects of gamma irradiation on the compressive strength of polyethylene waste forms containing 60 wt% sodium nitrate were examined. Waste form test specimens were exposed to a ^{60}Co gamma source at a dose rate of 3.6×10^6 rad/hr, for a total dose of 10^8 rad. Compressive strength of irradiated waste forms (Table 1) increased about 18%, thus confirming the effects of crosslinking.

Evolution of hydrogen gas from irradiated polyethylene is low due to the small number of bond scissions that occur. The potential impact of hydrogen gas generation on polyethylene waste forms was estimated based on a "G value" (number of H_2 molecules produced per 100 eV) of 4.1 μmoles/g/Mrad. Typical boiling water reactor waste forms with absorbed dose rates ranging from 10^7 Bq/m^3 to 10^{11} Bq/m^3 (0.01 Ci/ft^3 to 100 Ci/ft^3) were assumed. Estimated hydrogen gas evolution from 208-liter (55-gallon) polyethylene waste forms containing 60 wt% sodium nitrate ranges from about 4×10^{-6} l/day to 4×10^{-2} l/day, corresponding to dose rates of 0.03 - 300 gray/day (3 - 3 x 10^4 rad/day), as shown in Figure 2. Typical mild steel drums used for waste disposal are not gas-tight (especially to hydrogen). Therefore, this gas generation is not expected to cause any measurable pressure increase within the drum. Quantification of a pressure increase would require empirical data on the diffusion rate of hydrogen from the waste container and the effective volume of gas generated (based on the porosity of polyethylene to hydrogen, void volume within the waste form, and the fraction of hydrogen reabsorbed to the solid phase). Assuming a gas-tight container and an effective volume of 25% of the total volume, however, yields a conservative bounding estimate for pressure increases between 8×10^{-9} and 8×10^{-5} MPa (10^{-6} to 10^{-2} psi). This conservative estimate confirms the negligible impact of hydrogen gas generation due to radiolysis of polyethylene.

Leachability of Radioactive and Hazardous Constituents. Leachability is a measure of the waste form's ability to retard the release of radioactive and hazardous constituents into the accessible environment. Leach testing of polyethylene waste forms containing 30 to 70 wt% sodium nitrate was conducted in distilled water according to the ANS 16.1 Standard. (16) This is a semi-dynamic leach test in which the leachant is replaced and the leachate is sampled at prescribed intervals for a period of 90 days. Results are expressed in terms of a leach index, a dimensionless figure-of-merit that quantifies the relative leachability for a given waste type-solidification agent combination. Since the leach index is inversely proportional to the log of effective diffusivity, an incremental increase in leach index represents an order of magnitude reduction in leachability. Leach index values (Figure 3) ranged from a minimum of 7.8 (70 wt% sodium nitrate loading) to a maximum of 11.1 (30 wt% sodium nitrate). The minimum leach index specified by NRC is 6.0.

Sodium nitrate waste often contain both traces of toxic metals (in levels above those established by the EPA defining characteristic hazardous wastes) and radioactive contaminants. Thus, they are classified as mixed waste and are subject to regulations promulgated by EPA contained in 40 CFR 261 in support of the Resource Conservation and Recovery Act (RCRA). (17) Leaching of hazardous constituents (e.g., chromium) was conducted for waste forms containing 60 wt% sodium nitrate, in accordance with the Toxicity Characteristic Leaching Procedure (TCLP). Leachate concentrations of the encapsulated waste contained 3.6 mg/l chromium, compared with 9.0 mg/l for the unencapsulated waste, The EPA maximum allowable concentration for chromium in TCLP leachates is $\leq$ 5.0 ppm.

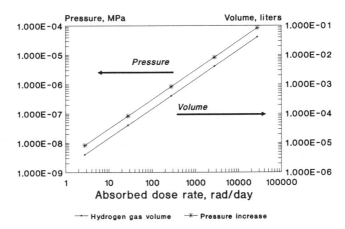

Figure 2.   Conservative estimates of hydrogen gas evolution and resultant gas pressure for self-irradiated polyethylene waste forms containing 60 wt% sodium nitrate.

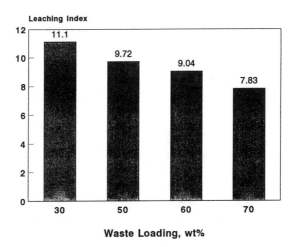

Figure 3.   Leaching index determined according to the ANS 16.1 Leach Test as a function of sodium nitrate waste loading for polyethylene waste forms.

Resistance to Chemical Attack. The ability of a waste form to resist attack from aggressive chemical environments, internally from the waste itself and externally from the disposal site, is a key factor in maintaining long-term durability. Polyethylene's ability to resist chemical attack is a primary reason for its widespread use in diverse applications. It is insoluble in virtually all organic solvents (at ambient temperature) and is resistant to many acids and alkaline solutions. (*18*) A summary of polyethylene suitability for use in specific chemical environments is given in Reference 2.

Flammability. Flammability of waste forms is not a concern for existing shallow land disposal facilities and advanced designs that incorporate subterranean disposal, since oxygen to support combustion is unavailable. However, for above-ground disposal and during processing, temporary storage, and transportation, conditions to sustain combustion may be present. The flash ignition temperature is the lowest temperature at which a material will ignite if exposed to a small flame under specified conditions. The self-ignition temperature is the lowest temperature at which a material will ignite without the presence of a flame. Samples of low density polyethylene used for waste encapsulation were tested in accordance with ASTM D-1929, "Standard Method of Test for Ignition Property of Plastics." (*19*) Flash ignition temperature was 409°C and self-ignition temperature was 430°C. The extrusion processing temperature of polyethylene is about 120°C, so process-related fires are unlikely. If ignited, polyethylene burns slowly in a controlled manner, at about 1.0 in/min under conditions specified in ASTM D-635. When cooled below the flash-ignition temperature, the material is self-extinguishing. The National Fire Protection Association (NFPA) has given polyethylene a flammability rating of 1 (slight), on a scale of 0 (least) to 4 (extreme). (*20*)

Environmental Stress Cracking and Photodegradation. Low density polyethylene can become embrittled if exposed to conditions that lead to environmental stress cracking (ESC). ESC can occur under conditions of stress, in the presence of chemical agents such as soaps, detergents, wetting agents, or oils and results in mechanical failure by cracking at stresses below those that would cause cracking in the absence of these agents. This phenomenon should have little or no impact on the long-term durability of polyethylene waste forms, since they are generally not exposed to significant concentrations of chemical agents required to initiate crack propagation and extruded waste forms would not be under high stresses that result from complex molds shapes.

Polyethylene exposed to sunlight (ultraviolet radiation) and oxygen for extended periods is subject to deterioration of mechanical properties, cracking, and molecular weight loss by a process known as photodegradation. Several researchers have found that these effects are limited to outer surface regions (on the order of tens of angstroms) since insufficient oxygen is available in deeper layers. (21,22) Containerized polyethylene waste forms that are stored or disposed will not likely encounter solar irradiation that could lead to photodegradation. Even if waste forms were exposed to sunlight and oxygen for prolonged periods, photodegradation impacts would be minimal since the phenomenon is limited to surface effects.

Summary/Conclusions

Review of existing property data and studies of material behavior, supplemented by the results of waste form performance testing, confirm the long-term durability of polyethylene for encapsulation of low-level radioactive, hazardous, and mixed wastes. Polyethylene is highly resistant to microbial degradation and attack by aggressive chemicals. Radiation doses through 10^8 rad increase crosslinking and improve strength and other physical properties. Flash- and self-ignition points are relatively high, and NFPA has rated polyethylene flammability as "slight." Other failure mechanisms including environmental stress cracking and photodegradation should have little effect on long-term durability. Polyethylene waste forms containing sodium nitrate exceed minimum performance standards established by NRC and EPA for commercial LLW and hazardous wastes, respectively. Waste form testing for other waste types (sodium sulfate, boric acid, incinerator ash, ion exchange resins) reported previously (1) also indicated performance well above minimum regulatory standards. While confirmatory testing is needed for any new waste stream application, data generated for this effort indicate that polyethylene encapsulation of mixed wastes yields stable, durable waste forms capable of withstanding anticipated conditions of storage, transport and disposal.

Acknowledgments

The authors gratefully acknowledge the efforts of Sandra G. Lane for providing editorial assistance and Patricia Durcan for help in preparation of this manuscript.

This work was sponsored by the U.S. Department of Energy under contract No. DE-AC02-76CH00016.

Literature Cited

1. Kalb, P.D., and P. Colombo, "Polyethylene Solidification of Low-Level Wastes," BNL-51867, Brookhaven National Laboratory, Upton, NY, October 1984.
2. Kalb, P.D., J.H. Heiser, III, and P. Colombo, "Polyethylene Encapsulation of Nitrate Salt Wastes: Waste Form Stability, Process Scale-up, and Economics," BNL-52293, Brookhaven National Laboratory, Upton, NY, July 1991.

3. U.S. Nuclear Regulatory Commission, "Technical Position on Waste Form (Revision 1)," Low-Level Waste Management Branch, United States Nuclear Regulatory Commission, Washington, D.C., January 1991.

4. U.S. Environmental Protection Agency, Toxicity Characteristic Leaching Procedure, 40 CFR 261, Appendix II, Fed. Reg. 55, 11863, March 29, 1990.

5. ASTM, Standard Method of Test for Compressive Properties of Rigid Plastics, D-695, American Society for Testing and Materials, Philadelphia, PA.

6. ASTM, Standard Method of Test for Thermal Cycling of Electroplated Plastics, D-695, American Society for Testing and Materials, Philadelphia, PA.

7. Albertsson, A., S.O. Andersson, and S. Karlsson, "The Mechanism of Biodegradation of Polyethylene," *Polymer Degradation and Stability*, 18, pp. 73-87, 1987.

8. Potts, J.E., R.A. Clendinning, and W.B. Ackart, "An Investigation of the Biodegradability of Packaging Plastics," EPA-R2-72-046, U.S. Environmental Protection Agency, Washington, D.C., August 1972.

9. Rodriguez, F., "The Prospects for Biodegradable Plastics," *Chemical Technology*, p. 409, July 1971.

10. Barua, P.K., et al., "Comparative Utilization of Paraffins by a Trichosporon Species," *Applied Microbiology*, pp. 657-661, November 1970.

11. ASTM, Standard Practice for Determining Resistance of Synthetic Polymeric Materials to Fungi, ASTM G-21, American Society for Testing and Materials, Philadelphia, PA.

12. ASTM, Standard Practice for Determining Resistance of Plastics to Bacteria, ASTM G-22, American Society for Testing and Materials, Philadelphia, PA.

13. Makhlis, F.A., *Radiation Physics and Chemistry of Polymers*, John Wiley and Sons, New York, (1975).

14. de Hollain, G., "Physical Properties of Irradiated Polymers," *Plastics and Rubber: Materials and Applications*, Vol. 5, (3) August 1980.

15. Barlow, A., J. Biggs, and M. Maringer, "Radiation Processing of Polyolefins and Compounds," *Radiat. Phys. Chem.*, Vol. 9, pp. 685-699, 1977.

16. ANS Standards Committee, Working Group 16.1, "Measurement of the Leachability of Solidified Low-Level Wastes," American Nuclear Society, June 1984.

17. U.S. Environmental Protection Agency, "Identification and Listing of Hazardous Waste," 40 CFR 261, Fed. Reg. 45, 33119, May 19, 1980.

18. Raff, R.A.V., and J.B. Allison, *Polyethylene*, Interscience Publishers, New York, (1956).

19.   ASTM, Standard Method of Test for Ignition Property of Plastics, ASTM-1929, American Society for Testing and Materials, Philadelphia, PA.

20.   "Material Safety Data Sheet, Gulf Polyethylene (All Grades)," MSDS Number: PL0001, Chevron Chemical Co., Houston, TX, (1989).

21.   La Manitia, F.P., "Natural Weathering of Low Density Polyethylene - Mechanical Properties," *Eur. Polym. J.*, Vol. 20, No. 10, pp. 993-995, 1984.

22.   Benachour, D. and C.E. Rogers, "Strain-Enhanced Photodegradation of Polyethylene," American Chemical Society, Division of Organic Coating and Plastics Chemistry, *Preprints*, Vol. 42, pp. 767-769, 1980.

RECEIVED November 9, 1992

INDEXES

Author Index

Affiliation Index

New Jersey Institute of Technology, 358
North Carolina State University, 18
The Pennsylvania State University, 203
Polytechnic University, 343
Purus Inc., 380
SRI International, 106
Tennessee Eastman Company, 219
U.S. Department of Agriculture, 278

U.S. Environmental Protection Agency, 159, 203
University of Cincinnati, 159
University of Kentucky, 308
University of Pittsburgh, 1
University of Sriwijaya, 308
Utah State University, 191
Virginia Polytechnic Institute and State
  University, 119

Subject Index

A

Acetic acid, supercritical water oxidation, 59
Activated sludge processes, effluent
  treatment during bioremediation, 153
Activity coefficients, determination, 282
Adsorption edges, description, 287
Adsorptive and chemical pretreatment of
  reactive dye discharge
  chemical reduction, 126,129–135
  color removal by dye adsorption,
    120,122–125
  experimental materials, 119–121
  flocculation studies, 126,128–129t
  sorption studies, 120,122–125
  total organic carbon removal by dye
    adsorption, 120,122–123t,124
Advanced electronic reactor process,
  description, 344
Advanced oxidation process(es)
  applications, 17
  chemistries
    bicarbonate inhibition, 24,27t,28,29f
    ease of implementation, 30,31t
    heterogeneous photocatalysis, 23
    high organic compound concentrations,
      28,31t
    oxidant cost, 28,31t
    ozone oxidation mechanism, 18,19–20f
    ozone–peroxide mechanism, 21,23
    ozone–UV mechanism, 18,20f,21,22f
    pH sensitivity, 24,25–26f,27t
    reactor design, 30,32
    semiconductor-mediated photooxidation, 23
    UV absorbance and scattering, 28,31t
    UV cost, 28,30,31t
    UV–peroxide mechanism, 21,22f
  description, 5

Aeration of contaminated soils
  advantages, 260–261
  comparison to other technologies,
    270,271–272t,273
  disadvantages, 261
  performance, 260
  technology description, 259–260
Aerobic sludge, industrial activated,
  anaerobic digestion, 219–233
Ag(II)
  mechanism of aliphatic destruction,
    431,432f
  observed vs. predicted time-dependent
    ethylene glycol destruction, 431,432f
  use as oxidizing agent, 431
Airborne volatiles, treatment, 9–10
Ammonia, supercritical water oxidation, 58
Ammonia–methanol mixtures, supercritical
  water oxidation, 59
Anaerobic digestion of industrial
  activated aerobic sludge
  advantages, 219
  ammonia concentration determination
    procedure, 224
  anaerobic reactors, 222–224
  batch cultures at different initial
    solids concentration, 221–222
  enrichment cultures and reactors, 220–221
  enrichment cultures at different pH
    levels, 221
  experimental description, 220
  GC procedure, 224
  initial solids concentration effect,
    225,228f,229
  interaction of bacterial groups in
    metabolic steps, 220
  operation, 229–233
  pH effect, 224–225,226–227f

Production: Anne Wilson
Indexing: Deborah H. Steiner
Acquisition: Rhonda Bitterli
Cover design: Neal Clodfelter

Printed and bound by Maple Press, York, PA

Bestsellers from ACS Books

The ACS Style Guide: A Manual for Authors and Editors
Edited by Janet S. Dodd
264 pp; clothbound ISBN 0–8412–0917–0; paperback ISBN 0–8412–0943–X

The Basics of Technical Communicating
By B. Edward Cain
ACS Professional Reference Book; 198 pp;
clothbound ISBN 0–8412–1451–4; paperback ISBN 0–8412–1452–2

Chemical Activities (student and teacher editions)
By Christie L. Borgford and Lee R. Summerlin
330 pp; spiralbound ISBN 0–8412–1417–4; teacher ed. ISBN 0–8412–1416–6

Chemical Demonstrations: A Sourcebook for Teachers,
Volumes 1 and 2, Second Edition
Volume 1 by Lee R. Summerlin and James L. Ealy, Jr.;
Vol. 1, 198 pp; spiralbound ISBN 0–8412–1481–6;
Volume 2 by Lee R. Summerlin, Christie L. Borgford, and Julie B. Ealy
Vol. 2, 234 pp; spiralbound ISBN 0–8412–1535–9

Chemistry and Crime: From Sherlock Holmes to Today's Courtroom
Edited by Samuel M. Gerber
135 pp; clothbound ISBN 0–8412–0784–4; paperback ISBN 0–8412–0785–2

Writing the Laboratory Notebook
By Howard M. Kanare
145 pp; clothbound ISBN 0–8412–0906–5; paperback ISBN 0–8412–0933–2

Developing a Chemical Hygiene Plan
By Jay A. Young, Warren K. Kingsley, and George H. Wahl, Jr.
paperback ISBN 0–8412–1876–5

Introduction to Microwave Sample Preparation: Theory and Practice
Edited by H. M. Kingston and Lois B. Jassie
263 pp; clothbound ISBN 0–8412–1450–6

Principles of Environmental Sampling
Edited by Lawrence H. Keith
ACS Professional Reference Book; 458 pp;
clothbound ISBN 0–8412–1173–6; paperback ISBN 0–8412–1437–9

Biotechnology and Materials Science: Chemistry for the Future
Edited by Mary L. Good (Jacqueline K. Barton, Associate Editor)
135 pp; clothbound ISBN 0–8412–1472–7; paperback ISBN 0–8412–1473–5

For further information and a free catalog of ACS books, contact:
American Chemical Society
Distribution Office, Department 225
1155 16th Street, NW, Washington, DC 20036
Telephone 800–227–5558